PLANT GENOTYPING
The DNA Fingerprinting of Plants

PLANT GENOTYPING
The DNA Fingerprinting of Plants

Edited by

R.J. HENRY

Centre for Plant Conservation Genetics
Southern Cross University
Lismore
Australia

CABI *Publishing*

CABI *Publishing* is a division of CAB *International*

CABI Publishing
CAB International
Wallingford
Oxon OX10 8DE
UK

CABI Publishing
10 E 40th Street
Suite 3203
New York, NY 10016
USA

Tel: +44 (0)1491 832111
Fax: +44 (0)1491 833508
Email: cabi@cabi.org
Web site: http://www.cabi.org

Tel: +1 212 481 7018
Fax: +1 212 686 7993
Email: cabi-nao@cabi.org

© CAB *International* 2001. All rights reserved. No part of this publication may be reproduced in any form or by any means, electronically, mechanically, by photocopying, recording or otherwise, without the prior permission of the copyright owners.

A catalogue record for this book is available from the British Library, London, UK.

Library of Congress Cataloging-in-Publication Data

Plant genotyping : the DNA fingerprinting of plants / edited by R.J. Henry.
    p. cm.
  Includes bibliographical references.
  ISBN 0-85199-515-2 (alk. paper)
    1. DNA fingerprinting of plants. I. Henry, Robert J.

QK981.45 .P56 2001
581.3′5--dc21

00-045428

ISBN 0 85199 515 2

Typeset in 10/12pt Photina by Columns Design Ltd, Reading
Printed and bound in the UK by Biddles Ltd, Guildford and King's Lynn

Contents

Contributors ix

Preface xiii

Molecular Markers Available for Use in Plant Genotyping

1 Plant Genotyping by Analysis of Single Nucleotide
  Polymorphisms 1
  K.J. Edwards and R. Mogg

2 Plant Genotyping by Analysis of Microsatellites 15
  T.A. Holton

3 Plant Genotyping Using Arbitrarily Amplified DNA 29
  G. Caetano-Anollés

4 Plant Genotyping Based on Analysis of Single Nucleotide
  Polymorphisms Using Microarrays 47
  B. Lemieux

Genotyping Plant Genetic Resource Collections

5 Genotyping in Plant Genetic Resources 59
  B.V. Ford-Lloyd

6 Applications of Molecular Marker Techniques to the Use
  of International Germplasm Collections 83
  M. Warburton and D. Hoisington

Genotyping Cultivated and Wild Germplasm

7 Molecular Analysis of Wild Plant Germplasm: the Case of
  Tea Tree (*Melaleuca alternifolia*) 95
  L.S. Lee, M. Rossetto, L. Homer and R.J. Henry

8   Genotyping Pacific Island Taro (*Colocasia esculenta* (L.)
    Schott) Germplasm   109
    I.D. Godwin, E.S. Mace and Nurzuhairawaty

9   Molecular Marker Systems for Sugarcane Germplasm Analysis   129
    G.M. Cordeiro

10 Microsatellite Analysis in Cultivated Hexaploid Wheat and
    Wild Wheat Relatives   147
    A. McLauchlan, R.J. Henry, P.G. Isaac and K.J. Edwards

11 Comparison of RFLP and AFLP Marker Systems for
    Assessing Genetic Diversity in Australian Barley Varieties
    and Breeding Lines   161
    K.J. Chalmers, S.P. Jefferies and P. Langridge

Development of Molecular Markers for Use in Plant Genotyping

12 Discovery and Application of Single Nucleotide
    Polymorphism Markers in Plants   179
    D. Bhattramakki and A. Rafalski

13 Producing and Exploiting Enriched Microsatellite Libraries   193
    T.L. Maguire

14 Sourcing of SSR Markers from Related Plant Species   211
    M. Rossetto

15 Microsatellites Derived from ESTs, and their Comparison with
    those Derived by Other Methods   225
    K.D. Scott

Technical Developments and Issues in Plant Genotyping

16 Plant DNA Extraction   239
    R.J. Henry

17 Collection, Reporting and Storage of Microsatellite
    Genotype Data   251
    N. Harker

18 Commercial Applications of Plant Genotyping   265
    L.S. Lee and R.J. Henry

19	Non-gel Based Techniques for Plant Genotyping *R. Kota*	275
20	Using Molecular Information for Decision Support in Wheat Breeding *H.A. Eagles, M. Cooper, R. Shorter and P.N. Fox*	285
21	Application of DNA Profiling to an Outbreeding Forage Species *J.W. Forster, E.S. Jones, R. Kölliker, M.C. Drayton, M.P. Dupal, K.M. Guthridge and K.F. Smith*	299

Index   321

Contributors

Dinakar Bhattramakki, Pioneer Hi-Bred International Inc., Reid 33C, 7300 N.W. 62nd Avenue, Johnston, IA 50131, USA

Gustavo Caetano-Anollés, Laboratory of Molecular Ecology and Evolution and Division of Molecular Biology, Department of Biology, University of Oslo, Oslo N-0316, Norway

Ken J. Chalmers, CRC for Molecular Plant Breeding, Department of Plant Science, Waite Campus, University of Adelaide, SA 5064, Australia

Mark Cooper, School of Land and Food Sciences, The University of Queensland, Brisbane, QLD 4072, Australia

Giovanni M. Cordeiro, Centre for Plant Conservation Genetics, Southern Cross University, Lismore, NSW 2480, Australia

Michelle C. Drayton, Plant Biotechnology Centre, Agriculture Victoria, La Trobe University, Bundoora, VIC 3083, Australia

Mark P. Dupal, Plant Biotechnology Centre, Agriculture Victoria, La Trobe University, Bundoora, VIC 3083, Australia

Howard A. Eagles, CRC for Molecular Plant Breeding, Department of Natural Resources and Environment, VIDA, PB 260, Horsham, VIC 3400, Australia

Keith J. Edwards, IACR-Long Ashton Research Station, Department of Agricultural Sciences, University of Bristol, Long Ashton, Bristol BS41 9AF, UK

Brian V. Ford-Lloyd, School of Biosciences, University of Birmingham, Edgbaston, Birmingham B15 2TT, UK

John W. Forster, Plant Biotechnology Centre, Agriculture Victoria, La Trobe University, Bundoora, VIC 3083, Australia

Paul N. Fox, CIMMYT, Lisboa 27, Apdo. Postal 6-641, 06600 México DF, México

Ian D. Godwin, School of Land and Food Sciences, The University of Queensland, Brisbane, QLD 4072, Australia

Kathryn M. Guthridge, Plant Biotechnology Centre, Agriculture Victoria, La Trobe University, Bundoora, VIC 3083, Australia

Natalie Harker, Centre for Plant Conservation Genetics, Southern Cross University, PO Box 157, Lismore, NSW 2480, Australia

Robert J. Henry, Cooperative Research Centre for Molecular Plant Breeding, Centre for Plant Conservation Genetics, Southern Cross University, PO Box 157, Lismore, NSW 2480, Australia

Dave Hoisington, Applied Biotechnology Center, International Maize and Wheat Improvement Center, CIMMYT, Apdo. Postal 6-641, 06600 México DF, México

Timothy A. Holton, Centre for Plant Conservation Genetics, Southern Cross University, Lismore, NSW 2480, Australia

Laura Homer, Centre for Plant Conservation Genetics, Southern Cross University, PO Box 157, Lismore, NSW 2480, Australia

Peter G. Isaac, Agrogene, 620 rue Blaise Pascal Z.I., 77550 Moissy-Cramayel, France

Stephen P. Jefferies, CRC for Molecular Plant Breeding, Department of Plant Science, Waite Campus, University of Adelaide, SA 5064, Australia

Elizabeth S. Jones, Plant Biotechnology Centre, Agriculture Victoria, La Trobe University, Bundoora, VIC 3083, Australia

Roland Kölliker, Swiss Federal Research Station for Agroecology and Agriculture, Reckenholzstrasse 191, 8046 Zürich, Switzerland

Raja Kota, Plant Genome Research Centre, Institute for Plant Genetics and Crop Plant Research (IPK), Corrensstra 3, D-06466 Gatersleben, Sachsen-Anhalt, Germany

Peter Langridge, CRC for Molecular Plant Breeding, Department of Plant Science, Waite Campus, University of Adelaide, SA 5064, Australia

L. Slade Lee, Centre for Plant Conservation Genetics, Southern Cross University, PO Box 157, Lismore, NSW 2480, Australia

Bertrand Lemieux, Department of Plant and Soil Sciences, University of Delaware, Newark, DE 19717-1303, USA

Emma S. Mace, School of Land and Food Sciences, The University of Queensland, Brisbane, QLD 4072, Australia

Tina L. Maguire, Department of Botany, The University of Queensland, Brisbane, QLD 4072, Australia

Anne McLauchlan, Cooperative Research Centre for Molecular Plant Breeding, Centre for Plant Conservation Genetics, Southern Cross University, PO Box 157, Lismore, NSW 2480, Australia

Rebecca Mogg, IACR-Long Ashton Research Station, Department of Agricultural Sciences, University of Bristol, Long Ashton, Bristol BS41 9AF, UK

Nurzuhairawaty, School of Land and Food Sciences, The University of Queensland, Brisbane, QLD 4072, Australia

Antoni Rafalski, DuPont Agricultural Products – Genomics, Delaware Technology Park, Suite 200, 1 Innovation Way, Newark, DE 19711, USA

Maurizio Rossetto, Centre for Plant Conservation Genetics, Southern Cross University, PO Box 157, Lismore, NSW 2480, Australia

Kirsten D. Scott, Centre for Plant Conservation Genetics, Southern Cross University, PO Box 157, Lismore, NSW 2480, Australia

Ray Shorter, CSIRO Tropical Agriculture, 120 Meiers Road, Indooroopilly, QLD 4068, Australia

Kevin F. Smith, Pastoral and Veterinary Institute, Agriculture Victoria, Hamilton, VIC 3300, Australia

Marilyn Warburton, Applied Biotechnology Center, International Maize and Wheat Improvement Center, CIMMYT, Apdo. Postal 6-641, 06600 México DF, México

Preface

Plant genotyping, or DNA fingerprinting of plants, is a technology that has matured and is poised for very widespread practical application. Plant genotype analysis has application in the identification of plants in commerce, plant breeding and research. Commercial applications include the protection of plant breeders' rights and patents, quality control in plant production and processing and labelling of plant-derived foods and other products. Plant breeding applications range from marker-assisted selection to the confirmation of identity of parents and progeny in breeding populations. Research applications include analysis of evolutionary relationships and population genetics. Comparison of the fingerprints of large numbers of genotypes from different locations or laboratories will require the establishment of international standard formats for the collection, storage and analysis of data. This will be especially important in fingerprinting the large international collections of germplasm of the major crops to support the conservation and exploitation of available genetics resources. This book examines the technologies available and their application in the analysis of wild plant populations, germplasm collections and plant breeding. Microsatellite analysis is a state-of-the-art technology and a major focus of this book. Single nucleotide polymorphism markers have great potential for application in plants and have been covered in several chapters. New technologies, such as microarrays and non-gel based analysis of markers, are also described.

R.J. Henry
Cooperative Research Centre for Molecular Plant Breeding
Centre for Plant Conservation Genetics
Southern Cross University

Chapter 1

Plant Genotyping by Analysis of Single Nucleotide Polymorphisms

K.J. EDWARDS AND R. MOGG

IACR-Long Ashton Research Station, Department of Agricultural Sciences, University of Bristol, Bristol, UK

Introduction

Before the advent of molecular markers, plant breeders based their selection on phenotypic agronomic traits such as plant height and grain yield. Even today, such an approach has considerable merit. However, with the development of marker technology, plant breeders were provided with tools capable of monitoring the whole genome in the absence of a phenotype (Helentjaris *et al.*, 1986). Given the advantages of molecular markers, it is not surprising that in the years that followed, molecular marker-based genotyping was adopted by all the major plant breeding companies and academic laboratories (Welsh and McClelland, 1990; Vos *et al.*, 1995; Struss and Plieske, 1998). Today, laboratories involved in plant genotyping have a multitude of marker systems to call upon. Unfortunately, the methods for detecting most molecular markers (restriction fragment length polymorphisms (RFLPs), random amplified polymorphic DNA (RAPDs), amplified fragment length polymorphisms (AFLPs) and microsatellites) rely upon electrophoretic separation of DNA fragments in agarose or polyacrylamide gels. Developments in fluorescent DNA fragment analysis have made it practical to analyse multiple loci simultaneously (Heyen *et al.*, 1997); however, despite this and the advent of semi-automated systems such as capillary gel electrophoresis (Gonen *et al.*, 1999), gel-based technology is labour intensive and time-consuming for the large-scale genotyping required in experimental genome analysis, DNA fingerprinting and marker-assisted breeding programmes.

Molecular markers are polymorphic when there is DNA sequence variation between the individuals under study. Molecular markers are, therefore, simply an indicator of sequence polymorphism. Sequence polymorphism between individuals can take many forms; for instance, it can be due to the insertion or

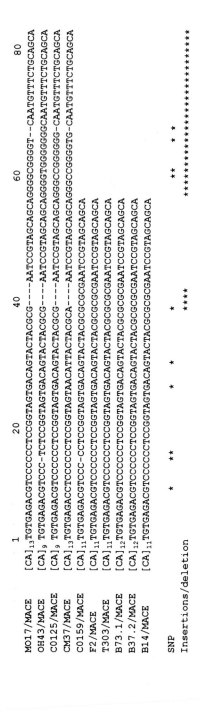

Fig. 1.1. Identification of SNPs in the 3′-flanking region of the maize microsatellite MACE01F07. SNPs associated with the flanking regions of maize microsatellites were identified as follows. Following amplification using the published primer sequences (http://www.agron.missouri.edu), PCR products from ten maize inbred lines were sequenced using the reverse primer. Sequence comparisons were carried out using the CLUSTALW program and the location of either SNPs or insertions/deletions identified (*). The microsatellite motif is located at the 5′ end of the sequence ([CA]).

deletion of multiple bases, or it can be due to single nucleotide polymorphisms (SNPs, but pronounced SNiPs: Brookes, 1999; Fig. 1.1). Insertions, deletions and SNPs are important in determining sequence variation between individuals; for the purpose of this study, only SNPs will be discussed further. However, it must be emphasized that a number of the properties attributed to SNPs also apply to insertions or deletions.

SNPs in plants

Characterization of SNPs in humans suggests that, when comparing two unrelated individuals, one SNP can be found on average every 1 kbp of sequence (Li and Sadler, 1991). Limited work has been carried out to examine the occurrence of SNPs in plants; Bryan and co-workers found that the sequence variation present between different wheat RFLP alleles was insufficient to design SNPs (Bryan *et al.*, 1999). However, Germano and Klein (1999) have shown that SNPs are present in the nuclear and chloroplast DNA of both *Picea rubens* and *Picea mariana*. Moreover, they can be used to genotype individuals more efficiently than either RFLPs or RAPDs. In soybean, Coryell *et al.* (1999) identified two SNPs in 400 bp of sequence from the nuclear RFLP locus A519-1. In recent work, Mogg *et al.* (1999), have shown that by sequencing the flanking regions of maize microsatellites, a SNP could be found every 40 bp. Given that the maize genome is estimated to be 2.5×10^9 bp in size, this means that there is the potential for up to 62 million SNPs! Further work in the laboratory of Powell *et al.* (http://www.scri.sari.ac.uk) suggested that the flanking regions of barley microsatellites were also a rich source of SNPs. In 1998, the National Science Foundation (NSF) awarded US$2.2 million to a consortium of American university laboratories to develop SNPs for genes involved in quality traits in maize (http://www.nsf.gov/bio/pubs/awards/genome98.html). In 1999, the NSF awarded a further US$2.9 million to Iowa State University to develop complementary SNP-related technology (http://www.nsf.gov/bio/pubs/awards/genome99.html). Recent industrial experiences with SNPs are described in Chapter 12 (Bhattramakki and Rafalski).

This current activity, therefore, clearly indicates that SNPs are abundant in plant genomes. Given this, and the fact that SNPs have the potential to provide the basis for a superior and highly informative genotyping assay, it is surprising that they are not already in regular use within plant genotyping laboratories. SNPs are, however, used for genotyping human populations for certain genetic diseases (Saiki *et al.*, 1989). The reason for this difference becomes clear when one considers the enormous cost of developing SNPs; for each locus, a mapped single copy probe has to be sequenced and suitable PCR primers designed. The primers must then be used to amplify the corresponding fragment from all the other possible genotypes. These fragments must then be sequenced and the sequences compared with one another to determine the SNPs for each haplotype (Brown, 1999). The term 'haplotype' is used in the context of SNPs instead of the term 'allele'. Until recently, an allele was defined as a single gene having a specific sequence, while a haplotype is defined as a collection of alleles which, due to their proximity, are usually inherited together. With SNPs it is possible for several alleles to share some SNPs but not others. In this case the current consensus is that each defined sequence type is a single haplotype and not a single allele.

A number of reviews have highlighted the difficulties of using a biallelic SNP-based marker system in place of multi-allelic systems such as RFLPs and microsatellites. Xiong and Jin (1999) demonstrated that, whereas the biallelic nature of SNPs makes them less informative per locus examined than multi-allelic microsatellites, this limitation is easily overcome by using more loci. For instance, Kruglyak (1997) determined that a 4 cM map of 750 SNP-based markers was equivalent in the information content to a 10 cM map of 300 microsatellite markers. Given the greater abundance of SNPs, such a requirement should not prove excessive. The relatively high frequency of SNPs in the genome brings about the possibility of cloning genes via linkage disequilibrium (Brookes, 1999). In essence linkage disequilibrium is the non-random association of alleles at different loci. Logically this will occur when two loci occur close together on the same chromosome. The dense genetic maps (up to one marker per 1 kb) possible with SNPs allow genome scans for linkage disequilibrium (of SNPs) to be studied in association with complex phenotypes (Xiong and Jin, 1999). In the future it may be possible to use SNPs in combination with linkage disequilibrium to screen directly the coding polymorphisms of genes derived from complex populations to identify directly candidates for the agronomic trait under investigation (Brookes, 1999).

Clearly, the exploitation of SNPs relies not just on their application, but also on their initial development. Therefore, for the rest of this chapter I will examine some of the recent developments that could enable researchers to identify and utilize SNPs efficiently for large-scale DNA fingerprinting.

Identification of SNPs

There are a number of methods for identifying SNPs within a genetic locus. Whatever method is chosen to detect the polymorphism, the initial step is almost always to determine the sequence of the locus for a reference genotype. Once determined, this sequence is used to design oligonucleotide primers for use in the PCR, which forms the cornerstone of all subsequent SNP-based technology (Erlich, 1989).

Direct sequencing

Sequence analysis is the most direct way of identifying SNPs; however, it is also the most time-consuming and costly. When one considers that an average gene is 5 kbp, it would take more than ten separate sequence reactions to cover it all via the single pass sequencing of PCR products. This highlights one of the main problems with the use of direct sequencing to identify SNPs; that of sequencing errors. A sequencing error rate of just one base per 100 would equal the rate at which SNPs are thought to occur in most plant species. If these sequencing errors are not detected at an early stage, they would result in a considerable waste of resources in both designing allele-specific oligonucleotides (ASOs) and carrying out the SNP assay. In addition, a detailed search for SNPs would examine many genotypes, compounding the sequencing error. A further significant problem exists with direct sequencing: the occurrence in many plant species of either heterozygotes (for instance sugarbeet) or ploidy levels higher than diploid (for instance wheat). In both cases the direct sequencing of PCR products would lead to the sequencing of multiple products derived from either homologous or homoeologous loci. In these cases, direct sequencing would only highlight the base differences, it would not identify the exact bases that were changed. This would be especially true if the polymorphism was based on insertions or deletions. Therefore, when dealing with either heterozygotes or higher levels of ploidy, it is necessary first to clone the PCR fragments before sequencing numerous clones (to avoid errors due to mis-incorporation by the *Taq* DNA polymerase). Given this complex procedure one would not be surprised to discover that there are relatively few examples of direct sequencing of PCR products to produce SNPs in agronomically important crops beyond those mentioned previously.

With the recent initiation of large-scale plant expressed sequence tag (EST) sequencing programmes (Shen *et al.*, 1994; Sasaki, 1998; see also http://wheat.pw.usda.gov/genome; http://wheat.pw.usda.gov/wEST/insf/title.html; http://www.zmdb.iastate.edu/zmdb/EST_project.html) a new and potentially rich source of SNPs has been uncovered. There are two reasons why ESTs might provide a large number of SNPs for genotyping. First, the sequencing of ESTs provides sequence data for expressed genes. It is reasonable to expect that at least some of these ESTs will be responsible for the observed agronomic traits. Therefore, unlike anonymous markers such as microsatellites, the SNPs derived from ESTs could underlie the traits being examined. Secondly, the independent sequencing of ESTs derived from specific and different genotypes will allow the *in silico* search for SNPs to take place. However, while this latter approach is likely to be the method of choice for the future discovery of plant SNPs, it too is not without problems; for instance, the occurrence of multigene families and the presence of large introns may interfere with the sequence analysis. Furthermore, it is clear that not all genotypes will be the subject of large-scale EST sequencing and, therefore, further targeted sequencing will be necessary to cover all useful genotypes.

Single-strand conformation polymorphism (SSCP)

An excellent review of SSCP has already been written in relation to its application to agriculture (Prosser, 1993). Briefly, SSCP is based on the variation in mobility of small polymorphic single-stranded DNA fragments in non-denaturing acrylamide gels. The technique is based on the assumption that the mobility of any single strand fragment will vary according to its exact sequence, and that even single base changes can modify mobility. SSCP works well with fragments of 100–400 bp which can be generated by PCR. A simple SSCP test would require amplification of a small fragment independently from numerous genotypes, followed by denaturation, polyacrylamide gel electrophoresis and silver staining (Jordan *et al.*, 1998). Any change in mobility across the genotype range would indicate a sequence change which could be targeted by direct sequencing. There have been several modifications and some improvements to the SSCP technique, the most widely used of which are denaturing gradient gel electrophoresis and temperature gradient gel electrophoresis (Etscheid and Riesner, 1998).

Chemical cleavage of mismatches (CCM)

CCM relies on the ability of certain chemicals, for instance piperidine (Prosser, 1993), to cleave a chemically modified DNA heteroduplex precisely at mismatched bases. Practically, the analysis involves a reference sample which is labelled with radioactivity, and the sample under investigation. For most purposes, both samples can be generated by PCR. Following denaturation, the two samples are brought together and allowed to anneal to form both homoduplexes and heteroduplexes. When the two samples have sequence differences and a heteroduplex is formed, this is subsequently chemically modified (for instance using osmium tetraoxide) and then cleaved by further treatment with piperidine. Cleaved and uncleaved products are then visualized on a denaturing polyacrylamide gel following autoradiography. CCM does have two significant disadvantages: first, it is time-consuming and, secondly, it requires the use of a number of hazardous reagents. However, CCM is capable of scanning regions of up to 3 kb with almost 100% accuracy.

Enzyme mismatch cleavage (EMC)

EMC has recently gained favour as a method for detecting mismatches in heteroduplexes. Acknowledging this, Nycomed Amersham recently announced the launch of an enzyme mismatch detection (EMD^{TM}) assay based on their existing Passport^{TM} kit (http://194.239.190.159/product/product_news/201099b.html). EMC is very similar in its principle to CCM, where a resolvase enzyme replaces the osmium tetraoxide. An excellent review of EMC has been published on the Internet (http://www.ich.ucl.ac.uk/cmgs/resolve.html). Briefly, amplified DNA from either an individual plant or two individual plants is heat denatured and allowed to re-anneal to form both homo- and heteroduplexes. This DNA is then labelled, with either radioactivity or a fluorescent marker. Any heteroduplex structures present are then subjected to cleavage by the resolvase enzyme at the mismatch site. The products of the reaction can be examined on a denaturing polyacrylamide gel. As for CCM, EMC is capable of scanning several kilobases of DNA. However, EMC is also time-consuming to perform. Unlike CCM, EMC does not require the use of hazardous chemicals. As a result of this, and the fact that the components are available in kit form, EMC is likely to supersede CCM as the method of choice for detection of base pair mismatches in genomic DNA.

Using SNPs in DNA fingerprinting

In previous sections several methods have been described for identifying SNPs in two or more genotypes. These same techniques can also be used to DNA fingerprint previously uncharacterized material. However, in this section I will focus on those technologies which have been developed specifically to fingerprint commercial or semi-commercial material.

DNA chips

Numerous reports and reviews have been written on the subject of DNA chips (Ramsay, 1997; reviewed in the *Nature Genetics* Supplement, 1999, 21, 1–60). However, when used in relation to SNPs we need to clarify that we are dealing with genotyping chips and not the more commonly discussed EST-based expression chips. At the heart of all genotyping chips is the ASO. As its name suggests, ASOs are oligonucleotides whose sequence is specific for either an allele or a portion of an allele (Ala-Kokko *et al.*, 1990). Therefore, the presence or absence of at least one SNP is a prerequisite for the design of an ASO. Genotyping chips can consist of just a few ASOs covering a single locus or many tens of thousands covering hundreds or thousands of loci (Service, 1998). The choice of using either a simple or a complex chip is usually dictated by the availability of multiple allele sequences covering multiple loci. Currently, for all plant species, multiple allele sequence information is very limited. It would, therefore, be impossible to design a plant genotyping chip with more than 20–50 ASOs covering more than five loci. However, in the near future it is likely that for the major crops such as maize, wheat, barley and soybean, genotyping chips will become commercially available.

There are two major strategies for coupling ASOs to the solid matrix that comprises the genotyping chip. The simplest approach is to attach the ASO to the matrix, after synthesis (Guo *et al.*, 1994). This is usually achieved by synthesizing the ASO with a 12-carbon spacer arm and an amino group at the 5′ end of the oligonucleotide. The amino group can then be covalently linked to the surface (for instance a glass microscope slide or a glass coverslip) following its activation to yield reactive aldehyde groups. This 'low-tech' approach is amenable to the coupling of any number of ASOs, from just a few to several thousand. Once coupled, the ASOs are sterically free to hybridize to DNA probes using conventional hybridization technology. The second and more complex approach is to synthesize the ASOs directly on to a specially prepared solid support (Fodor, 1997). This approach is very similar to the approach taken when generating computer chips, as individual photolithographic masks are required for base addition (Fodor, 1991). While this latter approach is clearly more resource intensive, requiring highly specialized facilities, it is capable of generating chips consisting of several hundred thousand ASOs covering an area of less than 1 cm^2 (Pirrung et al., 1998). As might be expected, in situ synthesis of ASOs is also more amenable to the large-scale synthesis of highly consistent genotyping chips (Beecher et al., 1997).

Once generated, the purpose of the genotyping chip is to detect which SNPs are present in the individuals under study. Normally this is achieved by labelling the amplified product (derived from the locus or loci under examination) and hybridizing this to the chip. Following hybridization, it is then necessary to remove not only the non-hybridizing probe, but also any probe which has hybridized to form a heteroduplex. Unfortunately, although numerous computer programs have been designed to predict the hybridization kinetics of ASOs, these are rarely accurate under field conditions. While the hybridization properties of a single or even a few ASOs can be determined empirically, it is virtually impossible to design ASOs for more than a handful of SNPs that allow only homoduplex formation and exclude heteroduplex formation. Currently, this practical problem threatens the future of all but simple genotyping chips. Work by my laboratory and others has suggested that the use of tetramethylammonium chloride in the washing steps makes it possible to increase the discrimination between homo- and heteroduplexes, and removes the effect of base composition on the melting profile of the various ASOs (Melchior and von Hippel, 1973). To date, my laboratory has used this procedure with 40 ASOs covering four maize loci with encouraging results. In the future we intend to try to increase the complexity of our genotyping chip to include ~350 ASOs from ~50 loci (Mogg et al., 1999). However, it is clear that without a solution to this problem we cannot be optimistic about the outlook for genotyping chips for DNA fingerprinting.

Matrix-assisted laser desorption/ionization time of flight mass spectrometry (MALDI-TOF MS)

MALDI-TOF MS is a powerful tool for the rapid analysis of biomolecules based on the intrinsic mass-to-charge ratio of their ions (Hillenkamp et al., 1991). MALDI-TOF MS-based assays are only a recent development for the analysis of SNPs (Ross et al., 1998). There are numerous modifications to the basic technique; however, for the purpose of this study, only the 'Invader™ technology' (Griffin et al., 1999) will be discussed in detail.

When used for the analysis of SNP-derived ASOs, MALDI-TOF MS is not affected by DNA secondary structure which is problematical when using genotyping chips. However, in common with the other technologies discussed, current MALDI-TOF MS procedures rely on the faithful hybridization of amplification DNA to the appropriate ASO (Ross *et al.*, 1998). Following hybridization, the bound ASO can then be eluted and analysed by MALDI-TOF MS. In the Invader™ protocol (Griffin *et al.*, 1999; Mein *et al.*, 2000), two sequence-specific oligonucleotides (one called the Invader oligonucleotide, and the other called the probe oligonucleotide) are allowed to hybridize to the DNA target (i.e. a DNA fragment, usually prepared via PCR, representing the allele under investigation). The two oligonucleotides are designed so that the 3′ end of the Invader oligonucleotide 'invades' one or more nucleotides into the downstream duplex which is formed by the probe oligonucleotide and the DNA target. This forms a sequence overlap at that position. The assay itself makes use of the ability of the 5′-nuclease domains of Eubacteria PolA DNA polymerase and homologous DNA repair proteins called flap endonucleases to recognize and cleave the unpaired regions on the 5′ end of the probe oligonucleotide, producing a 3′-hydroxyl terminating DNA cleavage product. Any sequence variation (for instance the presence of a SNP) between the probe, the invader and the target would result in no hybridization and therefore no cleavage (Lyamichev *et al.*, 1999). The use of thermostable variants of the flap endonuclease permits the reaction to be run at or near the melting temperature of the duplex between the probe and target DNA. In this case cleaved and uncleaved probe oligonucleotides will cycle off and on the target strand. Therefore, when excess probe oligonucleotide is present, cleaved products will be replaced by uncleaved probe which in turn will be cleaved, and so on. Numerous cycles of cleavage, replacement, cleavage, result in a linear accumulation of cleavage products. An improvement in the invader assay, termed the Invader Squared Assay, takes the cleaved product for the standard invader assay and uses it as an Invader oligonucleotide in a secondary invasive cleavage reaction. As Griffin *et al.* (1999) state 'the use of two sequential stages of cleavage reaction approximately squares the amount of amplification of cleavage products compared with a single-stage Invader reaction'. To analyse the results of the Invader™ assay, the products (or non-products if there is insufficient hybridization between probe/Invader and target DNA) are designed to produce biotinylated oligonucleotide molecules suitable for MALDI-TOF MS analysis. Given that the Invader technology is: (i) highly amenable to automation; (ii) open to multiplexing (Mein *et al.*, 2000); and (iii) rapid (in so much as samples can be processed via MALDI-TOF MS every few seconds), then it becomes clear that Invader might become the method of choice for genotyping plant samples. Finally, although current MALDI-TOF MS-based protocols rely on PCR-generated target DNA, it is possible that when used in combination with both the Invader and Invader square assay, nanogram quantities of genomic DNA, without pre-amplification, could be used.

Minisequencing

As for MALDI-TOF MS, numerous methods exist for assaying SNPs via minisequencing (Pastinen *et al.*, 1997; Syvanen, 1999). Essentially all minisequencing protocols rely on the extension of an oligonucleotide primer, which is designed to anneal immediately adjacent to the site of the SNP. Primer extension is then monitored via the inclusion of tagged nucleotides (Nikiforov *et al.*, 1994). The main differences between the various protocols are in the method of detecting the tagged nucleotides; for instance, Syvanen *et al.* (1990) reported the use of a fully automated solid-phase assay using microtitre plates. However, Piggee *et al.* (1997) used capillary-based electrophoresis and fluorescently tagged nucleotides to analyse the extended minisequencing primers. In these and other cases, the protocols have been further modified to allow the multiplexing of one or both of the minisequencing reactions and the detection procedure (Pastinen *et al.*, 1997).

A significant advantage of minisequencing is that in most cases it uses standard laboratory equipment to carry out the procedure and analyse the results. Today, most genotyping laboratories have access to large-scale sequencing facilities; therefore, it is possible to analyse several thousand minisequence reactions per day with the minimum of human labour.

Comparison of SNPs with other marker systems

When considering which marker system to use for DNA fingerprinting, a number of factors are usually considered:

1. What information is already available for the species under consideration? If there is very little information available and resources are limited, then either RAPDs or AFLPs could be the methods of choice, certainly SNPs should not be considered as a viable alternative.

2. What type of information is required? For instance, if single locus, co-dominant, multi-allelic information is required, microsatellites or even RFLPs might be considered. However, if biallelic information is sufficient and there is scope and funds for a high degree of automation, then using SNPs represents a viable alternative to either microsatellites or RFLPs.

3. How many genotypes will be fingerprinted and for how long will the work continue? Current plant-based genotyping assays are often required to genotype tens of thousands of samples over tens of loci within a single breeding season. However, plant breeding is largely a numbers game. Therefore, technology which increases either (or both) the number of samples analysed or the number of loci used could provide the user with a significant advantage. Through their ability to be automated, once generated SNPs could provide such an advantage. Right through from the DNA preparation stage to the scoring of the results, SNPs are highly suitable for automation. This is the great strength of SNPs when compared with other marker systems.

Conclusions

Although relatively new in their concept, SNPs are well on their way to becoming the dominant marker system in commercial plant breeding. Given the large resources required to develop and utilize SNPs, it is probable that their use in academic laboratories will require more time (perhaps 15–20 years). However, given the significant practical advantages, and the ability to examine polymorphisms within ESTs' underlying traits, there is no doubt that SNPs will become the method of choice for DNA fingerprinting.

Acknowledgements

IACR-Long Ashton receives grant-aided support from the Biotechnology and Biological Sciences Research Council of the United Kingdom. Part of the work described in this report was funded by a BBSRC-GAIT award to K.J.E.

References

Ala-Kokko, L., Baldwin, C.T., Moskowitz, R.W. and Prockop, D.J. (1990) Single base mutation in the type II procollagen gene (*COL2A1*) as a cause of primary osteoarthritis associated with a mild chondrodysplasia. *Proceedings of the National Academy of Sciences USA* 87, 6565–6568.

Beecher, J.E., McGall, G.H. and Goldberg, M.J. (1997) Chemically amplified photolithography for the fabrication of high density oligonucleotide arrays. *Polymeric Materials Sciences and Engineering* 76, 597–598.

Brookes, A.J. (1999) The essence of SNPs. *Gene* 234, 177–186.

Brown, T.A. (1999) *Genomics*, 1st edn. BIOS Scientific Publishers, Oxford.

Bryan, G.J., Stephenson, P., Collins, A., Kirby, J., Smith, J.B. and Gale, M.D. (1999) Low levels of DNA sequence variation among adapted genotypes of hexaploid wheat. *Theoretical and Applied Genetics* 99, 192–198.

Coryell, V.H., Jessen, H., Schupp, J.M., Webb, D. and Keim, P. (1999) Allele specific hybridization markers for soybean. *Theoretical and Applied Genetics* 98, 690–696.

Erlich, H.A. (1989) *PCR Technology*, 1st edn. Macmillan Publishers, Basingstoke, UK.

Etscheid, M. and Riesner, D. (1998) TGGE and DGGE. In: Karp, A., Isaac, P.G. and Ingram, D.S. (eds) *Molecular Tools for Screening Biodiversity*. Chapman & Hall, London, pp. 133–156.

Fodor, S.P.A. (1991) Light-directed, spatially addressable parallel chemical synthesis. *Science* 251, 767–773.

Fodor, S.P.A. (1997) Massively parallel genomics. *Science* 277, 393–395.

Germano, J. and Klein, A.S. (1999) Species-specific nuclear and chloroplast single nucleotide polymorphisms to distinguish *Picea glauca*, *P-mariana* and *P-rubens*. *Theoretical and Applied Genetics* 99, 37–49.

Gonen, D., VeenstraVanderWeele, J., Yang, Z., Leventhal, B.L. and Cook, E.H. (1999) High throughput flourescent CE-SSCP SNP genotyping. *Molecular Psychiatry* 4, 339–343.

Griffin, T.J., Hall, J.G., Prudent, J.R. and Smith, L.M. (1999) Direct genetic analysis by matrix-assisted laser desorption/ionization mass spectrometry. *Proceedings of the National Academy of Sciences USA* 96, 6301–6306.

Guo, Z., Guilfoyle, R.A., Thiel, A.J., Wang, R. and Smith, L.M. (1994) Direct fluorescence analysis of genetic polymorphisms by hybridization with oligonucleotide arrays on glass support. *Nucleic Acids Research* 22, 5456–5465.

Helentjaris, T., Slocum, M., Wright, S., Schaefer, A. and Nienhuis, J. (1986) Construction of genetic linkage maps in maize and tomato using restriction fragment length polymorphism. *Theoretical and Applied Genetics* 72, 761–769.

Heyen, D.W., Beever, J.E., Da, Y., Evert, R.E., Green, C., Bates, S.R.E., Ziegle, J.S. and Lewin, H.A. (1997) Exclusion probabilities of 22 bovine microsatellite markers in fluorescent multiplexes for semiautomated parentage testing. *Animal Genetics* 28, 21–27.

Hillenkamp, F., Karas, M., Beavis, R. and Chait, B. (1991) Matrix-assisted laser desorption-ionization mass spectrometry of biopolymers. *Analytical Chemistry* 63, 1193–1203.

Jordan, W.C., Foley, K. and Bruford, M.W. (1998) Single-strand conformation polymorphism (SSCP) analysis. In: Karp, A., Isaac, P.G. and Ingram, D.S. (eds) *Molecular Tools for Screening Biodiversity*. Chapman & Hall, London, pp. 152–156.

Kruglyak, L. (1997) The use of a genetic map of biallelic markers in linkage studies. *Nature Genetics* 17, 21–24.

Li, W. and Sadler, L.A. (1991) Low nucleotide diversity in man. *Genetics* 129, 513–523.

Lyamichev, V., Mast, A.L., Hall, J.G., Prudent, J.R., Kaiser, M.W., Takova, T., Kwiatowski, R.W., Sander, T.J., de Arruda, M. and Arco, D.A. (1999) Polymorphism identification and quantitative detection of genomic DNA by invasive cleavage of oligonucleotide probes. *Nature Biotechnology* 17, 292–296.

Mein, C.A., Barratt, B.J., Dunn, M.G., Siegmund, T., Smith, A.N., Esposito, L., Nutland, S., Stevens, H.E., Wilson, A.J., Phillips, M.S., Jarvis, N., Law, S., de Arruda, M. and Todd, J.A. (2000) Evaluation of single nucleotide polymorphism typing with Invader on PCR amplicons and its automation. *Genome Research* 10, 330–343.

Melchior, W.B. and von Hippel, P.H. (1973) Alteration of the relative stability of dA·dT and dG·dC base pairs in DNA. *Proceedings of the National Academy of Sciences USA* 70, 298–302.

Mogg, R., Hanley, S. and Edwards, K.J. (1999) Generation of maize allele specific oligonucleotides from the flanking regions of microsatellite markers. *Plant and Animal Genome VII Conference Abstract Guide*, Scherago International Inc., P491.

Nikiforov, T.T., Rendle, R.B., Goelet, P., Rogers, Y.H., Kotewicz, M.L., Anderson, S., Trainor, G.L. and Knapp, M.R. (1994) Genetic bit analysis: a solid phase method for typing single nucleotide polymorphisms. *Nucleic Acids Research* 22, 4167–4175.

Pastinen, T., Kurg, A., Metspalu, A., Peltonen, L. and Syvanen, A.C. (1997) Minisequencing: a specific tool for DNA analysis and diagnostics on oligonucleotide arrays. *Genome Research* 7, 606–614.

Piggee, C.A., Muth, J., Carrilho, E. and Karger, B.L. (1997) Capillary electrophoresis for the detection of known point mutations by single-nucleotide primer extension and laser induced fluorescence detection. *Journal of Chromatography* 781, 367–375.

Pirrung, M.C., Fallon, L. and McGall, G. (1998) Proofing of photolithographic DNA synthesis with 3′5′-dimethoxybenzoinyloxycarbonyl-protected deoxynucleoside phosphoramidites. *Journal of Organic Chemistry* 63, 241–246.

Prosser, J. (1993) Detecting single-base mutations. *Trends in Biochemistry* 11, 238–246.

Ramsay, G. (1997) DNA chips: state-of-the art. *Nature Biotechnology* 16, 40–45.

Ross, P., Hall, L., Smirnov, I. and Haff, L. (1998) High level multiplex genotyping by MALDI-TOF mass spectrometry. *Nature Biotechnology* 16, 1347–1351.

Saiki, R.K., Walsh, P.S., Levenson, C.H. and Erlich, H.A. (1989) Genetic analysis of amplified DNA with immobilized sequence-specific oligonucleotide probes. *Proceedings of the National Academy of Sciences USA* 86, 6230–6234.

Sasaki, T. (1998) The rice genome project in Japan. *Proceedings of the National Academy of Sciences USA* 95, 2027–2028.

Service, R.F. (1998) Microchip arrays put DNA on the spot. *Science* 282, 396–399.

Shen, B., Carneiro, N., Torresjerez, I., Stevenson, B., McCreery, T., Helentjaris, T., Baysdorfer, C., Almira, E., Ferl, R.J., Habben, J.E. and Larkins, B. (1994) Partial sequencing and mapping of clones from 2 maize cDNA libraries. *Plant Molecular Biology* 26, 1085–1101.

Struss, D. and Plieske, J. (1998) The use of microsatellite markers for detection of genetic diversity in barley populations. *Theoretical and Applied Genetics* 97, 308–315.

Syvanen, A.C. (1999) From gels to chips: 'minisequencing' primer extension for analysis of point mutations and single nucleotide polymorphisms. *Human Mutations* 13, 1–10.

Syvanen, A.C., Aalto-Setala, K., Harju, I., Kontula, K. and Soderlund, H. (1990) A primer-guided nucleotide incorporation assay in the genotyping of apolipoprotcin E. *Genomics* 8, 684–692.

Vos, P., Hogers, R., Bleeker, M., Rijans, M., Van de Lee, T., Hornes, M., Frijters, A., Pots, J., Peleman, J., Kuiper, M. and Zabeau, M. (1995) AFLP: a new technique for DNA fingerprinting. *Nucleic Acids Research* 23, 4407–4414.

Welsh, J. and McClelland, M. (1990) Fingerprinting genomes using PCR with arbitrary primers. *Nucleic Acids Research* 18, 7213–7218.

Xiong, M. and Jin, L. (1999) Comparison of the power and accuracy of biallelic and microsatellite markers in population-based gene-mapping methods. *American Journal of Human Genetics* 64, 629–640.

Chapter 2
Plant Genotyping by Analysis of Microsatellites

T.A. HOLTON

*Centre for Plant Conservation Genetics,
Southern Cross University, Lismore, Australia*

Introduction

What are microsatellites?

Microsatellites are simple sequence repeats (SSRs) of 1–6 nucleotides. They appear to be ubiquitous in higher organisms, although the frequency of microsatellites varies between species. They are abundant, dispersed throughout the genome and show higher levels of polymorphism than other genetic markers. These features, coupled with their ease of detection, have made them useful molecular markers. Their potential for automation and their inheritance in a co-dominant manner are additional advantages when compared with other types of molecular markers. SSRs have recently become important genetic markers in cereals, including wheat and barley.

Isolation of SSRs

Isolation of useful SSR loci can be a time-consuming and expensive process. Many approaches have been used to isolate SSRs and their flanking sequences. Genomic clones containing SSRs can be isolated by screening with labelled oligonucleotides containing the desired repeat sequences. SSRs may be obtained by screening sequences in databases or by screening libraries of clones. Standard methods for the isolation of SSRs from clones (Powell *et al.*, 1996a) involve:

- the creation of a small insert genomic library;
- library screening by hybridization;
- DNA sequencing of positive clones;

- primer design and PCR analysis; and
- identification of polymorphisms.

To improve the efficiency of SSR isolation, SSR-enriched libraries have been developed using a variety of methods with selection either before or after library construction (see Chapter 13).

In species where many gene sequences exist due to expressed sequence tag (EST) or genomic sequencing efforts, it is possible to identify SSRs by examination of DNA sequence databases. Identification of $(AT)_n$ and $(GC)_n$ microsatellites from sequence databases is also possible. The isolation of such SSRs by hybridization screening with labelled probes or enrichment is problematic due to the palindromic nature of these sequences. An additional advantage of using SSRs present in ESTs is that genes of known function can be mapped.

Methods of detection

Agarose, polyacrylamide gel electrophoresis (PAGE), denaturing PAGE and capillary electrophoresis have been used to determine the size of SSR amplification products and identify size polymorphisms. Ethidium bromide staining is commonly used to detect PCR products in non-denaturing gels. However, accurate sizing is difficult with agarose and polyacrylamide and these gel matrices do not allow single base pair resolution. The resolution of MetaPhor agarose gels (2% MetaPhor and 1% ultra Pure agarose) makes it possible to distinguish allelic differences as small as 3 bp (Becker and Heun, 1995).

Single nucleotide resolution of DNA fragments requires the use of denaturing PAGE or capillary electrophoresis. SSR amplification products are detected in PAGE by silver staining, radiolabelling (primer or PCR product) or fluorescence labelling (primer or PCR product). The use of fluorescently labelled primers, combined with an automated electrophoretic system, greatly simplifies the analysis of SSR allele sizes. The use of a fluorescent primer enables PCR products to be detected without any post-electrophoresis treatments. Sample throughput can also be increased through multiplexing by the use of different fluorescent tags on different PCR products. PCR amplification can be performed separately for each SSR locus and then the products run in the same lane. The use of an internal size standard in each lane also allows for accurate automated sizing of bands. Multiplexing of PCR reactions is generally not performed because this has been shown to produce poor results (Donini *et al.*, 1998).

Allele sizing artefacts

Stutter peaks

PCR amplification of dinucleotide SSRs often leads to the production of numerous bands referred to as stutter peaks (see Fig. 2.1). These bands, which differ

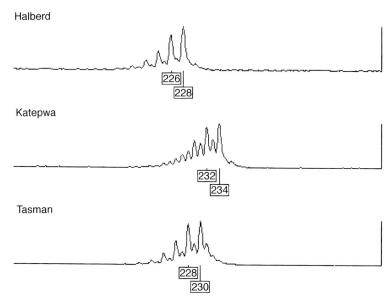

Fig. 2.1. Artefacts due to 'stutter' and non-template A-addition to PCR amplification products. Electropherogram of *wmc112* amplification products generated by the Applied Biosystems 373A automated DNA sequencer. Amplification of *wmc112* from the wheat varieties 'Halberd', 'Katepwa' and 'Tasman' show major stutter peaks, which differ by 2 bp increments (e.g. 226 and 228). Additional peaks between the stutter peaks are due to non-template addition of a single A nucleotide to the 3' end of the PCR products by *Taq* DNA polymerase.

in size by 2 bp increments, are caused by slippage of the DNA polymerase. The presence of such peaks can sometimes lead to difficulties in identifying the true allele size and can make it difficult to discriminate between two alleles differing by 2 bp. If the two alleles are sufficiently different in size, stutter peaks do not cause a problem for genetic mapping. The use of SSRs containing trinucleotide or higher order repeats eliminates this problem. However, such markers may be less polymorphic than dinucleotide SSRs (Bryan *et al.*, 1997).

Non-template A-addition

Thermostable DNA polymerases can catalyse non-templated addition of an extra A to the 3' end of PCR products (Clark, 1988; Smith *et al.*, 1995). The extent of A-addition is dependent on primer sequences and PCR conditions (see Fig. 2.1). This represents a potential source of error in genotyping studies using PCR amplification of SSR loci. Modification of PCR protocols (Smith *et al.*, 1995) or redesign of PCR primers via 'PIG-tailing' (Brownstein *et al.*, 1996) can increase the proportion of products with an extra A to simplify the banding pattern. Post-PCR treatment with T4 DNA polymerase will remove the non-template added A; however, the additional sample manipulation required makes this approach less amenable to high throughput genotyping.

Gel system

The type of gel system used for SSR analysis can lead to differences in measured sizes of PCR amplification products. Williams *et al.* (1999) compared the sizing of CAG repeats using radioactive and fluorescent PCR amplification, and the subsequent separation of these products by slab gel and capillary electrophoresis. The assays were performed on both cloned and sequenced CAG repeats, as well as genomic DNA from Huntington's disease patients with a wide range of repeat lengths. The mobility of the CAG repeat amplification products was greater using capillary electrophoresis compared with slab gel electrophoresis.

Physical mapping of SSRs

For genome analysis, bread wheat (*Triticum aestivum*) has a disadvantage compared with other cereals such as barley, due to its triple number of chromosomes. However, the allohexaploid status of the genome and the ability of homeologous chromosomes to compensate each other provided the opportunity to develop aneuploid genetic tester stocks such as monosomic, nulli-tetrasomic or ditelosomic lines. Nullisomic and ditelosomic lines are widely used to determine the chromosomal and chromosome arm assignment of genes and molecular markers. Plaschke *et al.* (1996) describe the chromosomal assignment of 64 PCR-amplified SSR loci, and 29 additional fragments amplified by the same primer pairs are described for bread wheat (*T. aestivum*). Figure 2.2 shows an example of the use of nulli-tetrasomic lines to map the physical location of the SSR marker *wmc110* in hexaploid wheat. Röder *et al.* (1998a) used 25 homozygous group 2 chromosome deletion stocks of 'Chinese Spring' to physically map 31 wheat SSR markers to deletion intervals.

Detection of polymorphisms

Variations in the length of tandem repeats can be identified by amplification of the region containing the repeat via PCR using primers designed to the regions flanking individual SSRs. Polymorphisms are detected based on size differences (see Fig. 2.3), which result from differences in the number of repeats. SSR loci are believed to evolve in a step-wise manner by the addition or subtraction of a single repeat.

PCR primer sequences flanking microsatellite repeats are generally designed to produce primers 17–22 nucleotides long, with a GC content of approximately 50% and a T_m about 60°C (McCouch *et al.*, 1997). The ideal size for analysis of SSR amplification products is approximately 100–250 bp.

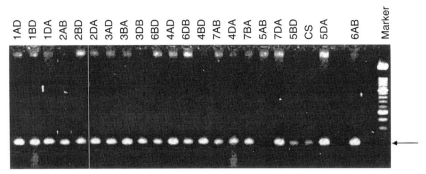

Fig. 2.2. Physical mapping of SSRs. Nulli-tetrasomic lines were used to determine the physical location of the microsatellite *wmc110* in the hexaploid wheat genome. PCR amplification products were detected in 'Chinese Spring' (CS) and all nulli-tetrasomic lines except 5AB, indicating that *wmc110* was located on chromosome 5A. The position of the amplified products is shown by the arrow.

Advantages and disadvantages over other markers

SSRs are easier to use than restriction fragment length polymorphisms (RFLPs) owing to the smaller amount of DNA required, higher polymorphism and the ability to automate assays. SSR markers can easily be exchanged between researchers because each locus is defined by the primer sequences. SSR assays are more robust than random amplified polymorphic DNA (RAPDs) and more transferable than amplified fragment length polymorphisms (AFLPs). SSRs are now replacing RFLPs in genetic mapping of crop plants. A combination of SSRs with AFLPs is used to produce detailed genetic maps. The co-dominant nature of SSRs is also an advantage for genetic mapping.

Powell *et al.* (1996b) examined the ultility of RFLP, RAPD, AFLP and SSR markers for soybean germplasm analysis by evaluating information content (expected heterozygosity), number of loci simultaneously analysed per experiment (multiplex ratio) and effectiveness in assessing relationships between accessions. SSR markers have the highest expected heterozygosity (0.60), while AFLP markers have the highest effective multiplex ratio.

Use of SSRs in genetic mapping

Their ease of use and high information content has ensured that SSRs have largely replaced RFLPs as a mapping technology in humans (Dib *et al.*, 1996). The development of SSRs in plants is accelerating, and SSR loci are now being incorporated into established genetic maps of all the major cereals (Liu *et al.*, 1996; Korzun *et al.*, 1997; Smith *et al.*, 1997; Stephenson *et al.*, 1998). The dinucleotide repeats $(AC)_n$ and $(AG)_n$ have been most commonly used for construction of genetic maps in crop plants.

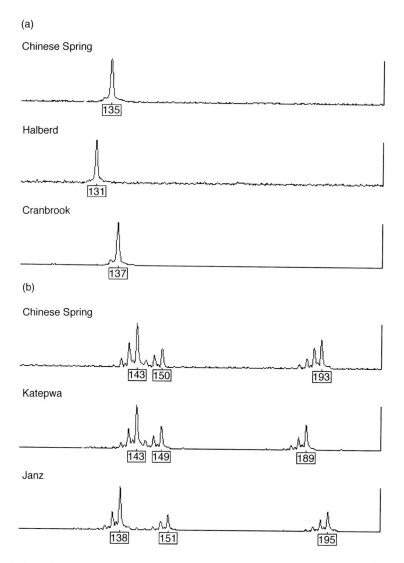

Fig. 2.3. Screening of wheat varieties for polymorphisms. Wheat varieties were screened for polymorphisms using: (a) *wmc047* and (b) *wmc048*. The *wmc047* primers amplify a single locus (4D), whereas the *wmc048* primers amplify three loci (4A, 4B and 4D).

The construction of high-resolution linkage maps based exclusively on SSRs may be an unrealistic goal for most crop plants due to the great effort and cost involved in isolation of markers from each species. However, the creation of 'skeletal' genetic maps with SSRs is an achievable goal, where SSRs can provide anchor points for specific regions of the genome. Gaps between the SSR markers may be 'filled in' with other markers such as AFLPs.

In contrast to RFLP markers, most wheat SSR markers are genome-specific and amplify only one specific locus containing an SSR in the A, B or D genome of bread wheat (Röder et al., 1998b). Cytogenetic data suggest that in some cases SSRs are not randomly distributed in plant genomes. For sugar beet, this is true for the dinucleotide repeat CA, which, besides the GA repeat, is the repeat most frequently used for SSR analysis. Such distribution precludes genome-wide coverage and may reduce the value of SSRs as genetic markers.

High throughput genotyping

There is a recognized need to distinguish varieties of crop plants reliably and to establish their purity. This need is due to the introduction of plant variety rights, seed certification, and associated requirements to protect proprietary germplasm. Morphological characters have traditionally been used to evaluate distinctiveness, uniformity and stability and to establish a description of genotype. Such methods are labour-intensive, subjective and are often influenced by environmental conditions. Therefore, a great deal of attention has been paid to the development of laboratory-based methods for variety identification. SSRs have been shown to provide a powerful methodology for discrimination between genotypes (Yang et al., 1994; Russell et al., 1997).

Isolation of SSRs from barley and wheat

Barley

Liu et al. (1996) isolated SSRs from a barley genomic library and from public databases. They estimated that the barley genome contains one GA repeat every 330 kb and one CA repeat every 620 kb. A total of 45 SSRs were identified and mapped to seven barley chromosomes using doubled-haploid lines and/or wheat–barley addition-line assays. Struss and Plieske (1998) screened a barley genomic library with GA and GT probes for developing SSR markers. The frequency of dinucleotide repeats in the barley genome was estimated among 39,000 plaques from a lambda genomic library. A total of 482 clones (1.2%) hybridized with GA and 285 clones (0.7%) hybridized with GT. Ramsay et al. (1999) characterized 290 dinucleotide repeat-containing clones from SSR-enriched libraries and revealed that a high percentage is associated with cereal retrotransposon-like and other dispersed repetitive elements.

Wheat

It has proved to be more difficult to identify polymorphisms in wheat than in many species, but recently there has been a rapid increase in identification of polymorphic SSR markers (Röder et al., 1998b; Stephenson et al., 1998; the wheat microsatellite consortium (WMC) coordinated by Agrogene, http://www.microsatellites.agrogene.com/).

Röder *et al.* (1995) screened a wheat genomic library with an average insert size of ~20 kb with synthetic GT and GA oligomers. Among 18,000 plaques, 511 positive clones were found for GT and among 27,000 plaques, 1228 clones hybridized with GA. Based on these results, the wheat genome contains on average one GT block every 704 kb and one GA block every 440 kb. The average distance between any of the two dinucleotide blocks is approximately 271 kb. With a genome size of 16×10^6 kb per haploid genome for wheat, the total number of microsatellite loci is 3.6×10^4 for GA and 2.3×10^4 for GT. For large-scale production of SSR markers, a phage library with an average insert size of 1 kb was constructed. A total of 70 SSRs were isolated after screening with GT and GA and sequencing.

Ma *et al.* (1996) found on average that there was one AC SSR every 292 kb and one AG SSR every 212 kb. Trinucleotide repeats were about ten times less common than the two dinucleotide tandem repeats tested, and tetranucleotide repeats were rare. Among 32 pairs of SSR primers tested on 18 genotypes, seven produced polymorphic products in the expected size range and these loci were mapped using a hexaploid wheat mapping population or aneuploid stocks.

Bryan *et al.* (1997) probed M13 libraries with six different SSR sequences: $(CA)_{16}$, $(GA)_{16}$, $(CAA)_{10}$, $(GAA)_{10}$, $(ACG)_{10}$ and $(CAG)_{10}$. Two thousand positive plaques were identified from 700,000 clones. From the positive clones, 222 clones containing SSRs were isolated from the wheat genome and 153 primer pairs were tested for genetic polymorphism using a panel of ten wheat varieties. A high proportion (approximately two-thirds) of primer pairs designed to detect SSR variation in wheat did not generate the expected amplification products and often generated unresolvable PCR products. Only 49 of the 153 clones amplified a PCR product of the predicted size from the variety 'Chinese Spring'. Stephenson *et al.* (1998) mapped 51 SSR loci, using these markers, in a 'Chinese Spring' $\times$ SQ1 wheat population.

Röder *et al.* (1998b) sequenced 1380 clones, and primer pairs were designed for 720 SSR clones. A total of 294 primer pairs (41%) yielded a discrete fragment of the expected fragment size. Eighty per cent of the primer pairs amplifying a fragment of the expected size detected polymorphism between Opata 85 and the synthetic wheat W7984. A total of 279 loci amplified by 230 primer sets were placed on to a genetic framework map composed of RFLPs previously mapped in an Opata 85 $\times$ W7984 cross. Only 37 of 230 primer sets produced more than one mappable locus. The majority of the 193 SSR markers constitute genome-specific markers. Of the 279 SSRs, 93 mapped to the A genome, 115 to the B genome and 71 to the D genome.

Variety screening

Many studies have been undertaken to examine the level of polymorphism of SSR markers in a large number of barley varieties and *Hordeum* species (Table

Table 2.1. Screening of barley varieties for SSR polymorphisms.

Publication	No. of accessions	SSRs screened	No. of alleles
Saghai Maroof et al., 1994	207	4	71
Struss and Plieske, 1998	163	15	130
William et al., 1997	11	38	40
Russell et al., 1997	24	11	48
Donini et al., 1998	135	21	106
Becker and Heun, 1995	11	15	

Table 2.2. Screening of wheat varieties for SSR polymorphisms.

Publication	No. of accessions	SSRs screened	No. of alleles
Röder et al., 1995	18	15	69
Plaschke et al., 1995	40	23	142
Bryan et al., 1997	10	49	169
Bohn et al., 1999	11	13	33

2.1). Levels of polymorphism are high; up to 37 different alleles for the one SSR locus were found in barley (Saghai Maroof et al., 1994). A number of similar studies have also been undertaken in wheat, which shows a lower level of polymorphism than barley (Table 2.2).

DNA fingerprinting and variety identification

SSRs are particularly attractive for distinguishing between cultivars because the level of polymorphism detected at SSR loci is higher than that detected with any other molecular marker assay (Saghai Maroof et al., 1994; Powell et al., 1996b). A study by Olufowote et al. (1997) showed that a selection of six SSR markers was sufficient to discriminate between 71 related lines of rice.

Polymorphic RAPD and SSR markers were used for distinguishing commercial barley cultivars and for comparison of differentiation capability of RAPD and SSR techniques (Kraic et al., 1998). All 23 cultivars were distinguished from each other by SSR markers. The differentiation effectiveness of SSRs was more than seven times higher in comparison with RAPDs.

Transferability between species

The ability to use the same SSR primers in different plant species depends on the extent of sequence conservation in the primer sites flanking the SSRs and the stability of the SSR during evolution. Transferability of SSR primers across species would obviously increase the value of such markers. However, wheat SSR primer sets generally do not amplify SSRs in barley (Roder et al., 1995), suggesting the need to develop separate SSR primer sets for each species. In contrast, sequence-tagged sites (STS) derived from the conversion of RFLP markers are readily transferable between wheat and barley (Erpelding *et al.*, 1996).

Data storage/analysis

In order to make the greatest use of SSR markers and other molecular markers it is important that data are stored in a readily accessible form. Publicly available data on cereal molecular markers are accessible from the GrainGenes web site (http://wheat.pw.usda.gov/). We have also developed dedicated databases to handle SSR information (see Chapter 19) for use in the Australian wheat (http://cpcg.scu.edu.au/genetics/) and barley (http://cpcg.scu.edu.au/barley_microsat/) breeding programmes.

The future

EST and genomic sequencing

The increasing availability of EST and genomic sequences from wheat and barley is providing a potentially valuable source of new SSR markers. Using *in silico* discovery of SSRs, it is also possible to identify polymorphic loci directly when sequences of the same gene are available from more than one variety. SSR markers derived from expressed gene sequences will provide 'perfect' markers for those genes. This may allow direct identification of alleles which cause important phenotypic traits. Different repeat sequences, such as AT and GC, may also be isolated. Many trinucleotide and higher repeats are also present within ESTs and these markers could be useful additions to the predominantly dinucleotide repeats that have been used so far in plant genotyping. These markers may be more transferable between species than dinucleotide repeats, and assays may be more easily automated owing to the absence of artefacts such as 'stutter'. Mapping of SSR markers in genes will allow more detailed comparative mapping between related species and allow correlation between RFLP, STS and SSR markers and gene locations.

ITEC (International Triticeae EST Cooperative)

In 1998, at the 9th International Wheat Genetics Symposium, a proposal was developed to establish a public database of ESTs from species of the *Triticeae*. Phase One of this project was to have at least 40,000 ESTs available publicly from 1 July 2000. Phase Two of this project plans to have 300,000 ESTs sequenced for wheat and 300,000 ESTs sequenced for barley. Sequences will be available at http://wheat.pw.usda.gov/genome/. Initial screening of the ITEC wheat and barley EST sequences for SSRs by our lab has identified a large number of potential SSR markers, which are currently being screened for polymorphisms.

Rice genome project

The rice genome project is a consortium of researchers that aims to sequence the rice genome completely by 2004. However, the recent release of a 'working draft' of the rice genome sequence by Monsanto should greatly accelerate the completion of a fully annotated version of the rice genome sequence. The availability of the complete sequence of a cereal genome should facilitate construction of genetic maps in other cereals by comparative mapping and increased opportunities to develop transferable markers.

Automation of assays

Large-scale genetic mapping and genetic fingerprinting of plants using SSR markers requires efficient methods for setting up PCR, gel electrophoresis and allele calling. There is a need to increase automation of these procedures through the use of robotics and the adoption of high throughput capillary electrophoresis systems with fluorescence detection. To increase throughput and decrease costs, multiplexing of reactions is necessary. Multiplexing can be performed using primers labelled with different fluorophors or based on size. Multiplexing of electrophoresis of SSRs up to 20-fold has been achieved with analysis of human SSRs. The recent development of new 96-capillary automated sequencers and increasing numbers of fluorescent dyes should enable even greater automation and increased levels of multiplexing.

References

Becker, J. and Heun, M. (1995) Barley microsatellites: allele variation and mapping. *Plant Molecular Biology* 27, 835–45.
Bohn, M., Utz, H.F. and Melchinger, A.E. (1999) Genetic similarities among winter wheat cultivars determined on the basis of RFLPs, AFLPs, and SSRs and their use for predicting progeny variance. *Crop Science* 39, 228–237.
Brownstein, M.J., Carpten, J.D. and Smith, J.R. (1996) Modulation of non-templated nucleotide addition by *Taq* DNA polymerase – primer modifications that facilitate genotyping. *Biotechniques* 20, 1004–1010.
Bryan, G.J., Collins, A.J., Stephenson, P., Orry, A., Smith, J.B. and Gale, M.D. (1997) Isolation and characterisation of microsatellites from hexaploid bread wheat. *Theoretical and Applied Genetics* 94, 557–563.
Clark, J.M. (1988) Novel non-templated nucleotide addition reactions catalyzed by procaryotic and eucaryotic DNA polymerases. *Nucleic Acids Research* 16, 9677–9686.

Dib, C., Faure, S., Fizames, C., Samson, D., Drouot, N., Vignal, A., Millasseau, P., Marc, S., Hazan, J., Seboun, E., Lathrop, M., Gyapay, G., Morissette, J. and Weissenbach, J. (1996) A comprehensive genetic map of the human genome based on 5,264 microsatellites. *Nature* 380, 152–154.

Donini, P., Stephenson, P., Bryan, G.J. and Koebner, R.M.D. (1998) The potential of microsatellites for high throughput genetic diversity assessment in wheat and barley. *Genetic Resources and Crop Evolution* 45, 415–421.

Erpelding, J.E., Blake, N.K., Blake, T.K. and Talbert, L.E. (1996) Transfer of sequence tagged site PCR markers between wheat and barley. *Genome* 39, 802–810.

Korzun, V., Borner, A., Worland, A.J., Law, C.N. and Roder, M.S. (1997) Application of microsatellite markers to distinguish inter-varietal chromosome substitution lines of wheat (*Triticum aestivum* L.). *Euphytica* 95, 149–155.

Kraic, J., Zakova, M. and Gregova, E. (1998) Comparison of differentiation capability of RAPD and SSR markers in commercial barley (*Hordeum vulgare* L.) cultivars. *Cereal Research Communications* 26, 375–382.

Liu, Z.-W., Biyashev, R.M. and Saghai Maroof, M.A. (1996) Development of simple sequence repeat DNA markers and their integration into a barley linkage map. *Theoretical and Applied Genetics* 93, 869–876.

Ma, Z.Q., Roder, M. and Sorrells, M.E. (1996) Frequencies and sequence characteristics of di-, tri-, and tetra-nucleotide microsatellites in wheat. *Genome* 39, 123–30.

McCouch, S.R., Chen, X., Panaud, O., Temnykh, S., Xu, Y., Cho, Y.G., Huang, N., Ishii, T. and Blair, M. (1997) Microsatellite marker development, mapping and applications in rice genetics and breeding. *Plant Molecular Biology* 35, 89–99.

Olufowote, J.O., Xu, Y.B., Chen, X.L., Park, W.D., Beachell, H.M., Dilday, R.H., Goto, M. and McCouch, S.R. (1997) Comparative evaluation of within-cultivar variation of rice (*Oryza sativa* L.) using microsatellite and RFLP markers. *Genome* 40, 370–378.

Plaschke, J., Ganal, M.W. and Roder, M.S. (1995) Detection of genetic diversity in closely related bread wheat using microsatellite markers. *Theoretical and Applied Genetics* 91, 1001–1007.

Plaschke, J., Borner, A., Wendehake, K., Ganal, M.W. and Roder, M.S. (1996) The use of wheat aneuploids for the chromosomal assignment of microsatellite loci. *Euphytica* 89, 33–40.

Powell, W., Machray, G.C. and Provan, J. (1996a) Polymorphism revealed by simple sequence repeats. *Trends in Plant Science* 1, 215–222.

Powell, W., Morgante, M., Andre, C., Hanafey, M., Vogel, J., Tingey, S. and Rafalski, A. (1996b) The comparison of RFLP, RAPD, AFLP and SSR (microsatellite) markers for germplasm analysis. *Molecular Breeding* 2, 225–238.

Ramsay, L., Macaulay, M., Cardle, L., Morgante, M., degli Ivanissevich, S., Maestri, E., Powell, W. and Waugh, R. (1999) Intimate association of microsatellite repeats with retrotransposons and other dispersed repetitive elements in barley. *Plant Journal* 17, 415–425.

Röder, M.S., Plaschke, J., Konig, S.U., Borner, A., Sorrells, M.E., Tanksley, S.D. and Ganal, M.W. (1995) Abundance, variability and chromosomal location of microsatellites in wheat. *Molecular and General Genetics* 246, 327–33.

Röder, M.S., Korzun, V., Gill, B.S. and Ganal, M.W. (1998a) The physical mapping of microsatellite markers in wheat. *Genome* 41, 278–283.

Röder, M.S., Korzun, V., Wendehake, K., Plaschke, J., Tixier, M.-H., Leroy, P. and Ganal, M.W. (1998b) A microsatellite map of wheat. *Genetics* 149, 2007–2023.

Russell, J., Fuller, J., Young, G., Thomas, B., Taramino, G., Macaulay, M., Waugh, R. and

Powell, W. (1997) Discriminating between barley genotypes using microsatellite markers. *Genome* 40, 442–450.

Saghai Maroof, M.A., Biyashev, R.M., Yang, G.P., Zhang, Q. and Allard, R.W. (1994) Extraordinarily polymorphic microsatellite DNA in barley: species diversity, chromosomal locations, and population dynamics. *Proceedings of the National Academy of Sciences USA* 91, 5466–5470.

Smith, J.R., Carpten, J.D., Brownstein, M.J., Ghosh, S., Magnuson, V.L., Gilbert, D.A., Trent, J.M. and Collins, F.S. (1995) Approach to genotyping errors caused by nontemplated nucleotide addition by *Taq* DNA polymerase. *PCR Methods and Applications* 5, 312–317.

Smith, J.S.C., Chin, E.C.L., Shu, H., Smith, O.S., Wall, S.J., Senior, M.L., Mitchell, S.E., Kresovich, S. and Ziegle, J. (1997) An evaluation of the utility of SSR loci as molecular markers in maize (*Zea mays* L.): comparisons with data from RFLPs and pedigree. *Theoretical and Applied Genetics* 95, 163–173.

Stephenson, P., Bryan, G., Kirby, J., Collins, A., Devos, K., Busso, C. and Gale, M. (1998) Fifty new microsatellite loci for the wheat genetic map. *Theoretical and Applied Genetics* 97, 946–949.

Struss, D. and Plieske, J. (1998). The use of microsatellite markers for detection of genetic diversity in barley populations. *Theoretical and Applied Genetics* 97, 308–315.

William, M., Dorocicz, I. and Kasha, K.J. (1997) Use of microsatellite DNA to distinguish malting and nonmalting barley cultivars. *Journal of the American Society of Brewing Chemists* 55, 107–111.

Williams, L.C., Hegde, M.R., Herrera, G., Stapleton, P.M. and Love, D.R. (1999) Comparative semi-automated analysis of (CAG) repeats in the Huntington disease gene: use of internal standards. *Molecular and Cellular Probes* 13, 283–289.

Yang, G.P., Maroof, M.A., Xu, C.G., Zhang, Q. and Biyashev, R.M. (1994) Comparative analysis of microsatellite DNA polymorphism in landraces and cultivars of rice. *Molecular and General Genetics* 245, 187–194.

Chapter 3

Plant Genotyping Using Arbitrarily Amplified DNA

G. CAETANO-ANOLLÉS

Laboratory of Molecular Ecology and Evolution and Division of Molecular Biology, Department of Biology, University of Oslo, Oslo, Norway

Introduction

The survey, management and manipulation of biological resources require tools capable of measuring constitution, change and evolution of genetic material. In recent years, a number of molecular techniques have been recruited for the task. Many of them complement traditional methods for the evaluation of biodiversity, are based on the analysis of information-rich nucleic acid molecules, and provide reliable estimators of relatedness, phylogeny and inheritance of genetic material (Weising *et al.*, 1994; Caetano-Anollés and Gresshoff, 1998). Nucleic acid markers are by far the most powerful and widely used, because they portray genome sequence composition. They constitute distinguishing features obtained by confining analysis of the typically 10^6–10^{10} bp of a genome to selected nucleic acid regions representing only 1–10^4 bp of the overall sequence. Nucleic acid marker techniques generally use hydrogen-bonding interactions between nucleic acid strands ('hybridization') and oligonucleotide-driven enzymatic accumulation of specific nucleic acid sequences ('amplification') to uncover sequence polymorphism and genome variability, and have been divided into four groups according to experimental strategy: (i) hybridization-based analysis (e.g. restriction fragment length polymorphism (RFLP) analysis); (ii) amplification-based nucleic acid scanning (e.g. amplified fragment length polymorphism (AFLP analysis)) (iii) amplification-based nucleic acid profiling (e.g. PCR amplification of microsatellites); and (iv) sequence-targeted techniques (e.g. oligonucleotide arrays) (cf. Caetano-Anollés and Trigiano, 1997). Markers in these four categories have been widely used in plant genome analysis (Rafalski and Tingey, 1993), plant improvement (breeding and genetic engineering; Winter and Kahl, 1995), plant ecological research (Bachmann, 1994), evolutionary biology (Schierwater *et al.*, 1997), management of plant genetic

© CAB *International* 2001. *Plant Genotyping: the DNA Fingerprinting of Plants* (ed. R.J. Henry)

resources (Bretting and Wirdrlechner, 1995; Tanksley and McCouch, 1997), and evaluation of biodiversity (Karp et al., 1997). Molecular markers can identify genetic differences in plant populations, exotic varieties and wild species that, for example, can help guide the introgression of new and important traits into elite germplasm. Similarly, markers can establish the levels of heterozygosity and homozygosity in breeding stock. Their use in genetic mapping can pinpoint genes controlling important traits and identify gene combinations that are most favourable in breeding strategies. Finally, markers can facilitate the study of the distribution, diversity and evolution of plants, plant pathogens and pests, allowing the identification of interactions and co-evolution strategies that can aid in forecasting epidemics and in disease management.

The amplification-based scanning methods that use arbitrarily amplified DNA (AAD) characterize nucleic acids without prior knowledge of nucleotide sequence or cloned and characterized hybridization probes (Livak et al., 1992; Bassam et al., 1995; McClelland et al., 1996). These techniques are versatile and universal, as demonstrated by the many applications and wide range of organisms studied in recent years and covering all domains of life (reviewed in Caetano-Anollés, 1993, 1996; Rafalski and Tingey, 1993; McClelland et al., 1995; Schierwater 1995; Micheli and Bova, 1996). AAD methods produce characteristic amplification signatures from virtually any nucleic acid template and use at least one short oligonucleotide of arbitrary or semi-arbitrary sequence to target a multiplicity of anonymous sites. These signatures (fingerprints) are composed mainly of amplification products of varying length. Primers as short as 5-mers can fingerprint nucleic acids (Caetano-Anollés et al., 1991, 1992). However, primers can harbour an arbitrary sequence of only three nucleotides (nt), if an extraordinarily stable mini-hairpin is attached at their 5' termini (Caetano-Anollés and Gresshoff, 1994). These mini-hairpin primers have been used effectively in the fingerprinting of small templates, such as PCR products (0.2–1 kb), plasmids (2–5 kb), AAD signatures (15–25 kb), and cloned genomic fragments (50–250 kb) (e.g. Kolchinsky et al., 1993; Caetano-Anollés and Gresshoff, 1994; Starman and Abbitt, 1997; Trigiano et al., 1998; Caetano-Anollés, 1999).

Three techniques have been the most popular: random amplified polymorphic DNA (RAPD; Williams et al., 1990), arbitrarily primed PCR (AP-PCR; Welsh and McClelland, 1990) and DNA amplification fingerprinting (DAF; Caetano-Anollés et al., 1991). Following their inception, many others have introduced modifications in primer design, amplification and overall strategy (cf. Caetano-Anollés, 1996). One in particular, the AFLP technique (Vos et al., 1995), has been widely adopted in genetic mapping and marker-assisted breeding applications. In this method, DNA is digested with two endonucleases, restriction fragments are then ligated to double-stranded oligonucleotide adaptors, and PCR primers complementary to adaptor and restriction site but harbouring 2–3 nt of arbitrary sequence at their 3' ends are used to generate highly informative fingerprints. Other AAD methods can detect polymorphic DNA with high efficiency, permitting analysis of the same set of templates at different taxonomical levels. Examples include mini-hairpin-primed DAF (mhpDAF; Caetano-Anollés and Gresshoff, 1994), arbitrary signatures from amplification profiles (ASAP; Caetano-Anollés and Gresshoff, 1994), and the use of oligonucleotide arrays in nucleic acid scanning-by-hybridization (NASBH; Salazar and Caetano-Anollés, 1996). These techniques have pushed the limits of detection of nucleic acid scanning.

Reaction mechanism and primer–template interactions

Only a fraction of targeted sites is efficiently amplified

An 'amplicon' is a region defined by two primer annealing sites in opposite strands of a nucleic acid molecule (Mullis, 1991). The AAD amplification reaction is generally driven by the annealing of a single arbitrary primer to short and complementary inverted repeats. These repeats, which are scattered throughout a template nucleic acid in a more or less random manner (Kesseli *et al.*, 1994), must be in near proximity to define AAD amplicons. However, not every possible amplicon will or can be amplified. Annealed primers will be efficiently extended by the thermostable DNA polymerase in only those cases favoured by a 'contextual' nucleic acid landscape defined by amplicons, flanking sequences and combinatorial interactions of primer and templates. This landscape defines

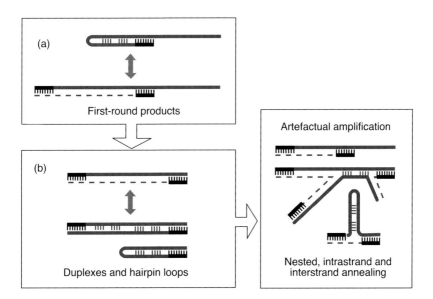

Fig. 3.1. Major molecular species establishing interactions during primer screening of first-round amplification products (a) and during the exponential phases of amplification (b). Hatched lines represent products expected to be formed by extension of annealed primers (in black).

a subset of successful amplicons capable of outcompeting all others for target and reagent resources during the first few amplification cycles.

The sequence and concentration of the arbitrary primer primarily determine the generation of AAD (Caetano-Anollés *et al.*, 1992). Amplification can accommodate extensive primer–template mismatching events (Parker *et al.*, 1991; Caetano-Anollés *et al.*, 1992; Venugopal *et al.*, 1993; Bertioli *et al.*, 1994), depending mainly on primer length and template complexity. Based on experiments that used sequence-related primers of different lengths and re-amplification of engineered amplification products, we demonstrated that only a fraction of template annealing sites was efficiently amplified and that primer–template mismatching could influence the competitive ability of many amplification products (Caetano-Anollés *et al.*, 1992). We proposed a model to explain these results (Caetano-Anollés *et al.* 1992; Caetano-Anollés, 1993, 1996) invoking the existence of primer–template and template–template interactions capable of curtailing or enhancing the amplification efficiency of individual amplicons. Because amplification products generated with a single primer have a region of terminal hairpin symmetry at least as long as the primer itself, it was proposed that the primer had to displace hairpin loop complexes before it could be extended efficiently by the DNA polymerase (Fig. 3.1). The hypothesis was tested by amplifying sets of DNA fragments with variable regions of hairpin terminal symmetry with homologous primers of increasing length (Caetano-Anollés *et al.*, 1992). Results showed that primer length and primer–template mismatching were crucial in determining the competitive ability of a primer and supported the hairpin loop formation of first-round amplification products. Simulation studies of the amplification of plasmids pUC18 and pBR322 using mini-hairpin primers with short arbitrary sequences allowed assignment of amplification products to expected regions and revealed physical interaction between annealing sites, probably during amplification of first-round products (Caetano-Anollés and Gresshoff, 1994). Results showed that hairpin-looped template duplexes served as preferential annealing sites during the amplicon 'screening' phase of amplification. Therefore, the ability of the primer to displace secondary structures in the template appeared to be a decisive factor in determining which rare primer–template duplexes were stabilized by primer extension and later transformed into accumulating amplification products.

Mismatching plays an important role

Primer–template mismatching plays an important role in amplification (reviewed in Caetano-Anollés, 1993). Extensive mismatching between the oligonucleotide primer and the template can be tolerated (Caetano-Anollés *et al.*, 1992). However, primer hybridization to perfect or partially complementary sites in the target nucleic acid requires that the first 5–6 nt from the 3′ terminus of the primer match faithfully those in the template. The successful amplification of an amplicon is driven by the interactions of the primer with its annealing site, followed by enzyme anchoring and primer extension. Inefficient extension of mismatched primers appears to be caused by failed DNA elongation rather than by a differential enzyme-binding affinity of matched versus mismatched termini (e.g. Huang et al., 1992). Successful AAD amplification is directed by a 3'-terminal primer domain of only 8 nt (Caetano-Anollés et al., 1992). This domain determines the spectrum of amplified products and is moderately influenced by the 5'-terminal region of the primer. Primer–template mismatching influences the competitive ability of many amplification products and the kinetics of the overall amplification reaction. The amplification of mixed samples of DNA in dilution series has shown the existence of competition phenomena between amplification products (Eskew et al., 1993; Williams et al., 1993; Tinker et al., 1994), which are, for example, important when searching for linked markers in bulked segregant analysis (Michelmore et al., 1991). The re-amplification of templates originally generated by amplification with sequence-related primers showed that primer mismatching curtailed the competitive ability of amplification products (G. Caetano-Anollés, unpublished). Similarly, the amplification of differentially mixed samples of DNA from several soybean cultivars, *Glycine soja* and bermudagrass showed that not all amplification products were equally competitive (Abbitt, 1996). Instead, products tolerated varying degrees of dilution, exhibiting different 'amplification potentials' (*vide* Tinker et al., 1994). An interesting observation was that, in general, bermudagrass and *G. soja* amplification products were more competitive than soybean products. This suggests that differences in competition resulted from variations in sequence diversity within templates and not from differences in genome size or number of targets. In this regard, a number of studies have identified *G. soja* as harbouring a panel of more diverse genetic loci (e.g. microsatellites) than soybean.

Template interactions are evident in AAD-derived sequence tags

The existence of template–template interactions during the nucleic acid scanning reaction was further analysed in sequence characterized amplified regions (SCARs) from soybean, rice, lettuce, bean, apple, strawberry, tomato, burrowing nematodes, zebrafish and mosquito. SCARs are PCR-based enhanced AAD markers produced by cloning, sequencing and PCR amplification of arbitrary sites in the genome (Paran and Michelmore, 1993). SCAR sequence analysis provides an estimate of the extent of duplex formation between AAD product termini. Figure 3.2 shows results obtained from sequence tags in mosquito. Chi-square analysis indicated that while base pair matching values fitted with a high level of confidence those expected from average GC or AT ratios for the individual nucleotide positions, there were significant departures that could only be explained by the selective interaction of sequence positions internal to the amplified products. In particular, complementarity of the nucleotide position immediately following the primer 3' termini was highly disfavoured, indicating

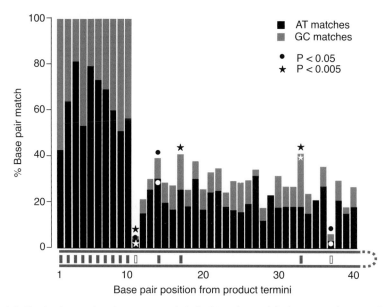

Fig. 3.2. Duplex interactions between termini of arbitrarily amplified DNA products used as sequence-tagged sites in the mosquito genetic map (Dimopoulos et al., 1996). Chi-square values for AT, GC and total base pair matches were calculated from a set of 34 sequence tags. Dots and stars show values that reject a null hypothesis of no significant differences in random duplex formation at the $P < 0.05$ and $P < 0.005$ confidence levels, respectively.

a crucial role for this position in secondary structure displacement by the primer. Selection of sequence-tagged sites depends on internal sequences within targeted regions and is, therefore, biased by factors other than those related to primer sequence and target specificity.

Interactions also occur during exponential amplification

A number of interactions can occur during the amplification phase that follows the initial screening of targets by the primer. These include nested priming and template duplex formation. Mathematical modelling and computer simulation have been used to predict the stochastic amplification of products produced from nested priming (Schierwater et al., 1996). These artefactual products have been observed in some RAPD studies but are usually rare (e.g. Smith et al., 1994). Heteroduplex molecules between allelic sequences that differ in length can also form during the exponential amplification phase, and can be revealed by nondenaturing electrophoresis (Ayliffe et al., 1994). Artefactual fragments resulting from template rearrangement were found following sequencing and Southern hybridization (to genomic DNA and RAPD profiles) of amplified fragments generated with primers containing restriction endonuclease sites (Rabouam et al., 1999). The study suggests that intra- and interstrand template interactions (Fig. 3.1) also occur in reactions driven by low primer concentration. It is important to note that similar artefacts have been rare in other RAPD homology studies (Smith *et al.*, 1994; Thormann *et al.*, 1994; Lannér *et al.*, 1996; Rieseberg, 1996).

Optimization and reproducibility

Reaction environment

Targeting of template sites occurs under a non-stringent reaction environment that is both adequate for annealing of short primers and specific enough to permit the exclusive amplification of legitimate amplicons (Caetano-Anollés, 1993). The mass ratio between primer and template adjusts the overall stringency of the amplification reaction and constitutes the single most important variable in the AAD reaction (Bassam *et al.*, 1992). Operationally, this ratio defines the different nucleic acid scanning techniques (cf. Caetano-Anollés, 1996). For example, RAPD uses primer-to-template ratios of less than one while DAF uses ratios generally ranging from 5 to 50,000. Increased ratios usually result in more complex DNA profiles and a more stable amplification reaction. This is why there has been a tendency during recent years to increase primer concentrations in optimized RAPD protocols.

Since AAD techniques depend on many amplification factors (reviewed by Tyler *et al.*, 1997), robust performance requires careful optimization. Measures of central tendency of experimental variables (reproducibility windows) have guided optimization exercises with great effectiveness (e.g. Bassam *et al.*, 1992; Bentley and Bassam, 1996; Bentley *et al.*, 1998). However, success depends on what you optimize and how the optimization exercise is set up. DNA amplification can be appropriately characterized by the parameters of *specificity*, *efficiency* (i.e. yield) and *fidelity*. The number, reproducibility and intensity of bands obtained by gel electrophoresis usually measure the first two of these parameters. However, amplification products need to be separated adequately and visualized in order to avoid borderline experimental conditions. The comparison of RAPD and DAF optimization illustrates this point. Only the most abundant amplification products are visualized by agarose gel electrophoresis in RAPD; optimization is tailored towards few and high intensity bands. Unfortunately, the generation of simple fingerprints favours amplification products competing disproportionately with each other for target and resources (the so-called 'context' effect; McClelland *et al.*, 1995). This usually results in an overall decrease of reproducibility that complicates optimization, explaining why different RAPD optimization exercises have yielded widely different optima. In contrast, the sensitive silver staining of numerous amplification products in DAF forces the operator to optimize for complex and uniform fingerprints, decreasing the context effect and increasing reproducibility. This is probably why independent DAF optimization exercises have always rendered similar protocols.

Optimization using Taguchi methods

Optimization is usually a laborious process that relies on the sequential investigation of individual variables in large experiments. The simplicity and proven success of Taguchi methods (Taguchi, 1986) in industrial process design offer a cost-effective alternative for optimization of DNA amplification (Cobb and Clarkson, 1994; Caetano-Anollés, 1998b). For example, L_9 (3^4) and L_{18} (3^8) orthogonal arrays have been used to study the interaction of amplification components and thermal cycling parameters in DAF with great success (Caetano-Anollés, 1998b; Caetano-Anollés *et al.*, 1999). Analysis of variance (ANOVA) decomposed the contribution of individual amplification factors to the responses of amplification yield or product number, while verification experiments established that optimum conditions were predictable and reproducible. Several amplification components (primer, magnesium and enzyme) controlled the amplification reaction. Conversely, annealing temperature and time were the only important thermal cycling contributing factors. The Taguchi approach defined a robust and transportable amplification protocol based on high annealing temperatures (typically 48°C) and primer concentrations (typically 8 µM), which can be applied to the fingerprinting of a wide range of DNA templates of plant and fungal origin. A worldwide web site offers protocols and information on AAD techniques and their optimization (DNA kaffe; http://biologi.uio.no/FellesAvdelinger/DNA_KAFFE/default.html).

Overcoming AAD marker limitations

AAD analysis presents practical problems such as band co-migration and uncertainties in band–locus assignment in the absence of preliminary marker segregation data. These difficulties can be overcome by proper experimental design (Hadrys *et al.*, 1992; Rieseberg, 1996). In contrast, the analysis of population genetic structure is hampered inherently by the dominant nature of most AAD markers. Parameters in population genetics, such as gene frequencies, heterozygosity and measures of population subdivision, can, nevertheless, be corrected using suitable estimators that diminish bias at the expense of larger sampling of individuals and pruning of loci with low-frequency null alleles (Lynch and Milligan, 1994). Finally, AAD markers can also be used to estimate nucleotide divergence, provided nine precautionary criteria are satisfied (Clark and Lanigan, 1993). While it may be unrealistic to expect compliance with all requisites, the methods should not be used when true nucleotide divergence exceeds 10%. The estimation of nucleotide diversity has been extended to AFLP (Innan *et al.*, 1999) and ASAP (Caetano-Anollés, 1999) data and, for example, has been used to measure DNA variation of *Arabidopsis thaliana* ecotypes distributed throughout the world (Miyashita *et al.*, 1999).

In genetic mapping, the conversion of AAD markers to SCAR loci has facilitated the transfer of marker information across species or between mapping progenies. The effectiveness of SCARs has, however, been hampered by the need to clone AAD fragments before sequencing them. Fortunately, an efficient method for the direct end-sequencing of purified AAD fragments using 3′-terminal extended primers has been proposed recently (Mitchelson *et al.*, 1999). This will surely facilitate the use of SCAR markers in a variety of applications.

Extending the limits of polymorphic DNA detection

Different applications in the plant sciences demand experimental approaches capable of differentiating the sampled material with efficiency and at different taxonomic levels. Profiling methods differ in their ability to detect DNA polymorphism in a given population (informativeness) and in the number of loci simultaneously targeted per experiment (multiplex ratio). DNA markers can also be dominant or co-dominant and can exhibit different expected heterozygosities (a measure of the number of alleles that are detected). Informativeness and co-dominance are especially important in genetic mapping and trait tagging. Desirable markers for these applications should be highly polymorphic and should exhibit multiple co-dominant alleles (e.g. microsatellite markers). Alternatively, robust multilocus-fingerprinting techniques with high multiplex ratios (e.g. DAF and AFLP) can be used efficiently despite low expected heterozygosities. When markers are used to estimate genetic diversity and build molecular phylogenies, the taxonomical level of analysis becomes relevant. Nucleic acid scanning techniques are useful here for distinguishing closely related organisms (usually at or below the species level). The challenges and limitations of AAD markers are well illustrated below in applications related to turfgrass science and floriculture.

AAD genotyping in turfgrass: off-types, cultivar instabilities and genome-wide mutation

DNA analysis has been widely used to characterize turfgrass species, cultivars and accessions (reviewed in Caetano-Anollés, 1998c). One interesting example is bermudagrass (*Cynodon*). The genus *Cynodon* comprises nine species and ten varieties that constitute a diverse group of important warm-season perennial sod-forming grasses, most of which have been used as turf, pasture and fodder throughout warm temperate and tropical regions of the world (Taliaferro, 1995). Only a few clonally propagated lines bred during the later part of the 20th century are currently being widely cultured. These include sterile triploids ($2n = 3x = 27$) resulting from the interspecific hybridization of tetraploid *Cynodon dactylon* var. *dactylon* and diploid *Cynodon transvaalensis*. Such is the case for cultivars 'Tifway', 'Tifgreen' and 'Tifdwarf'. The narrow genetic base of these cultivars poses a risk for extensive damage from virulent or introduced pests (Taliaferro, 1995), an apparent vulnerability that must be counteracted

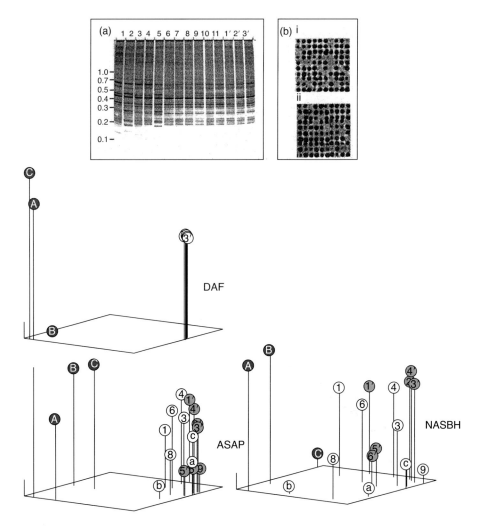

Fig. 3.3. Genetic relationships between 'Tifgreen' and 'Tifdwarf' bermudagrass accessions and off-types revealed by DNA amplification fingerprinting (DAF), arbitrary signatures from amplification profiles (ASAP) and nucleic acid scanning by hybridization (NASBH). Genetic relationships were analysed by principal coordinate analysis using the NTSYS PC program (Rohlf, 1992) using Dice similarity coefficients. Cultivar accessions and off-types (described in Caetano-Anollés, 1998a) are indicated with numbers and letters, respectively. Prime numbers correspond to 'Tifdwarf' accessions. Cultivars 'Tiffine' (A), 'Tifton 10' (B) and 'Tifway' (C) were defined as the outgroup. ASAP data are from Caetano-Anollés (1998a). In NASBH, DAF products generated with primer GCGACAGCTGC were hybridized to a 10 × 10 gridded panel of arbitrary oligonucleotides of sequence NX_9N (X being any one nucleotide and N representing degenerate positions; Salazar and Caetano-Anollés, 1996). Insert (a) shows representative DAF profiles generated with the mini-hairpin primer GCGAAAGCGGA from selected 'Tifgreen' and 'Tifdwarf' accessions. Molecular weight markers are given in kb. Insert (b) shows NASBH array profiles obtained from cultivar 'Tifgreen' accession 1 (i) and 'Tifway' (ii).

by assessing and broadening the genetic diversity of plant material used in breeding programmes. We used DAF with standard and mini-hairpin primers to establish the levels of genetic variation within and between selected species and interspecific hybrids of bermudagrass (Caetano-Anollés et al., 1995) and examined the origin of 'off-type' derivatives that exhibit variant morphology and performance (Caetano-Anollés et al., 1997; Caetano-Anollés, 1998a). Studies showed that 'Tifway' was intrinsically stable and that off-types originated from contamination were diverse and were probably interspecific hybrids. In contrast, off-types in cultivars 'Tifgreen' and 'Tifdwarf' resulted predominantly from genetic instabilities arising from somatic mutation but not from gross chromosomal rearrangements. The detection of genetic instabilities was particularly facilitated by the use of the ASAP cascade amplification strategy and the oligonucleotide array-based NASBH method (Fig. 3.3). In ASAP, a single arbitrary primer is used to produce a collection of AAD products which then serves as a template for a second round of DNA amplification that is generally directed by one or more mini-hairpin primers. This results in additional target sites being analysed and an overall increase in informativeness. The enhanced detection of polymorphic DNA in AAD profiles was particularly successful when oligonucleotide arrays were used to detect variation in first-round amplified fragments from bermudagrass accessions and off-types (Fig. 3.3).

An accumulation of point mutations within plants that are multiplying vegetatively at high rates by specialized means (stolons and rhizomes) could have important consequences on biological fitness and could be the cause of the high incidence of genetic off-types in bermudagrass sterile hybrids. Kondrashov (1994) has developed a general model whereby mutation load in a population exhibiting obligate vegetative reproduction appears substantially higher than under sexual or asexual reproduction, explaining the rarity of this reproductive mode throughout evolution. In line with this proposal, we were able to show that genetic instabilities detected by phenetic analysis in bermudagrass sterile hybrids resulted from increased genome-wide mutation levels in vegetative culture (Caetano-Anollés, 1999). In this study, genome-wide somatic mutation rates were measured directly using the ASAP strategy. Allelic signatures were produced by amplification of monomorphic DAF profiles. These signatures were characteristic of the sequence of the initial DAF-amplified products. The uncovered ASAP polymorphisms allowed measurement of sequence divergence in organisms as varied as bermudagrass, chrysanthemum and *Discula destructiva* fungi, and provided a direct estimation of mutation rate at the whole genome level. The approach was used to study mutations that were induced by irradiation in 'Tifway II', those arising from genetic instabilities in the 'Tifgreen—Tifdwarf' bermudagrass complex, and those appearing as the result of normal vegetative growth in 'Tifdwarf'. Mutations studied were essentially nucleotide substitutions and strand inversions that occurred within monomorphic DAF products. They were not the result of insertions or deletion events, the activity of transposable elements, major rearrangements in the genome, artefacts produced by the DNA amplification process itself, or the existence of non-orthologous bands in DAF products.

The calculated vegetative mutation rate (1.05×10^{-8} per nucleotide per generation) was strikingly congruent with a long-term rate measured across accessions and indicative of the accumulation of mutations in 'Tifgreen–Tifdwarf' populations (1.02×10^{-8} per nucleotide per generation), suggesting absence of evolutionary constraints in the sampled genomic regions. Therefore, most mutations detected by ASAP in bermudagrass accumulate freely, appearing populationally neutral. Rates were similar to those found in human, *Drosophila melanogaster*, *Caenorhabditis elegans* and the mouse (μ ranging from 0.4×10^{-8} to 2×10^{-8}; reviewed in Kondrashov, 1998), supporting the contention that sequence evolution in eukaryotes is cell-or-organism generation-dependent rather than time-dependent. Mutation rates calculated from across-accessions divergence estimates indicated that plant material was evolving 100 times faster (3.8×10^{-7} changes per nucleotide year) than a molecular clock rate estimate for grasses, probably resulting from the compound effect of clonal growth and life span of the hybrid plant material.

Mutation rates have never been measured 'directly' in plants before but were proposed to occur at levels of one mutation per diploid genome per generation (Kondrashov, 1998). The relatively high genomic mutation rate ($\mu_g = 10$ per triploid genome) during bermudagrass vegetative growth results in effective μ_{eg} rates ($U = 1$) higher than calculated deleterious mutation rates using chlorophyll-deficient lethals ($U = 0.003$–0.074), but consistent with rates in self-fertilizing annual plants ($U = 0.2$–0.9). The high deleterious mutation rate compares well with recent estimates in hominids ($U = 1.2$–1.7) (Eyre-Walker and Keightley, 1999). The high incidence of deleterious mutations, with rate estimates (U) comparable with those in plants subjected to inbreeding depression, casts doubt on the long-term success of the interspecific hybrids, if mutational effects on fitness were to combine in a multiplicative manner during clonal growth. Further research is, therefore, needed to evaluate the effect of mutation accumulation on vegetative culture.

AAD genotyping in floriculture: distinguishing closely related plants

Cultivar identification and pedigree verification are important for the protection of intellectual property and royalty income, and for the development of 'essentially derived varieties'. Varietal identification has been particularly important for the floriculture industry. The production of bedding plants is one of the fastest growing segments in horticulture, with annual sales of billions of dollars. Floricultural crops, such as petunia, contribute importantly to this growth. However, cultivars are closely related and are difficult to differentiate at the genetic level. We have used AAD markers successfully to characterize cultivars of petunia (Cerny *et al.*, 1996), chrysanthemum (Scott *et al.*, 1996; Trigiano *et al.*, 1998), geranium (Starman and Abbitt, 1997) and carnation (Trigiano *et al.* 1998), clarifying in some cases their origin or establishing genetic relationships. In particular, the ASAP technique (Caetano-Anollés and Gresshoff,

1996) permitted the clear identification of somatic mutants and radiation-induced sports that are genetically highly homogeneous (Starman and Abbitt, 1997; Trigiano et al., 1998), facilitating future marker-assisted breeding and protection of plant breeders' rights of varieties or cultivars. One interesting application relates to the analysis of rates of radiation-induced and somatic mutation in chrysanthemum. Currently cultivated chrysanthemum cultivars are usually developed from a single progenitor either spontaneously or by radiation-induced mutagenesis (sports). Because of their close genetic relationship, there is a need to differentiate vegetatively derived accessions. For example, cultivars 'Dark Charm', 'Salmon Charm', 'Coral Charm' and 'Dark Bronze Charm' are either radiation-induced mutants or spontaneous sports of cultivar 'Charm' and constitute a family or series of plants that primarily differ in flower colour. These cultivars, were difficult to differentiate genetically by DAF analysis (Scott et al., 1996), but were easily identified by ASAP analysis (Trigiano et al., 1998). Sequence diversity levels (0.03–4%) were comparable with those reported in *Drosophila*, *Tradescantia chiensis* and *Anopheles gambiae*. However, no differences were observed between the individual mutants (one-way ANOVA, $P > 0.963$), indicating that radiation-induced and somatic mutations occur at a similar pace in chrysanthemum.

Concluding remarks

A decade has elapsed from the inception of AAD markers. Despite controversies on their limitations and reproducibility, they can still be regarded as valuable tools in molecular biology. More stringent optimization exercises, better knowledge of the amplification mechanism, improved experimental design, development of more accurate data analysis tools and modifications in methodology have enhanced their use considerably. If handled correctly and matched to the right application, AAD marker methods should be considered robust and powerful. Moreover, they have been regularly used in novel applications, such as those recent in toxicology and mutation research (Atienzar et al., 1998; Shimada and Shima, 1998; Caetano-Anollés, 1999). Their novel use in new areas of science is expected to continue in the near future. AAD techniques are also evolving and benefit from recent developments in genomics. Their coupling to oligonucleotide arrays constitutes one first example (Salazar and Caetano-Anollés, 1996; Beattie, 1998). Ultimately, the versatility and universal nature of AAD markers should guarantee their continuous use.

References

Abbitt, S. (1996) Analysis of soybean non-nodulation mutants by DNA amplification fingerprinting and bulked segregant analysis. MSc thesis, University of Tennessee, Knoxville, USA.

Atienzar, F., Child, P., Evenden, A., Jha, A., Savva, D., Walker, C. and Depledge, M. (1998) Application of the arbitrarily primed polymerase chain reaction to the detection of DNA damage. *Marine Environmental Research* 46, 331–335.

Ayliffe, M.A., Lawrence, G.J., Ellis, J.G. and Pryor, A.J. (1994) Heteroduplex molecules formed between allelic sequences cause nonparental RAPD bands. *Nucleic Acids Research* 22, 1632–1636.

Bachmann, K. (1994) Molecular markers in plant ecology. *New Phytologist* 126, 403–418.

Bassam, B.J., Caetano-Anollés, G. and Gresshoff, P.M. (1992) DNA amplification fingerprinting of bacteria. *Applied Microbiology and Biotechnology* 38, 70–76.

Bassam, B.J., Caetano-Anollés, G. and Gresshoff, P.M. (1995) Method for profiling nucleic acids of unknown sequence using arbitrary oligonucleotide primers. *US Patent* 5,413,909.

Beattie, K.L. (1998) Genomic fingerprinting using oligonucleotide arrays. In: Caetano-Anollés, G. and Gresshoff, P.M. (eds) *DNA Markers*. Wiley-Liss Inc., New York, pp. 213–224.

Bentley, S. and Bassam, B.J. (1996) A robust DNA amplification fingerprinting system applied to analysis of genetic variation within *Fusarium oxysporum* f. sp. *cubense*. *Journal of Phytopathology* 144, 207–213.

Bentley, S., Pegg, K.G., Moore, N.Y., Davis, R.D. and Buddenhagen, I.W. (1998) Genetic variation among vegetative compatibility groups of *Fusarium oxysporum* f. sp. *cubense* analysed by DNA fingerprinting. *Phytopathology* 88, 1283–1293.

Bertioli, D.J., Schlichter, U.H.A., Adams, M.J., Burrows, P.R., Steinbiß, H.H. and Antoniw, J.F. (1994) An analysis of differential display shows a strong bias towards high copy number mRNAs. *Nucleic Acids Research* 23, 4520–4523.

Bretting, P.K. and Widrlechner, M.P. (1995) Genetic markers and plant genetic resource management. *Plant Breeding Reviews* 13, 11–86.

Caetano-Anollés, G. (1993) Amplifying DNA with arbitrary oligonucleotide primers. *PCR Methods and Applications* 3, 85–94.

Caetano-Anollés, G. (1996) Scanning of nucleic acids by *in vitro* amplification: new developments and applications. *Nature Biotechnology* 14, 1668–1674.

Caetano-Anollés, G. (1998a) Genetic instability of bermudagrass (*Cynodon*) cultivars 'Tifgreen' and 'Tifdwarf' detected by DAF and ASAP analysis of accessions and off-types. *Euphytica* 101, 165–173.

Caetano-Anollés, G. (1998b) DAF optimization using Taguchi methods and the effect of thermal cycling parameters on DNA amplification. *Biotechniques* 25, 472–480.

Caetano-Anollés, G. (1998c) DNA analysis of turfgrass genetic diversity. *Crop Science* 38, 1415–1424.

Caetano-Anollés, G. (1999) High genome-wide mutation rates in vegetatively propagated bermudagrass. *Molecular Ecology* 8, 1211–1221.

Caetano-Anollés, G. and Gresshoff, P.M. (1994) DNA amplification fingerprinting using arbitrary mini-hairpin oligonucleotide primers. *Bio/technology* 12, 1011–1026.

Caetano-Anollés, G. and Gresshoff, P.M. (1996) Generation of sequence signatures from DNA amplification fingerprints with mini-hairpin and microsatellite primers. *Biotechniques* 20, 1044–1056.

Caetano-Anollés, G. and Gresshoff, P.M. (1998) *DNA Markers*. Wiley-Liss Inc., New York.

Caetano-Anollés, G. and Trigiano, R.N. (1997) Nucleic acid markers in agricultural biotechnology. *AgBiotech News and Information* 9, 235N–242N.

Caetano-Anollés, G., Bassam, B.J. and Gresshoff, P.M. (1991) DNA amplification fingerprinting using very short arbitrary oligonucleotide primers. *Bio/technology* 9, 553–557.

Caetano-Anollés, G., Bassam, B.J. and Gresshoff, P.M. (1992) Primer–template interactions during *in vitro* amplification with short oligonucleotides. *Molecular and General Genetics* 235, 157–165.

Caetano-Anollés, G., Callahan, L.M., Williams, P.E., Weaver, K.R. and Gresshoff, P.M. (1995) DNA amplification fingerprinting analysis of bermudagrass (*Cynodon*): genetic relationships between species and interspecific crosses. *Theoretical and Applied Genetics* 91, 228–235.

Caetano-Anollés, G., Callahan, L.M. and Gresshoff, P.M. (1997) The origin of bermudagrass (*Cynodon*) off-types inferred by DNA amplification fingerprinting. *Crop Science* 37, 81–87.

Caetano-Anollés, G., Schlarbaum, S.E. and Trigiano, R.N. (1999) DNA amplification fingerprinting and marker screening for pseudo-testcross mapping of flowering dogwood (*Cornus florida* L.). *Euphytica* 106, 209–222.

Cerny, T.A., Caetano-Anollés, G., Trigiano, R.N. and Starman, T.W. (1996) Molecular phylogeny and DNA amplification fingerprinting of Petunia taxa. *Theoretical and Applied Genetics* 92, 1009–1016.

Clark, A.G. and Lanigan, C.M.S. (1993) Prospects for estimating nucleotide divergence with RAPDs. *Molecular Biology and Evolution* 10, 1096–1111.

Cobb, B.D. and Clarkson, J.M. (1994) A simple procedure for optimizing the polymerase chain reaction (PCR) using modified Taguchi methods. *Nucleic Acids Research* 22, 3802–3805.

Dimopoulos, G., Zheng, L., Kumar, V., della Torre, A., Kafatos, F.C. and Louis, C. (1996) Integrated map of *Anopheles gambiae*: use of RAPD polymorphisms for genetic, cytogenetic and STS landmarks. *Genetics* 143, 953–960.

Eskew, D.L., Caetano-Anollés, G., Bassam, B.J. and Gresshoff, P.M. (1993) DNA amplification fingerprinting of the *Azolla–Anabaena* symbiosis. *Plant Molecular Biology* 21, 363–373.

Eyre-Walker, A. and Keightley, P.D. (1999) High genomic deleterious mutation rates in hominids. *Nature* 397, 344–347.

Hadrys, H., Balick, M. and Schierwater, B. (1992) Applications of random amplified polymorphic DNA (RAPD) in molecular ecology. *Molecular Ecology* 1, 55–63.

Huang, M., Arnheim, N. and Goodman, M.F. (1992) Extension of base mispairs by *Taq* DNA polymerase: implications for single nucleotide discrimination in PCR. *Nucleic Acids Research* 20, 4567–4573.

Innan, H., Terauchi, R., Hahl, G. and Tajima, F. (1999) A method for estimating nucleotide diversity from AFLP data. *Genetics* 151, 1157–1164.

Karp, A., Edwards, K.J., Bruford, M., Funk, S., Vosman, B., Morgante, M., Seberg, O., Kremer, A., Boursot, P., Arctander, P., Tautz, D. and Hewitt, G.M. (1997) Molecular technologies for biodiversity evaluation: opportunities and challenges. *Nature Biotechnology* 15, 625–628.

Kesseli, R.V., Paran, I. and Michelmore, R.W. (1994) Analysis of a detailed genetic linkage map of *Lactuca sativa* (Lettuce) constructed from RFLP and RAPD markers. *Genetics* 136, 1435–1446.

Kolchinsky, A.M., Funke, R.P. and Gresshoff, P.M. (1993) DAF-amplified fragments can be used as markers for DNA from pulse field gels. *Biotechniques* 14, 400–403.

Kondrashov, A.S. (1994) Mutation load under vegetative reproduction and cytoplasmic inheritance. *Genetics* 137, 311–318.

Kondrashov, A.S. (1998) Measuring spontaneous deleterious mutation. *Genetica* 102/103, 183–197.

Lannér, C., Bryngelsson, T. and Gustafsson, M. (1996) Genetic validity of RAPD markers at the intra- and inter-specific level in wild *Brassica* species with n=9. *Theoretical and Applied Genetics* 93, 9–14.

Livak, K.J., Rafalski, J.A., Tingey, S.V. and Williams, J.G. (1992) Process for detecting polymorphisms on the basis of nucleotide differences. *US Patent* 5,126,239.

Lynch, M. and Milligan, B.G. (1994) Analysis of population genetic structure with RAPD markers. *Molecular Ecology* 3, 91–99.

McClelland, M., Mathieu-Daude, F. and Welsh, J. (1995) RNA fingerprinting and differential display using arbitrarily primed PCR. *Trends in Genetics* 11, 242–246.

McClelland, M., Welsh, J.T. and Sorge, J.A. (1996) Arbitrarily primed polymerase chain reaction method for fingerprinting genomes. *US Patent* 5,467,985.

Micheli, M.R. and Bova, R. (1996) *Fingerprinting Methods Based on Arbitrarily Primed PCR*. Springer Verlag, Berlin.

Michelmore, R.W., Paran, I. and Kesseli, R.V. (1991) Identification of markers linked to disease-resistance genes by bulked segregant analysis: a rapid method to detect markers in specific genomic regions by using segregating populations. *Proceedings of the National Academy of Sciences USA* 88, 9828–9832.

Mitchelson, K.R., Drenth, J., Duong, H. and Chaparro, J.X. (1999) Direct sequencing of RAPD fragments using 3′-extended oligonucleotide primers and dye terminator cycle-sequencing. *Nucleic Acids Research* 27, e28.

Miyashita, N.T., Kawabe, A. and Innan, H. (1999) DNA variation in the wild plant *Arabidopsis thaliana* revealed by amplified fragment length polymorphism analysis. *Genetics* 152, 1723–1731.

Mullis, K.B. (1991) The polymerase chain reaction in an anemic mode: how to avoid cold oligodeoxyribonuclear fusion. *PCR Methods and Applications* 1, 1–4.

Paran, I. and Michelmore, R.W. (1993) Development of reliable PCR-based markers linked to downy mildew resistance genes in lettuce. *Theoretical and Applied Genetics* 85, 985–993.

Parker, J.D., Rabonovich, P.S. and Burmer, G. (1991) Targeted gene walking polymerase chain reaction. *Nucleic Acids Research* 19, 3055–3060.

Rabouam, C., Comes, A.M., Bretagnolle, V., Humbert, J.-F., Prequet, G. and Bigot, Y. (1999) Features of DNA fragments obtained by random amplified polymorphic DNA (RAPD) analysis. *Molecular Ecology* 8, 493–503.

Rafalski, J.A. and Tingey, S.V. (1993) Genetic diagnostics in plant breeding: RAPDs, microsatellites and machines. *Trends in Genetics* 9, 275–279.

Rieseberg, L.H. (1996) Homology among RAPD fragments in interspecific comparisons. *Molecular Ecology* 5, 99–105.

Rohlf, F.J. (1992) *Numerical Taxonomy and Multivariate Analysis System* (NTSYS-pc), version 1.8. Exeter Software, Setauket, New York.

Salazar, N. and Caetano-Anollés, G. (1996) Nucleic acid scanning-by-hybridization of enterohemorrhagic *Escherichia coli* isolates using oligonucleotide arrays. *Nucleic Acids Research* 24, 5056–5057.

Schierwater, B. (1995) Arbitrarily amplified DNA in systematics and phylogenetics. *Electrophoresis* 16, 1643–1647.

Schierwater, B., Metzler, D., Krüger, K. and Streit, B. (1996) The effect of nested primer binding sites on the reproducibility of PCR: mathematical modeling and computer simulation studies. *Journal of Computational Biology* 3, 235–251.

Schierwater, B., Ender, A., Schroth, W., Holzmann, H., Diez, A., Streit, B. and Hadrys, H. (1997) Arbitrarily amplified DNA in ecology and evolution. In: Caetano-Anollés, G. and Gresshoff, P.M. (eds) *DNA Markers: Protocols, Applications and Overviews.* John Wiley & Sons, New York, pp. 313–330.

Scott, M.C., Caetano-Anollés, G. and Trigiano, R.N. (1996) DNA amplification fingerprinting identifies closely related Chrysanthemum cultivars. *Journal of the American Society of Horticultural Science* 121, 1043–1048.

Shimada, A. and Shima, A. (1998) Combination of genomic DNA fingerprinting into the medaka specific-locus test system for studying environmental germ-line mutagenesis. *Mutation Research* 399, 149–165.

Smith, J.J., Scott-Craig, J.S., Leadbetter, J.R., Bush, G., Roberts, D.L. and Fulbright, D.W. (1994) Characterization of random amplified polymorphic DNA (RAPD) products from *Xanthomonas campestris* and some comments on the use of RAPD products in phylogenetic analysis. *Molecular Phylogenetics and Evolution* 3, 135–145.

Starman, T.W. and Abbitt, S. (1997) Evaluating genetic relationships of geranium using arbitrary signatures from amplification profiles. *HortScience* 32, 1288–1291.

Taguchi, G. (1986) *Introduction to Quality Engineering.* Asian Productivity Organization, UNIPUB, New York.

Taliaferro, C.M. (1995) Diversity and vulnerability of bermuda turfgrass species. *Crop Science* 35, 327–332.

Tanksley, S.D. and McCouch, S.R. (1997) Seed banks and molecular maps: unlocking genetic potential from the wild. *Science* 277, 1063–1066.

Thormann, C.E., Ferreira, M.E., Camargo, L.E.A., Tivang, J.G. and Osborn, T.C. (1994) Comparisons of RFLP and RAPD markers to estimating relationships within and among cruciferous species. *Theoretical and Applied Genetics* 88, 973–980.

Tinker, N.A., Mather, D.E. and Fortin, M.G. (1994) Pooled DNA for linkage analysis: practical and statistical considerations. *Genome* 37, 999–1004.

Trigiano, R.N., Scott, M.C. and Caetano-Anollés, G. (1998) Genetic signatures from amplification profiles characterize DNA mutation in somatic and radiation-induced sports of chrysanthemum. *Journal of the American Society of Horticultural Science* 123, 642–646.

Tyler, K.D., Wang, G., Tyler, S.D. and Johnson, W.M. (1997) Factors affecting reliability and reproducibility of amplification-based DNA fingerprinting of representative bacterial pathogens. *Journal of Clinical Microbiology* 35, 339–346.

Venugopal, G., Mohapatra, S., Salo, D. and Mohapatra, S. (1993) Multiple mismatch annealing: basis for random amplified polymorphic DNA fingerprinting. *Biochemical and Biophysical Research Communications* 197, 1382–1387.

Vos, P., Hogers, R., Bleeker, M., Reijans, M., van de Lee, T., Hornes, M., Frijters, A., Pot, J., Peleman, J., Kuiper, M. and Zabeau, M. (1995) AFLP: a new technique for DNA fingerprinting. *Nucleic Acids Research* 23, 4407–4414.

Weising, K., Nybom, H., Wolff, K. and Meyer, W. (1994) *DNA Fingerprinting in Plants and Fungi.* CRC Press, Boca Raton, Florida.

Welsh, J. and McClelland, M. (1990) Fingerprinting genomes using PCR with arbitrary primers. *Nucleic Acids Research* 18, 7213–7218.

Williams, J.G.K., Kubelik, A.R., Livak, K.J., Rafalski, J.A. and Tingey, S.V. (1990) DNA polymorphisms amplified by arbitrary primers are useful as genetic markers. *Nucleic Acids Research* 18, 6531–6535.

Williams, J.G.K., Reiter, R.S., Young, R.M. and Scolnik, P.A. (1993) Genetic mapping of mutations using phenotypic pools and mapped RAPD markers. *Nucleic Acids Research* 21, 2697–2702.

Winter, P. and Kahl, G. (1995) Molecular marker technologies for plant improvement. *World Journal of Microbiology and Biotechnology* 11, 438–448.

Chapter 4

Plant Genotyping Based on Analysis of Single Nucleotide Polymorphisms Using Microarrays

B. LEMIEUX

Department of Plant and Soil Sciences, University of Delaware, Newark, Delaware, USA

SNP discovery

Estimates based on the detection of nucleotide substitutions in expressed sequence tag (EST) databases suggest that the average number of differences between any pair of chromosomes is 3 per 10,000 bp in coding regions (Garg *et al.* 1999). Considering that coding regions are subjected to selective pressure, this estimate is probably an underestimate of the total number of single nucleotide polymorphisms (SNPs) throughout the human genome. In maize, SNPs can be readily found in the untranslated regions of gene coding regions and estimates of 1/300 bp have been reported (Bongard *et al.*, 2000).

The most efficient approach for SNP discovery is to mine the existing sequence databases. A number of tools have been developed to automate: (i) base calling for genome sequencing; (ii) the assembly of these sequences; and (iii) viewing and editing the assembled sequence contigs, i.e. the PHRED, PHRAP and CONSED suite of programs. These bioinformatic tools provide a framework for SNP mining software that uses sequence trace information.

SNP discovery can be through the detection of heterozygosity in sequence trace files derived from PCR products. The program POLYPHRED has a demonstrated accuracy of > 99% with single-pass data obtained from pooled PCR products containing known sequence variants when the sequences are generated with fluorescent dye-labelled primers, and approximately 90% for those prepared with dye-labelled terminators (Nickerson *et al.*, 1997).

The high level of endoduplication of sequences within plant genomes, however, may make the use of POLYPHRED problematic. Moreover, one would prefer to use existing sequence data without having to rely on resequencing of pools of PCR products. Therefore, researchers have begun to extract SNPs from EST databases (Picoult-Newberg *et al.*, 1999). However, as much of this sequence

© CAB *International* 2001. *Plant Genotyping: the DNA Fingerprinting of Plants* (ed. R.J. Henry)

information is in uncharacterized regions of the genome, it can be difficult to distinguish true allelic variations from sequencing errors. POLYBAYES uses base quality values to discern true polymorphisms from sequencing errors and assigns a probability value that a given site is polymorphic (Marth *et al.*, 1999). This rigorous treatment of base quality permits completely automated evaluation of sequence traces without the limitations due to low depth of alignments in a contig.

The major weakness of using SNPs as markers for DNA fingerprinting in plants is that they are 'less informative' than simple sequence repeat (SSR) markers in that they generally have only two alleles. Despite this inherent limitation, SNPs have great potential because they are so abundant. The key to their implementation as tools for DNA fingerprinting is the development of flexible and high throughput assay formats.

Selecting an optimal SNP genotyping strategy

Once a set of SNPs has been identified, a number of strategies are available for genotyping these markers. This usually entails PCR amplification as a means of preparing the test sample. Each method has its strengths and limitations. Among the considerations for practical implementation of these technologies are robustness, cost and labour input. For marker-assisted selection (MAS), robustness and cost must meet more demanding standards than for the biological research laboratory. In every case, flexibility in redirecting the assay for new marker sets is essential.

Solid-phase assays

DNA ligase-based assays

The oligonucleotide ligation assay (OLA) (Landegren *et al.*, 1988) exploits the fact that: (i) DNA ligase cannot join two oligonucleotides unless the ligation reaction is 'guided' by a second DNA strand; and (ii) the 3' end of the first oligonucleotide has perfect complementarity with the 'guide' DNA strand. In this assay, two oligonucleotides that differ in their 3' ends are used (one for each allele of a locus). Although the original version of this assay was performed in a microtitre dish (Landegren *et al.*, 1988), multiplex OLA assays use automated DNA sequencers to display the reaction products (Eggerding, 1995). The mobility of the latter can be manipulated by incorporating a different number of non-nucleotide mobility modifiers into each allele-specific OLA primer.

Recently a DNA microarray and a flow cytometry method have been reported for the detection of OLA products (Gerry *et al.*, 1999; Iannone *et al.*, 2000). Instead of mobility tags, these methods use hybridization tags designed to hybridize to DNA sequences that have been coupled to OLA allele-specific primers. These sequences are called 'zip codes' and direct the sorting of the OLA products to specific addresses on the microarray or to microspheres that have been coupled to zip code complementary sequences. In the former assay format, OLA reaction products are hybridized to the microarray and the level of fluorescence at each address is a measure of the frequency of the allele at the site. The use of microspheres with different ratios of red and orange fluorescence allows for the identification of individual alleles using a flow cytometer.

Microsequencing

The microsequencing assay consists of using DNA polymerase to add a single dideoxynucleotide to an oligonucleotide primer that binds to a DNA fragment containing the SNP. As such, this method exploits the fact that DNA polymerase requires a template to incorporate nucleotides on to a nucleic acid primer and that this reaction can be stopped by dideoxynucleotides. In the genetic bit analysis developed by Molecular Tool Inc. (Nikiforov *et al.*, 1994), the primers are immobilized in microtitre plates and the reaction products are detected with a fluorescence microtitre plate reader.

Alternatively, the introduction of a 'mass tag' on the primer allows for a multiplex analysis via matrix-assisted laser desorption/ionization time of flight mass spectroscopy (MALDI-TOF MS) (Haff and Smirnov, 1997; Monforte and Becker, 1997). By keeping the DNA fragments small, very high sample throughputs are possible; however, the number of different markers per run is limited because DNA fragments longer than 100 bases cannot be resolved with enough resolution for accurate genotyping. This limitation makes the microsequencing assay more adaptable to MALDI-TOF MS than the OLA assay.

Genetic bit analysis has recently been adapted to a DNA microarray format (Fan *et al.*, 2000). In this assay, marker-specific primers are used in PCR amplifications of genomic regions containing SNPs. These amplicons are used as templates in single base extension (SBE) reactions using chimeric primers with 3' complementarity to the specific SNP loci and 5' complementarity to hybridization tags. The SBE primers, terminating one base before the polymorphic site, are extended in the presence of labelled dideoxynucleotide triphosphates, using a different label for each of the two SNP alleles, and hybridized to the tag array composed on the tag complementary sequences. The fluorescence intensity ratio of the two colours is then used to genotype the SNP. This method was used to genotype 44 individuals for 142 human SNPs. Because the hybridization results are quantitative, this method can also be used for allele frequency estimation in pooled DNA samples.

MutS technology

The use of the DNA repair mismatch binding protein MutS has recently been proposed in a number of SNP scoring methods. The highest throughput method proposed thus far is that of Bellanne-Chantelot *et al.* (1997). It uses a filtration-based assay to identify DNA fragments generated by PCR that have been bound by MutS. By introducing a reporter group (e.g. biotin) on the PCR primers, one can detect the DNA fragments immobilized on to the filter surface. By attaching fluorescent spheres to the PCR products, it might be possible to develop a multiplex analysis of MutS-enriched fragments using a bead sorting approach similar to that used for detecting OLA products (Iannone et al., 2000).

Homogeneous assays

Invasive probe assays

Third Wave Technologies Inc. has developed a SNP genotyping based on a 'structure-dependent cleavase' assay (Brow et al., 1996; Lyamichev et al., 1999). This assay is based on the digestion of overlapping oligonucleotides hybridized to genomic DNA by flap endonucleases (FENs). The downstream oligonucleotide probe is cleaved, and the precise site of cleavage is dependent on the amount of overlap with the upstream oligonucleotide. By using thermostable archaeal FENs, it is possible to perform these digestions at temperatures that promote probe turnover without the need for temperature cycling. The resulting amplification of the cleavage signal enables the detection of specific DNA targets at sub-attomole levels within complex mixtures. This cleavage is sufficiently specific to enable discrimination of single-base differences and can differentiate homozygotes from heterozygotes in single-copy genes in genomic DNA. Griffin et al. (1999) have adapted this assay format to MALDI-TOF-based detection of 'mass tags' located on oligonucleotides cleaved in a genotyping assay. Although this format does not allow for multiplexing, its throughput is very high, making it a suitable assay for direct association studies.

DNA melting-based approaches

Methods based on the difference in T_m of two alleles of a locus are conceptually simple (Germer and Higuchi, 1999). Homogenous hybridization assays for SNP typing have also been reported, for example molecular beacons (Marras et al., 1999). Molecular beacons are quenched fluorescent probes with a hairpin structure, which, upon hybridization to a target, melt to separate the quencher from the reporter. Applying this method to genotyping is straightforward. The region containing the SNP is amplified by PCR and the identity of the variant nucleotide is determined by observing which of four differently coloured molecular beacons binds to the amplification product. Each of the molecular beacons is perfectly complementary to one variant of the target sequence and each is labelled with a different fluorophore.

Restriction endonuclease-based assays

Researchers often need a flexible, yet low labour input assay format for SNPs that does not require expensive instrumentation; particularly academic laboratories. The dCAPS technique is a variant of the cut amplified polymorphic sequence (CAPS) strategy that uses oligonucleotides that are designed such as to create a new restriction site in a DNA fragment with one of the two alleles of a SNP (Neff *et al.*, 1998). Unlike denaturing HPLC, this method requires knowledge of the DNA sequence of the polymorphism for primer design. Its throughput is limited by two factors: (i) a multitude of different restriction endonucleases may be needed to score a large collection of dCAPS markers; and (ii) gel electrophoresis is needed to resolve alleles by length differences, hence limiting the sample throughput of the method. These characteristics suggest that this method may be ideal for the research laboratory that seeks easily to map a given sequence with a small mapping population without resorting to costly instrumentation. In combination with high throughput oligonucleotide synthesis technology (Rayner *et al.*, 1998), the dCAPS and IP-RP-HPLC methods can be used rapidly to map and validate candidate genes, respectively.

PNA-directed PCR clamping

Peptide nucleic acids (PNAs) are synthetic analogues of DNA with a N-2-aminoethyl glycine backbone with nucleoside bases attached to the backbone by methylene carbonyl groups. This non-charged backbone allows PNAs to recognize and bind to their complementary nucleic acid sequences with higher thermal stability than the corresponding DNA oligonucleotides. Therefore, PNA–DNA duplex formation can block the formation of PCR product when the PNA is targeted against one of the PCR primer sites. The key characteristic that makes PNAs useful for SNP typing is that PNA–DNA duplexes are more easily destabilized by mismatches than DNA–DNA duplexes. Therefore, it is possible to use PNAs selectively to amplify target sequences that differ by only one base pair (Orum *et al.* 1993). The main disadvantage of this technology is that each allele of a SNP must be assayed independently. However, it may be possible to multiplex loci in the same PCR reaction tube and, by incorporating fluorescent reporter tags on the allele-specific DNA primers, multiplex the analysis of the products. The ideal readout of such an assay would be by hybridization to a DNA microarray composed of all the loci.

DNA chip-based methods

DNA chip technology can also be used to genotype SNPs using two formats:

1. Those based on hybridization of single-stranded DNA to a probe array of oligonucleotides.
2. Those based on the hybridization of DNA fragments that survive *in vitro* selection directed against single base mismatches.

Hybridization assays

Assay formats based on hybridizations of PCR products containing SNPs to custom designed oligonucleotide probe arrays hold the most promise for typing large numbers of loci. Custom probe arrays are designed using the DNA sequence of the region containing the SNP. A 'tiling path' of oligonucleotides complementary to the region of the genome containing the SNP can be produced on a surface using any number of manufacturing methods. The best known method is light-directed synthesis (Pease et al., 1994); however, other methods ranging from in situ synthesis (Southern et al., 1992) to 'maskless photolithography' (Singh-Gasson et al., 1999) have been reported. The PCR products hybridized to these oligonucleotide arrays are labelled with fluorescent reporter groups during amplification. Figure 4.1 illustrates a tiling path used to design an oligonucleotide microarray for genotyping brassica SNPs. In this case, a 13 nucleotide window containing the SNP is scanned by hybridization to a set of 20-mer probes complementary to each strand. As the middle position of an oligonucleotide probe is most sensitive to the disruption due to a mismatch, the central base of these probes is substituted with a non-complementary base to create a 'homo-mismatch' (e.g. an 'A' is paired with an 'A'). Comparison of the relative intensity of fluorescence at these 'mismatch probes' (MM) with their corresponding 'match probes' (PM) can be used to estimate the specificity of hybridization to the latter. These comparisons can be used for two tests: (i) an intensity test (PM-MM); and (ii) a signal test (PM/MM) which must meet thresholds specified by the investigator in order to validate a given 'mini-block', composed of a PM and its corresponding MM. These indices can then be used to estimate a confidence value that a given individual has a specified genotype.

These probe arrays can also be used for polymorphism detection. For example, in the 4L tiling path (or variant detection) used by Chee et al. (1996), each nucleotide in a given region of the genome is queried by four oligonucleotides that differ by a single base in their middle position. This strategy allows for the detection of SNPs because a divergence from the reference sequence used to design the tiling path leads to a 'footprint' of lower signal in the hybridization pattern. The use of this tiling path has also been applied to genotyping biallelic markers. In this case, a window of several nucleotides flanking each allele of a SNP is used as a reference sequence to produce two detection blocks, one for each allele. This tiling path seems favoured for SNP typing in humans (Wang et al., 1998) but has also been applied to plant markers (Cho et al., 1999).

The DNA fragments hybridized to these DNA chips are generated by PCR using primers designed with sequence data from the genomic region immediately surrounding the SNPs. By keeping these fragments to a minimum length and equalizing the concentration of each primer, it is possible to minimize the loss of markers during the initial cycles of multiplex amplification. These primers also contain a 'universal domain' in order to facilitate the incorporation of a fluorescent reporter group and equalize the amount of DNA generated for each

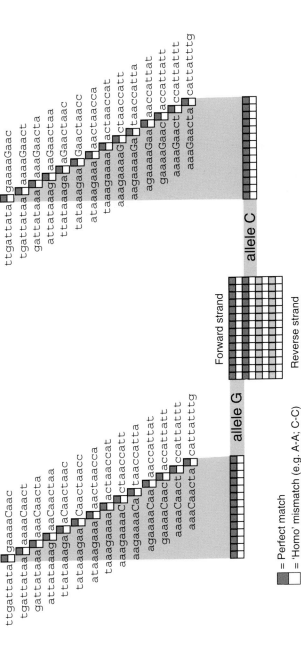

Fig. 4.1. Tiling pattern used for genotyping single nucleotide polymorphisms (SNPs). SNPs are genotyped by hybridization of short, fluorescent-labelled PCR products to custom designed oligonucleotide microarrays. The DNA sequence flanking a particular SNP is used to programme the design of photolithographic masks to allow for the production of a 'tiling pattern'. In this particular case, each allele is scored by 56 different molecular species of oligonucleotides (i.e. 28 per strand per allele). The grey squares in the tiling pattern represent the oligonucleotide tiles that are complementary to each of the alleles (the perfect matches (PM)). The tiling strategy also uses a purposeful 'homo' mismatch (shown as white squares) positioned in the centre of the oligonucleotides (mismatch (MM)). These are used to estimate the specificity of the hybridization signal at the matched tiles; the PM-MM must be greater than a minimum threshold value.

locus. Multiplex PCR of 25–100 fragments has been reported. The PCR products are then added to a hybridization cocktail and injected into the GeneChip hybridization chamber. After an overnight hybridization, the hybridization buffer is washed away and the GeneChip array read with a confocal scanner. The processing time of each GeneChip array is about 5 min and, depending on the GeneChip design, each hybridization reaction can harvest up to 4000 genotypes from one sample. As the density of GeneChip arrays is expected to increase by one order of magnitude in the next 5 years, this strategy has great potential. Its main limitation is economic; current GeneChip arrays cost several hundred dollars.

Genomic mismatch scanning (GMS)

Of all the SNP typing technologies, GMS potentially has the highest throughput of all in terms of the number of genotypes per organism. Indeed, the nucleic acid selection strategy used by this method enriches for all heteroduplexes that contain no mismatches and are thus identical by descent (Nelson *et al.*, 1993). The GMS procedure uses fragments generated by the digestion of genomic DNA with restriction endonucleases that yield protruding 3' ends. Heteroduplexes are formed by hybridization of DNA fragments from two samples (usually a patient and his or her grandparent). The heteroduplex fragments are selected by a series of enzymatic steps involving methylases as well as restriction enzymes that cut methylated and unmethylated DNA but not hemi-methylated duplexes. Mismatched bases in heteroduplexes are then nicked by the DNA repair enzymes MutSHL such that only fragments that contain perfectly matched strands survive after treatment with exonuclease III (an enzyme that cannot digest DNA with protruding 3' ends). The surviving fragments represent regions of 'identity by descent' in a given individual compared with one of its ancestors.

As eukaryotic genomes may contain millions of fragments, this technology potentially allows for the greatest number of markers per individual. In practice, the surviving fragments are labelled by PCR and hybridized to a DNA microarray. Therefore, the number of individual fragments printed on the DNA microarray places an upper limit on the number of loci that can be detected in one experiment. This technology is possibly the most powerful tool for indirect association studies because of the large number of loci that can be enriched in a single reaction. However, its applicability is severely handicapped by its complexity of implementation; therefore, its usefulness is likely to be confined to the research laboratory.

Summary

Of all the methods listed in Table 4.1, hybridization-based genotyping has the highest throughput if one considers both the total number of genotypes and the number of samples that can be processed in one person-day with a single instrument. As current GeneChip arrays can contain up to 400,000 different tiles, a chip capable of scoring 4000 SNP markers could be produced using the tiling path illustrated in Fig. 4.1.

MAS requires a robust, high throughput assay format for a limited number of SNPs. All of the aforementioned technologies have limitations to their applicability to MAS. Any post-PCR processing reduces the throughput and increases the cost of the markers. The only methods that do not require additional processing after PCR are denaturing HPLC (Underhill *et al.*, 1997) and hybridization to GeneChip arrays. The throughput of the former is limited to one genotype from one sample every 5 min for each US$80,000 instrument. The GeneChip assay format could generate ~400,000 genotypes day^{-1}; however, it requires a fixed chip design and is rather costly at present. Therefore, once these economic obstacles are eliminated, it is likely to become the dominant SNP typing technology for MAS. The DNA chip-based OLA and SBE assays (Gerry *et al.*, 1999; Fan *et al.*, 2000) would allow for the reuse of the DNA chip (the most expensive reagent in the assay) while eliminating the need for sample clean-up and chip redesign for new SNP markers.

SNP typing technology may be particularly attactive for germplasm identification. This opinion is not due to the relatively low throughput of SSR markers compared with SNP typing methods, but rather the greater precision of SNP scoring assays. Indeed, alleles of a given SSR locus with equal numbers of repeats do not always have the same nucleotide sequences embedded within the repeat (Di Rienzo *et al.*, 1994).

References

Bellanne-Chantelot, C., Beaufils, S., Hourdel, V., Lesage, S., Morel, V., Dessinais, N., Le Gall, I., Cohen, D. and Dausset, J. (1997) Search for DNA sequence variations using a MutS-based technology. *Mutation Research* 382, 35–43.

Bongard, D.K., Goodman, H.M., Mikkilineni, V., Rocheford, T.R., Farnworth, B., Peng, J. and Lemieux, B. (2000) SNP discovery using the maize EST database. *Maize Genetics Conference. Coeur d'Alene, Idaho.*

Brow, M.A., Oldenburg, M.C., Lyamichev, V., Heisler, L.M., Lyamicheva, N., Hall, J.G., Eagan, N.J., Olive, D.M., Smith, L.M., Fors, L. and Dahlberg, J.E. (1996) Differentiation of bacterial 16S rRNA genes and intergenic regions and *Mycobacterium tuberculosis katG* genes by structure-specific endonuclease cleavage. *Journal of Clinical Microbiology* 34, 3129–3137.

Chee, M., Yang, R., Hubbell, E., Berno, A., Huang, X.C., Stern, D., Winkler, J., Lockhart, D.J., Morris, M.S. and Fodor, S.P.A. (1996) Accessing genetic information with high-density DNA arrays. *Science* 274, 610–614.

Cho, R.J., Mindrinos, M., Richards, D.R., Sapolsky, R.J., Anderson, M., Drenkard, E., Dewdney, J., Reuber, T.L., Stammers, M., Federspiel, N., Theologis, A., Yang, W.H., Hubbell, E., Au, M., Chung, E.Y., Lashkari, D., Lemieux, B., Dean, C., Lipshutz, R.J., Ausubel, F.M., Davis, R.W. and Oefner, P.J. (1999) Genome-wide mapping with biallelic markers in *Arabidopsis thaliana*. *Nature Genetics* 23, 203–207.

Di Rienzo, A., Peterson, A.C., Garza, J.C., Valdes, A.M., Slatkin, M. and Freimer, N.B.

(1994) Mutational processes of simple-sequence repeat loci in human populations. *Proceedings of the National Academy of Sciences USA* 91, 3166–3170.

Eggerding, F.A. (1995) A one-step coupled amplification and oligonucleotide ligation procedure for multi-plex genetic typing. *PCR Methods and Applications* 4, 337–345.

Fan, J.-B., Chen, X., Halushka, M.K., Berno, A., Huang, X., Ryder, T., Lipshutz, R.J., Lockhart, D.J. and Chakravarti, A. (2000) Parallel genotyping of human SNPs using generic high-density oligonucleotide tag arrays. *Genome Research* 10, 853–860.

Garg, K., Green, P. and Nickerson, D.A. (1999) Identification of candidate coding region single nucleotide polymorphisms in 165 human genes using assembled expressed sequence tags. *Genome Research* 9, 1087–1092.

Germer, S. and Higuchi, R. (1999) Single-tube genotyping without oligonucleotide probes. *Genome Research* 9, 72–78.

Gerry, N.P., Witowski, N.E., Day, J., Hammer, R.P., Barany, G. and Barany, F. (1999) Universal DNA microarray method for multiplex detection of low abundance point mutations. *Journal of Molecular Biology* 292, 251–262.

Griffin, T.J., Hall, J.G., Prudent, J.R. and Smith, L.M. (1999) Direct genetic analysis by matrix-assisted laser desorption/ionization mass spectrometry. *Proceedings of the National Academy of Sciences USA* 96, 6301–6306.

Haff, L.A. and Smirnov, I.P. (1997) Single-nucleotide polymorphism identification assays using a thermostable DNA extraction MALDI-TOF mass spectrometry. *Genome Research* 7, 378–388.

Iannone, M.A., Taylor, J.D., Chen, J., Li, M.S., Rivers, P., Slentz-Kesler, K.A. and Weiner, M.P. (2000) Multiplexed single nucleotide polymorphism genotyping by oligonucleotide ligation and flow cytometry. *Cytometry* 39, 131–140.

Landegren, U., Kaiser, R., Sanders, J. and Hood, L. (1988) A ligase-mediated gene detection technique. *Science* 241, 1077–1080.

Lyamichev, V., Mast, A.L., Hall, J.G., Prudent, J.R., Kaiser, M.W., Takova T., Kwiatkowski, R.W., Sander, T.J., de Arruda, M., Arco, D.A., Neri, B.P. and Brow, M.A. (1999) Polymorphism identification and quantitative detection of genomic DNA by invasive cleavage of oligonucleotide probes. *Nature Biotechnology* 17, 292–296.

Marras, S.A., Kramer, F.R. and Tyagi, S. (1999) Multiplex detection of single-nucleotide variations using molecular beacons. *Genetic Analysis* 14, 151–156.

Marth, G.T., Korf, I., Yandell, M.D., Yeh, R.T., Gu, Z., Zakeri, H., Stitziel, N.O., Hillier, L., Kwok, P.Y. and Gish, W.R. (1999) A general approach to single-nucleotide polymorphism discovery. *Nature Genetics* 23, 452–456.

Monforte, J.A. and Becker, C.H. (1997) High-throughput DNA analysis by time-of-flight mass spectrometry. *Nature Medicine* 3, 360–362.

Neff, M.M., Neff, J.D., Chory, J. and Pepper, A.E. (1998) dCAPS, a simple technique for the genetic analysis of single nucleotide polymorphisms: experimental applications in *Arabidopsis thaliana* genetics. *Plant Journal* 14, 387–392.

Nelson, S.F., McCusker, J.H., Sander, M.A., Kee, Y., Modrich, P. and Brown, P.O. (1993) Genomic mismatch scanning: a new approach to genetic linkage mapping. *Nature Genetics* 4, 11–18.

Nickerson, D.A., Tobe, V.O. and Taylor, S.L. (1997) PolyPhred: automating the detection and genotyping of single nucleotide substitutions using fluorescence-based resequencing. *Nucleic Acids Research* 25, 2745–2751.

Nikiforov, T.T., Rendle, R.B., Goelet, P., Rogers, Y.H., Kotewicz, M.L., Anderson, S., Trainor, G.T. and Knapp, M.R. (1994) Genetic bit analysis: a solid phase method for typing single nucleotide polymorphisms. *Nucleic Acids Research* 22, 4167–4175.

Orum, H., Nielsen, P.E., Egholm, M., Berg, R.H., Buchardt, O. and Stanley, C. (1993) Single base pair mutation analysis by PNA directed PCR clamping. *Nucleic Acids Research* 21, 5332–5336.

Pease, A.C., Solas, D., Sullivan, E.J., Cronin, M.T., Holmes, C.P. and Fodor, S.P.A. (1994) Light-generated oligonucleotide arrays for rapid DNA sequence analysis. *Proceedings of the National Academy of Sciences USA* 91, 5022–5026.

Picoult-Newberg, L., Ideker, T.E., Pohl, M.G., Taylor, S.L., Donaldson, M.A., Nickerson, D.A. and Boyce-Jacino, M. (1999) Mining SNPs from EST databases. *Genome Research* 9, 167–174.

Rayner, S., Brignac, S., Bumeister, R., Belosludtsev, Y., Ward, T., Grant, O., O'Brien, K., Evans, G.A. and Garner, H.R. (1998) MerMade: an oligodeoxyribonucleotide synthesizer for high throughput oligonucleotide production in dual 96-well plates. *Genome Research* 8, 741–747.

Singh-Gasson, S., Green, R.D., Yue, Y., Nelson, C., Blattner, F., Sussman, M.R. and Cerrina, F. (1999) Maskless fabrication of light-directed oligonucleotide microarrays using a digital micromirror array. *Nature Biotechnology* 17, 974–978.

Southern, E.M., Maskos, U. and Elder, J.K. (1992) Analyzing and comparing nucleic acid sequences by hybridization to arrays of oligonucleotides: evaluation using experimental models. *Genomics* 13, 1008–1017.

Underhill, P.A., Jin, L., Lin, A.A., Mehdi, S.Q., Jenkins, T., Vollrath, D., Davis, R.W., Cavalli-Sforza, L.L. and Oefner, P.J. (1997) Detection of numerous Y chromosome biallelic polymorphisms by denaturing high-performance liquid chromatography. *Genome Research* 7, 996–1005.

Wang, D.G., Fan, J.-B., Siao, C.-J., Berno, A., Young, P., Sapolsky, R., Ghandour, G., Perkins, N., Winchester, E., Spencer, J., Kruglyak, L., Stein, L., Hsie, L., Topaloglou, T., Hubbell, E., Robinson, E., Mittmann, M., Morris, M.S., Shen, N., Kilburn, D., Rioux, J., Nusbaum, C., Rozen, S., Hudson, T.J., Lipshutz, R., Chee, M. and Lander, E.S. (1998) Large-scale identification, mapping, and genotyping of single-nucleotide polymorphisms in the human genome. *Science* 280, 1077–1082.

Chapter 5

Genotyping in Plant Genetic Resources

B.V. FORD-LLOYD

School of Biosciences, University of Birmingham, Birmingham, UK

Introduction

Over the last 60 or 70 years, scientists throughout the world have been engaged in developing better and higher yielding cultivars of crop plants to be grown on increasingly larger scales. This has inevitably involved the replacement of more variable, lower yielding, but locally adapted varieties (Fig. 5.1) grown in traditional agricultural systems. Increasingly every major crop has had a steadily decreasing genetic base and, under such situations, diversity within cultivated plants has been replaced by genetic uniformity. This loss of diversity in the crops themselves (genetic erosion) has also been accompanied by further loss of genes found in wild and weedy species related to the crop plants, because of improved crop husbandry and the destruction of natural ecosystems by human development. The paradox is that in order to be able to develop new cultivars of crop plants in the future, plant breeders will need to have access to the wealth of genes which have been or are being lost; hence the need for conservation.

Scientific development of genetic resource conservation took place in the 1960s and 1970s and this led to the establishment of the International Board for Plant Genetic Resources (now the International Plant Genetic Resources Institute). Techniques for collecting, sampling and storing were devised, along with data management procedures for maintaining data on characterization and evaluation of germplasm. Only much more recently has interest focused upon *in situ* genetic resource conservation techniques (Maxted *et al.*, 1997), partly as a result of the Convention on Biological Diversity, which supports the complementary nature of *in situ* and *ex situ* genetic resource conservation (UNCED, 1992).

Genetic resources can be described as the total genetic diversity of cultivated species and their wild relatives and more specifically take the form of: related

© CAB *International* 2001. *Plant Genotyping: the DNA Fingerprinting of Plants* (ed. R.J. Henry)

Fig. 5.1. A genetically diverse assemblage of potato tubers (*Solanum tuberosum*) representing landraces found in South America.

wild species which mainly occur in 'gene centres' of cultivated plants or outside them; weed races which occur as part of crop–weed complexes in gene centres; landraces (Fig. 5.1) which are the products of traditional agriculture rather than modern plant breeding, grown using traditional agricultural practices; modern breeding lines and genetic stocks; obsolete cultivars; and modern cultivars which will only be significant for conservation when they become obsolete.

Conservation methods

Seed storage

The most convenient way of maintaining most plant germplasm is by storing seeds. The main exceptions to this are plants where selected genotypes are vegetatively propagated (e.g. potatoes), or which may not produce viable seeds (e.g. banana), and crops where the seeds produced are very short-lived. The latter are often referred to as being 'recalcitrant' because they will not stand drying below some relatively high moisture content without very serious loss of viability (Roberts, 1973).

Fortunately, most crop plants do produce seeds, and these show 'orthodox' behaviour; they are tolerant of a decrease in moisture content coupled with temperature, allowing storage for relatively long periods at about 5% moisture content and $-18°C$ without damage (Frison and Bolton, 1994).

Ex situ *alternatives to seed storage*

The conservation of vegetatively propagated crops such as potato, cassava, yams, sweet potato, sugarcane and many temperate fruit trees, among others, presents special problems for *ex situ* conservation. Although some of these crops are sexually fertile, it is often not convenient to propagate them commercially from seed because of high levels of heterozygosity, and breeders and horticulturalists commonly require uniform clones. Many vegetatively propagated crops are, however, sexually sterile, or at the very least have reduced fertility which precludes the possibility of seed storage.

Field gene banks, botanical gardens, arboreta and plantations can be used to maintain germplasm as living collections. Much interest has also been focused on the application of tissue culture or *in vitro* techniques for plant genetic resource conservation (Ashmore, 1997; Engelmann, 1997). It is possible to store plants *in vitro* for short periods of time, or longer if subculturing is carried out after certain intervals. *In vitro* technology also opens up the possibility of ultra-low temperature storage of vegetative material, or cryopreservation at temperatures as low as $-196°C$. By this means, germplasm could be stored indefinitely.

Storing germplasm in the form of DNA is now a reality, despite the fact that current technology will allow for the recovery of single genes only, and not whole genomes, genome segments, gene complexes or sets of genes which control quantitative traits (quantitative trait loci or QTL). While such problems may be apparent now, the rapid advance of molecular technology means that well within the time span of medium term germplasm conservation (10–20 years), these problems may substantially be overcome, and there are, therefore, perceived advantages and usefulness in storing DNA in certain circumstances, as proposed in the DNA-BankNet programme (Adams, 1997).

In situ *methods*

Whereas *ex situ* conservation involves the maintenance of genetic resources away from their natural habitats, *in situ* conservation involves maintaining viable populations of target species in their natural environments (Maxted *et al.*, 1997). These environments can be 'wild' or 'agricultural' habitats. Genetic reserves are identified with the aim of maintaining natural populations of wild plants which may be relatives of crop plants and therefore of genetic resource significance. In contrast, conservation 'on-farm' can be attempted in order to maintain germplasm of traditionally cultivated landraces.

The use of genotyping and molecular markers in germplasm conservation

The whole process of conservation involves several sequential stages, ranging from the initial selection of target taxa and identification of conservation objectives, through field exploration and germplasm collection, the actual storage and maintenance of that germplasm over extended time periods, to the evaluation of genetic diversity generally, and for useful characters and eventual use or exploitation of the germplasm. A model identifying these stages is found in Fig. 5.2 (after Maxted *et al.*, 1997). It is possible to identify important uses for molecular markers and genotyping capability at all of these stages.

Molecular markers in use

The range of molecular markers that can now be relatively easily used is quite extensive; DNA-based markers have substantially overtaken those based on proteins and enzymes. Techniques applied to studying plant genetic resources variation include identifying polymorphism in the actual DNA sequence, the use of DNA hybridization methods to identify restriction fragment length polymorphisms (RFLPs), or the use of PCR-based technology to find polymorphism using random amplified polymorphic DNA (RAPD), simple sequence repeat (SSR) polymorphism, or combination techniques such as amplified fragment length polymorphism (AFLP). While reviews of these techniques are plentiful (Newbury and Ford-Lloyd, 1997; Westman and Kresovich, 1997; Karp *et al.*, 1998), because of the rapidity with which relevant technology is proceeding, these may not remain comprehensive for long. As an example, the uses of single nucleotide polymorphisms (SNPs) or microarrays for screening germplasm have yet to receive much attention (but see Chapters 1 and 4).

Nevertheless, it is possible to identify important criteria by which to judge the value of any particular marker system to a chosen application. For instance, for information on population history or phylogenetic relationships, sequence data or restriction site data may be most appropriate. At which taxonomic level is the

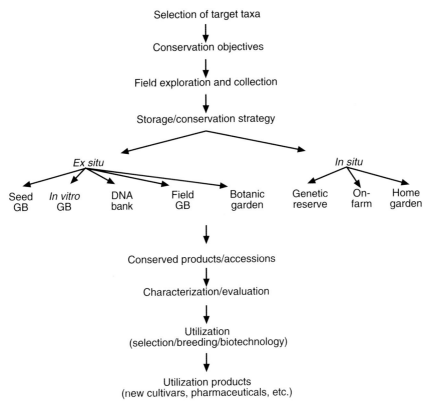

Fig. 5.2. A model scheme for plant genetic resource conservation (after Maxted *et al.*, 1997).

genetic variation being measured is an important question: within populations, between species, between genera? From how many loci will information be required, and how much allelic variation per locus is appropriate? To what extent are the methodologies robust and reproducible? Is cost per data point an important factor, and how important is the speed of analysis? How much DNA is available: most PCR-based methods require only tiny quantities of easily prepared DNA, RFLP analysis requires larger amounts, and sequencing may require the greatest quantities. Is it necessary to identify homozygotes and heterozygotes: are co-dominant markers needed, or will dominant markers suffice? All of these questions may demand answers before any study of germplasm is undertaken.

Using molecular markers at the different stages of germplasm conservation

Molecular markers may be employed to assist in most, if not all of the stages of germplasm conservation, as seen in Fig. 5.2, and including both *in situ* and *ex situ* approaches. The sorts of questions that can be answered may fall into several categories (Westman and Kresovich, 1997). First, there are those questions that involve determining how similar or distinct are any two or more genotypes, samples, accessions or populations. This assessment may be made by considering one or a few gene loci, several or many loci, or something approaching a complete sample of loci from the whole genome. These questions will be important when considering gene bank management and organization from the point of view of efficiency gains from identifying duplicates, ensuring accurate taxonomic identity, assessing material for entry into 'core collections' or the level of genetic stability in conserved germplasm. These questions will also be important for the design of genetic reserves, in terms of which populations to include, and in determining whether populations change genetically over time.

The second category of questions relates much more to the evaluation of germplasm and its use in plant breeding. Molecular markers can provide answers about whether there are particular DNA sequences, alleles, chromosome segments, whole chromosomes or whole genomes (for instance in interspecific hybrids) in one or other samples, accessions or populations of genetic resources. These answers will normally be revealed either because a particular marker is part of the DNA (gene) of interest, or because it is closely genetically linked to the target gene or sequence. One or a few diagnostic markers may serve these purposes.

Thirdly, there is an important set of questions which relate to the assessment of 'genetic diversity' or genetic variation that may exist among a set or sets of germplasm, without reference to any specific use for that variation. They will also include questions about genetic and evolutionary relationships, and will generally require larger sets of markers covering all chromosomes of a genome in order to be effective. Assessment in one way or another of genetic variation within and between samples, accessions or species can be important at almost every stage of the conservation process (Fig. 5.2) and certainly for both *in situ* and *ex situ* conservation methods.

Assessment of 'genetic diversity' within the whole genetic resource conservation process

The aim of plant genetic resources conservation should be to conserve, as far as possible, the broad genetic diversity which is found in the target species, with the expectation that this will maximize the chances of conserving potentially useful genes. Classical methods of estimating the genetic diversity among groups of plants have relied upon morphological characters. However, these characters can be influenced by environmental factors. Molecular markers avoid many of these complications by looking directly at the genetic material itself. Molecular markers, therefore, represent a powerful and potentially rapid method for characterizing diversity *per se* within *in situ* and *ex situ* conservation. Depending upon which molecular marker is used, direct and accurate measurements of a range of genetic diversity indicators may be made, although it should be appreciated that different molecular marker systems may give contrasting or even contradictory results (Parsons *et al.*, 1997; Virk *et al.*, 2000a), presenting problems which need to be resolved on a species by species basis. On the other hand, good agreement can be found when studying some species, even when contrasting techniques such as RAPD and DNA sequencing are used (Shen *et al.*, 1998). Additionally, while there may be increasing emphasis on choosing genetically mapped markers for studying diversity, it needs to be appreciated that a distorted or inaccurate view of diversity may be obtained (Fig. 5.3), and that markers chosen arbitrarily may be more valid (Virk *et al.*, 2000b).

The concept of 'diversity' may seem simple, but is not, and this can lead to the unnecessary and inappropriate use of some measures of diversity. There are two conventional views, the first being in terms of 'richness': the total number of genotypes or alleles present within germplasm regardless of their frequencies. The second is in relation to 'evenness' of the frequencies of different alleles or genotypes. Where richness is used to measure diversity, germplasm with more (and different) alleles or genotypes will be more diverse. In contrast, where evenness is considered important, a germplasm sample where the alleles or genotypes, albeit fewer, are all roughly equal in frequency will be more diverse than one where there may be greater numbers of alleles or genotypes, but where they are very unequal in frequency.

Estimations of allelic richness are basically the number of distinct alleles at a locus (A), and estimations of diversity (H) are measures based upon the frequencies of variants, allelic variants in the case of Nei's index of diversity (Nei, 1973) or phenotypic variants in the case of the Shannon–Weaver index (H'). These relatively simple estimators can be used to assess diversity occurring not only within populations, but also within different geographical regions or different germplasm collections. Other measures which measure population structure (Wright's F_{st}; Nei's G_{st}; Wright's fixation index) might be applied to looking at the way genetic variation is distributed among populations or samples within collections, as too might the analysis of molecular variation (Excoffier *et al.*, 1992).

The question of what analysis to choose is a difficult one, and useful guidance is given by Gonzalez-Candelas and Palacios (1997). However it seems unlikely that estimates of evenness will provide the desired answer as far as *ex situ* germplasm conservation is concerned. It is important to know how frequent alleles are so that care can be taken to conserve those which are less frequent, during seed rejuvenation for instance. But of greatest importance to the conservationist when considering the potential future use of germplasm is what genes or alleles can be found in which accessions and gene banks, to some extent irrespective of their frequencies. For this reason, the preference should be for the cluster analysis approach (Fig. 5.3) and the calculation of either similarity or genetic distance.

There are numerous examples of studies of the diversity of plant genetic resources. These have focused on wild populations, landraces and more

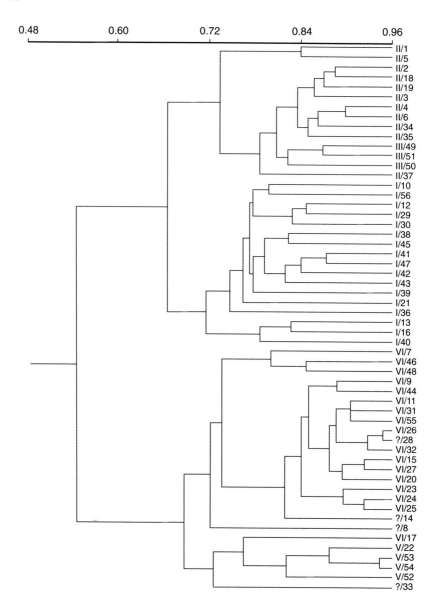

Fig. 5.3. Dendrograms of 56 accessions of rice generated by UPGMA cluster analysis of (a) 299 umapped markers generated using 14 primer combinations and (b) (*opposite*) 93 hypothetically mappable AFLP markers defined from a potential cross between accessions from isozyme groups I and II (after Virk *et al.*, 2000b).

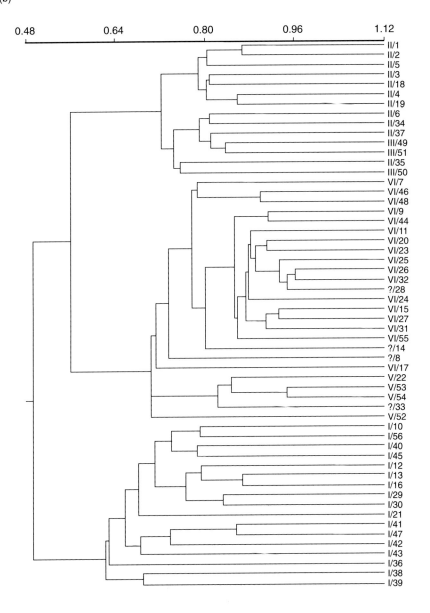

modern cultivars. For instance, Nevo *et al.* (1998) were able to associate high genetic diversity indices generated from RAPD data with stressful environments in which wild barley grew, while Ferguson *et al.* (1998b) related RAPD and isozyme diversity to geographical distribution of lentil landraces. Rice ecotype diversity was also partitioned geographically across Bangladesh using RAPD and the geographical information system by Parsons *et al.* (1999). High levels of intraspecific and within-population variation in Spanish cedar (*Cedrela odorata*) were recorded by Gillies *et al.* (1997) using RAPD data analysed by various methods including analysis of molecular variation, Shannon's diversity index and cluster analysis. In maize breeding, there is good reason to maintain genetic diversity for the improvement of the crop, and Senior *et al.* (1998) have used a range of markers including RFLPs, isozymes and SSRs to estimate similarity and divergence among diverse maize breeding material.

Bjornstad *et al.* (1997) used various markers to study diversity in Ethiopian landraces of barley in comparison with modern Western varieties. Their results were contrasting in that the Ethiopian material appeared to be less diverse, but genotypically very distinct. This contrast between richness and diversity was also highlighted by Ferguson *et al.* (1998a) when using a novel approach to study how genetic variation in lentils is partitioned geographically. The approach utilized cluster analysis to map geographical distribution of genetic variation in different *Lens* taxa, and to locate areas of high diversity and unique variation in each of the taxa. Allelic composition and richness in *ex situ* conserved germplasm were used to identify areas appropriate for *in situ* conservation, and where further *ex situ* germplasm collection was most likely to yield novel genetic variation. When compared with those produced using Nei's index of diversity, the results were sometimes conflicting, and this was almost certainly due to the greater emphasis placed on alleles of average frequency using Nei, and its inability to discriminate between situations where different alleles were present. Areas of both high diversity and unique diversity could be located using the cluster analysis approach.

Selection of target taxa and populations, exploration and collection

Molecular marker data can assist in taxonomic evaluation, particularly the accurate identification of germplasm. Our understanding of the relationships between crop gene pools has been substantially assisted by using molecular markers. Gepts (1995) surveyed this use, and described how molecular markers have been used to confirm teosinte as the actual and immediate progenitor of maize (Doebly *et al.*, 1984; Zimmer *et al.*, 1988; Doebly, 1990), while in rice, a species receiving increased attention in terms of its utilization potential and therefore conservation, *Oryza longistaminata*, has been shown to be genetically close to the common ancestor of the A genome species including *Oryza sativa* (Akagi *et al.*, 1998). SSR markers have been used to assess diversity among coconut ecotypes worldwide and to confirm how the dissemination of coconut has taken place historically (Perera *et al.*, 2000). The implications of this research are important in terms of selecting coconut germplasm for conservation: the fact that all alleles found in American and African samples are found in all the other regions indicates which material should receive most conservation effort.

Improving our understanding of crop gene pools, relationships with wild relatives, and particularly where crops were domesticated and display most genetic variation is important when attempting to capture that variation for conservation. This can be seen in the examples for maize, rice and coconut above. Biomolecular evidence has been assessed for several domesticates in the Sahul (Lebot, 1999). These include bananas, breadfruit, sugarcane, taro and yam, and evidence of patterns of genetic diversity will inevitably inform future genetic resources programme strategies and future surveys. *Brassica nigra* is a very widely distributed wild species which has known useful traits for *Brassica* crop improvement. However, little is known about the patterns and distribution of its diversity worldwide, and so SSRs are being used to plan strategies for its conservation (Westman and Kresovich, 1999). Petit *et al.* (1998) have considered the question of how best to assess diversity when planning a conservation strategy. They have studied an endangered tree species of Morocco (the argan tree), and using isozyme and chloroplast DNA data, developed a new method which emphasizes allelic richness rather than allele frequency (see above) to identify priority areas for conservation. Similarly, the conservation prospects for another endangered medicinal tree species (*Prunus africana*) have been improved across five countries by partitioning RAPD variation between populations and countries to enable selection of the most appropriate material for conservation (Dawson and Powell, 1999).

Improved taxonomic identification can increase the effectiveness of targeting material for collection. AFLP fingerprinting of Lima bean and wild relatives from South America has provided useful taxonomic information which will help in future germplasm collecting (Caicedo *et al.*, 1999). Improvement in our understanding of the taxonomic relationships between *Vicia* species has also led to a better understanding of the distribution of diversity, so aiding future exploration and collection (Potokina *et al.*, 1999).

Storage/conservation management

The management of increasingly larger collections of germplasm demands that procedures are developed which utilize fast and reliable methods for the identification of material, the measurement of diversity or the determination of redundancy; these will facilitate the organization and prioritization of germplasm (Virk *et al.*, 1995; Ford-Lloyd *et al.*, 1997; Jackson, 1997; Smith *et al.*, 2000).

The correct taxonomic identification of germplasm already being conserved can be achieved using molecular markers. In Asian rice, six crossability groups have been recognized comprising the bulk of the primary gene pool. Breeders require unambiguous identification of *indica* and *japonica* rices because these are regularly used for crosses (Ford-Lloyd *et al.*, 1997). This is not always possible using morphological characteristics, and breeders have turned to isozyme markers for this purpose. They can now also be unambiguously defined using other markers such as RAPD (Virk *et al.*, 1995), RFLP (Wang and Tanksley, 1989) and AFLP (Zhu *et al.*, 1998). Wild conspecific relatives of the cultivated sugar beet have been notoriously difficult to identify taxonomically, but with the use of several different marker systems this can be quite effectively achieved (Shen *et al.*, 1996).

Other examples where misidentification of germplasm can reduce the potential for utilization can be found, and molecular markers can be useful for correcting such mistakes in gene banks. SSR markers have been developed which have allowed the identification of misnamed accessions within a *Malus* germplasm collection (Hokanson *et al.*, 1998) and correct varietal identification (Gianfranceschi *et al.*, 1998). In *Capsicum* it is difficult to score the morphological traits which enable identification of the different species. Rodriguez *et al.* (1999) have, therefore, used RAPD to produce species-specific diagnostic markers. In *Oryza*, and particularly within the *O. sativa* complex of AA genome species, there is often uncertainty with regard to the allocation of germplasm to several of the species. *Oryza rufipogon* and *Oryza nivara* from Asia, and even *Oryza glumaepatula* have posed problems of identification using morphological characters. Much more precise identification can be achieved using various molecular markers (Martin *et al.*, 1997).

In terms of gene bank management, the estimation of genetic similarity using molecular markers can be taken further with regard to the identification of duplicates within collections. Gene banks have a finite capacity, and may often hold more than one sample of the same genotype. The scale of this problem cannot easily be estimated within seed gene banks, yet there are distinct advantages in being able to identify duplicate accessions, thereby focusing most effort on unique genetic materials for conservation. In the absence of molecular markers, identification of duplicate accessions has relied on comparison of morphological characters together with passport data. However, at the International Potato Center, duplicates of accessions have been routinely identified by comparison of tuber proteins coupled with morphological characters (Huaman *et al.*, 2000). In a more recent approach involving rice germplasm from the International Rice Research Institute (Jackson, 1997), tests have been carried out that use RAPD molecular markers for identifying duplicates, and schemes for using a combination of markers and passport data have been proposed (Virk *et al.*, 1995).

While most of the examples described above have general applications to the actual conservation process, and to the management of germplasm collections regardless of whether they are *in situ* or *ex situ*, and involving seeds, *in vitro* plants or field grown collections, there are some uses for molecular markers and genotyping which are more specifically related to one or other method of conservation. Some of these examples are covered below.

Seed storage

Seed gene banks represent the largest forms of germplasm collections in existence and, because of their size, present problems in terms of all aspects of management. These problems include the maintenance of the collections on the one hand, such as ensuring that seed samples are sufficiently viable and of acceptable quantity, to characterization and evaluation of the material to enhance its usefulness in terms of exploitation in crop improvement. A concept which is not new, but is receiving increased attention is that of the 'core collection'. Originally proposed by Frankel in 1984, core collections are supposed to be composed of a limited number of accessions specially chosen to be a representative sample of the total genetic variation of a crop species and its wild relatives (Brown and Spillane, 1999). There are many problems to overcome in their development, and many solutions, partly a result of the multiplicity of views about what can and should be achieved by developing core collections. Despite this, there are clearly some roles that molecular markers can play in their development. It seems likely that geographical (passport) data and morphological characterization data (Prasada Rao and Ramanathan Rao, 1995; Tohme *et al.*, 1995, 1999) will normally be most important for the effective construction of a core collection, while molecular markers can be used to good effect to test the extent to which diversity has been captured (Gepts, 1995; Ferguson *et al.*, 1998a; Divaret *et al.*, 1999; Parsons *et al.*, 1999; Huaman *et al.*, 2000).

The question of how many samples should be present in a core collection is one that can be answered to a great extent using molecular marker data. For instance, a figure derived by Lawrence *et al.* (1995), by surveying published isozyme marker data, points to the need for remarkably few accessions, and Glaszmann *et al.* (1996) used a sample of only 261 rice accessions to represent the whole array of variation in rice for several parameters, including geographical origin, culture type and particularly the classification based on isozymes.

The question of the maintenance of genetic integrity during seed conservation in terms of sample regeneration, rejuvenation and storage has received much attention, and is one where answers can be obtained by applying basic knowledge of population genetics (Hayward and Sackville-Hamilton, 1997), and without necessarily resorting to using molecular markers. However, some important questions have been raised about the accumulation of deleterious mutations and the accompanying recurrent regeneration of seed material in gene banks (Schoen *et al.*, 1998). Results based on modelling suggest that accumulating mutations have the potential to reduce the viability of stored germplasm, and more importantly to reduce the value of that germplasm when used in breeding programmes. Once again while these results have been obtained using theoretical calculations, it is suggested that more detailed studies of mutation rates in germplasm collections are needed which clearly would be assisted by using molecular marker genotyping.

In vitro *storage*

In vitro storage techniques such as slow growth storage and cryopreservation demand that the genetic stability of germplasm is considered and monitored, since somaclonal variation can arise when using plant tissue culture (Scowcroft, 1984; Israeli *et al.*, 1994), resulting in a change in the genotype of the material being conserved. Molecular marker genotyping of such material represents one way in which the genetic integrity of germplasm can be monitored (Karp and Edwards, 1997). However, the major problem with any form of assessment of genetic stability is that only a very small portion of any plant genome can be assessed for genetic change, and therefore no method can guarantee that no genetic change has taken place. Nevertheless, studies on genetic stability have been undertaken using molecular markers.

Assessment of the genetic stability of cassava shoot cultures maintained in an *in vitro* gene bank for 10 years revealed no apparent change when isozyme, DNA and morphological markers were used (IPGRI/CIAT, 1994). The molecular stability of rRNA genes varied depending upon the method of *in vitro* storage. Some changes were detected in potatoes in slow growth storage, but none in material that was cryopreserved (Harding, 1991, 1994). Isozymes and RFLPs gave inconclusive results in terms of possible genetic changes in sugarcane grown under field conditions versus cryopreservation (Glaszmann *et al.*, 1996). Isozyme, RFLP and RAPD markers each gave similar results when somaclonal variation was assessed among tissue culture regenerants of beets (*Beta vulgaris*) (Sabir *et al.*, 1992; Munthali *et al.*, 1996). Frequency per data point was found to be 0.05%. Perhaps more significantly it was highlighted that at this rate, given the size of the beet genome, on average 50 genes would be somaclonally variant in every regenerant, even though most were not detectable pheno-typically.

Banana and plantain germplasm is a focus for *in vitro* conservation, despite the fact that high levels of somaclonal variation can arise from tissue culture. Variants are often found among field-grown plants as well (Vuylsteke, 1989). RAPD markers have been used to assess how such variation might arise, and among *in vitro*-derived material has been found to be substantially pre-existing in the original shoot tips and of chimeral organization (Newbury *et al.*, 2000). Dwarf somaclones have been identified using RAPD (Damasco *et al.*, 1996). It seems likely that similar high levels of cultivar-dependent somaclonal variation can arise in garlic where *in vitro* conservation is fairly essential, and this has also been measured in terms of frequency using RAPD markers (Al-Zahim *et al.*, 1999).

Field gene banks and living collections

For germplasm maintained as living collections in field gene banks, as with *in vitro* collections, many of the same arguments in support of the use of molecular markers apply. Even if somaclonal variation is not a problem, questions concerning the genetic integrity of the germplasm and the reliability of its genotypic composition can be addressed using molecular genotyping. The very extensive United States Department of Agriculture collection of apple germplasm comprising 2500 accessions grown as trees presents many problems in terms of maintenance (Hokanson *et al.*, 1998), some of which are being solved by using SSR markers. Their use is streamlining the management of the collection by contributing to the development of a core subset, by allowing accurate identification of genotypes, and in the assessment of diversity and relatedness. Similarly, the largest cassava germplasm collections are maintained in the field and *in vitro*. The collection at CIAT contains about 5500 accessions, and the maintenance costs are quite substantial (Chavarriaga-Aguirre *et al.*, 1999). Semi-automated genotyping using fluorescently labelled SSR primers, coupled with more traditional use of isozymes and AFLPs has enabled the selection of unique genotypes for a core collection to be established, eliminating costly redundancy and ensuring adequate representation of diversity.

In situ *conservation of genetic resources*

Inevitably, when considering the conservation of wild species relatives of crops as populations in genetic reserves, it is important to be able to measure a number of population genetic parameters such as effective population size, minimum viable population and gene flow, as well as making estimates of genetic diversity, allele richness and so on (see above) to ensure that the most appropriate populations are being targeted (Hayward and Sackville-Hamilton, 1997; Lawrence and Marshall, 1997). Without molecular markers, such studies are virtually impossible. Hokanson *et al.* (1998) suggest that SSR markers can be used to manage *in situ* conservation of wild apple species. Improved genetic reserve design and the more effective identification of candidate populations of the argan tree have resulted from the use of both isozymes and chloroplast DNA markers (Petit *et al.*, 1998), while similar studies using RAPD markers have focused on the endangered Spanish cedar (Gillies *et al.*, 1997).

In situ on-farm conservation of landraces has not received quite the same attention, as yet, although there are questions that need addressing via the use of molecular markers. Some research has been undertaken by Dje *et al.* (1999) who have been able to use isozyme and SSR markers to propose that individual fields of sorghum landraces constitute valuable units of conservation.

Characterization/evaluation and utilization of germplasm

Molecular markers are appropriate for identifying useful genes within germplasm collections, and for their introduction into commercially important material. Gene mapping is important for genetic analysis of plant material (Kearsey, 1997), and maps provide a framework within which important genes can be located. They are of major significance for the detection of genes within closely or more distantly related germplasm. Hence they will be important when more distantly related species or even genera are being evaluated.

The location of genes of major effect can employ molecular markers, and the technique of bulked segregant analysis (Michelmore *et al.*, 1991) can be used for evaluating widely differing germplasm. Molecular markers have also become invaluable for analysing variation in quantitative traits, and the detection and location of QTL (Kearsey, 1997). While it will be necessary in many situations to have access to segregating populations derived from specific crosses to undertake QTL analysis, an approach allowing the detection but not the mapping of QTL, involving the analysis of 'raw' germplasm collections, has also been demonstrated (Virk *et al.*, 1996).

The transfer of 'alien' genes in 'wide crosses' can be assisted with molecular markers (marker-assisted gene transfer) and is achieved by using markers for a particular chromosome segment within which the gene of interest has been located. The transfer of the segment can then be monitored and traced during successive backcrosses to ensure that the target gene is still present in subsequent generations. Molecular markers can also be used to assist in the actual utilization process within plant breeding programmes, taking the form of marker-assisted selection (Ribaut and Hoisington, 1998). Linked markers help to trace alleles of target genes through the different generations of crosses in a breeding programme, making the production of new cultivars quicker and more efficient (Ribaut and Betran, 1999).

More appropriate selection of parental material in breeding programmes for *Phaseolus* using AFLP is suggested by Caicedo *et al.* (1999), and similarly for maize using RFLP (Dubreuil and Charcosset, 1999). Improved heterosis from the use of marker identification of suitable germplasm is proposed by Brummer (1999) for forage crops. Selection of superior genetic resources for the improvement of turfgrass using DNA profiling is suggested by Caetano-Anolles (1998), and similarly for many other crop–marker combinations (Ghesquiere *et al.*, 1997; Liu and Date, 1997; Pickersgill, 1997; Lanham and Brennan, 1998; Perera *et al.*, 1998; Senior *et al.*, 1998; Von Braun and Virchow, 1998).

The future

Greater availability of molecular markers makes the prospect for accurately defining genetic resources, in terms of the genetic diversity that they represent, much more achievable. Hence it becomes increasingly possible to answer the question 'are we conserving enough germplasm for future potential use'. Large collections already exist (particularly of the major cereal crops), where it is arguable that sufficient numbers of samples of the actual crop species are already in store. What of the numbers of lesser crops, and wild species? What proportion of the total diversity for these is being conserved, and how much remains to be conserved? These are questions which can now be addressed by way of molecular markers, and which need to be addressed before further collections are added to gene banks in the future. On the other hand, there would appear to be vast gaps in collections as far as wild species are concerned, and with increasing emphasis on their significance in crop improvement, such gaps should be filled. As examples, out of over 420,000 accessions of rice conserved worldwide, only 1% represents wild species, and for potato, 5% of 31,000 accessions are wild species (FAO, 1998). For the minor crops, where there are relatively fewer accessions being conserved worldwide, there are also often negligible numbers of wild relatives being conserved. What germplasm to choose in order to fill these gaps, and where to obtain it can be answered by using molecular markers to characterize the diversity already being conserved, and to determine what new variation would be contributed by adding any new accessions. Geographical surveying using molecular markers would assist in identifying new regions for collecting such novel germplasm.

Improvements in methods of identifying which germplasm to use for crop improvement may come about in a number of ways. The rapidly growing understanding of molecular genetics will affect the way that we utilize and exploit conserved genetic diversity. At the simplest level, molecular markers will allow us to perform marker-assisted evaluation of germplasm; perhaps even to locate genes of interest without having to screen the germplasm for the phenotypic trait of interest. This can most easily be done for major genes conferring pest and disease resistance, but will be extended to genes contributing to quantitative traits as well. At a more complex level, an increasing understanding of comparative genomics will allow us to use the information that we have about one species (crop or otherwise) to predict and locate genes present in a completely different species, leading to its increased exploitation. As an example, rice is regarded as a model crop species among the cereals because of its small genome size. Because of the high level of synteny existing between cereal species, we can use the genetic map of rice as a reference point for exploring the larger and more complex genomes of wheat and maize and even other grasses (Jackson et al., 1999). The pay-off in other crops may be even greater. Among the legumes, there are many different crop species which, although of local importance, are relatively minor, and will be unlikely to attract funding in the future for major molecular genetic study and exploitation. If, because of synteny among these legume species, mapping information from those that are more important (such as soybean) can be usefully shared among lesser legumes, then more rapid improvements may be made. This could not be achieved without an adequate supply of molecular markers.

Further advances involving molecular genetics and the evaluation of genetic resources are likely to arise from new marker technologies. The development of libraries of expressed sequence tags may lead much more directly to the gene of interest within a germplasm collection. Such technology, combined with even more advanced systems involving microarrays and DNA chips, will undoubtedly take genetic resources evaluation into a different era. The potential for evaluating germplasm on a much larger scale by being able to place 100,000 samples of DNA representing a whole crop or species collection on to one microscope slide, and then analysing for particular polymorphisms is enormous. With increasing amounts of DNA sequence information becoming publicly available, the potential for developing microarrays for screening SNPs among whole germplasm collections is becoming feasible. If large collections by definition are difficult and expensive to maintain and exploit, then the use of biotechnological advances surrounding microarrays and SNPs will improve the efficiency of germplasm management by allowing for the identification of redundant samples, increasing the accuracy of taxonomic identification and the targeting of subsamples of a collection to maximize genetic diversity and useful genes within core collections.

References

Adams, R.P. (1997) Conservation of DNA: DNA banking. In: Callow, J.A., Ford-Lloyd, B.V. and Newbury, H.J. (eds) *Biotechnology and Plant Genetic Resources: Conservation and Use*. CAB International, Wallingford, UK, pp. 163–174.

Akagi, H., Yokozeki, Y., Inagaki, A. and Fujimura, T. (1998) Origin and evolution of twin microsatellites in the genus *Oryza*. *Heredity* 81, 187–197.

Al-Zahim, M.A., Ford-Lloyd, B.V. and Newbury, H.J. (1999) Detection of somaclonal variation in garlic (*Allium sativum* L.) using RAPD and cytological analysis. *Plant Cell Reports* 18, 473–477.

Ashmore, S.A. (1997) *Status Report on the Development and Application of in vitro Techniques for the Conservation and Use of Plant Genetic Resources*. IPGRI, Rome.

Bjornstad, A., Demissie, A., Kilian, A. and Kleinhofs, A. (1997) The distinctness and diversity of Ethiopian barleys. *Theoretical and Applied Genetics* 94, 514–521.

Brown, A.H.D. and Spillane, C. (1999) Implementing core collections – principles, procedures, progress, problems and promise. In: Johnson, R.C. and Hodgkin, T. (eds) *Core Collections for Today and Tomorrow*. IPGRI, Rome, pp. 1–9.

Brummer, E.C. (1999) Capturing heterosis in forage crop cultivar development. *Crop Science* 39, 943–954.

Caetano-Anollés, G. (1998) DNA analysis of turfgrass genetic diversity. *Crop Science* 38, 1415–1424.

Caicedo, A.L., Gaitan, E., Duque, M.C., Chica, O.T., Debouck, D.G. and Tohme, J. (1999) AFLP fingerprinting of *Phaseolus lunatus* L. and related wild species from South America. *Crop Science* 39, 1497–1507.

Chavarriaga-Aguirre, P., Maya, M.M., Tohme, J., Duque, M.C., Iglesias, C., Bonierbale, M.W., Kresovich, S. and Kochert, G. (1999) Using microsatellites, isozymes and AFLPs to evaluate genetic diversity and redundancy in the cassava core collection and to assess the usefulness of DNA-based markers to maintain germplasm collections. *Molecular Breeding* 5, 263–273.

Damasco, O.P., Graham, G.C., Henry, R.J., Adkins, S.W., Smith, M.K. and Godwin, I.D. (1996) Random amplified polymorphic DNA (RAPD) detection of dwarf off-types in micropropagated Cavendish (*Musa* ssp. AAA) bananas. *Plant Cell Reports* 16, 118–123.

Dawson, I.K. and Powell, W. (1999) Genetic variation in the Afromontane tree *Prunus africana*, an endangered medicinal species. *Molecular Ecology* 8, 151–156.

Divaret, I., Margale, E. and Thomas, G. (1999) RAPD markers on seed bulks efficiently assess the genetic diversity of a *Brassica oleracea* L. collection. *Theoretical and Applied Genetics* 98, 1029–1035.

Dje, Y., Forcioli, D., Ater, M., Lefebvre, C. and Vekemans, X. (1999) Assessing population genetic structure of sorghum landraces from North-western Morocco using allozyme and microsatellite markers. *Theoretical and Applied Genetics* 99, 157–163.

Doebly, J. (1990) Molecular systematics of *Zea* (Gramineae). *Maydica* 35, 143–50.

Doebly, J.F., Goodman, M.M. and Stuber, C.W. (1984) Isoenzymatic variation in *Zea* (Gramineae). *Systematic Botany* 9, 203–218.

Dubreuil, P. and Charcosset, A. (1999) Relationships among maize inbred lines and populations from European and North-American origins as estimated using RFLP markers. *Theoretical and Applied Genetics* 99, 473–480.

Engelmann, F. (1997) *In vitro* conservation methods. In: Callow, J.A., Ford-Lloyd, B.V. and Newbury, H.J. (eds) *Biotechnology and Plant Genetic Resources*. CAB International, Wallingford, UK, pp. 119–161.

Excoffier, L., Smouse, P. and Quattro, J. (1992) Analysis of molecular variance inferred from metric distances among DNA haplotypes: application to human mitochondrial DNA restriction data. *Genetics* 131, 479–491.

FAO (1998) *The State of the World's Plant Genetic Resources for Food and Agriculture*. FAO, Rome.

Ferguson, M.E., Ford-Lloyd, B.V., Robertson, L.D., Maxted, N. and Newbury, H.J. (1998a) Mapping the geographical distribution of genetic variation in the genus *Lens* for the enhanced conservation of plant genetic diversity. *Molecular Ecology* 7, 1743–1755.

Ferguson, M.E., Robertson, L.D., Ford-Lloyd, B.V., Newbury, H.J. and Maxted, N. (1998b) Contrasting genetic variation among lentil landraces from different geographical origins. *Euphytica* 102, 265–273.

Ford-Lloyd, B.V., Jackson, M.T. and Newbury, H.J. (1997) Molecular markers and the management of genetic resources in seed genebanks: a case study of rice. In: Callow, J.A., Ford-Lloyd, B.V. and Newbury, H.J. (eds) *Biotechnology and Plant Genetic Resources: Conservation and Use*. CAB International, Wallingford, UK, pp. 103–118.

Frankel, O.H. (1984) Genetic perspectives in germplasm conservation. In: Arber, W.K., Llimensee, K., Peacock, W.J. and Starlinger, P. (eds) *Genetic Manipulation: Impact on Man and Society*. Cambridge University Press, Cambridge, pp. 161–170.

Frison, E.A. and Bolton, M. (1994) *Proceedings of a Joint FAO/IPGRI Workshop on ex situ Germplasm Conservation, 7–9 October 1993, Prague, Czech Republic*. IPGRI, Rome.

Gepts, P. (1995) Genetic markers and core collections. In: Hodgkin, T., Brown, A.H.D., van Hintum, Th.J.L. and Morales, E.A.V. (eds) *Core Collections of Plant Genetic Resources*. John Wiley & Sons, Chichester, UK, pp. 127–146.

Ghesquiere, A., Sequier, J., Second, G. and Lorieux, M. (1997) First steps towards a rational use of African rice, *Oryza glaberrima*, in rice breeding through a 'contig line' concept. *Euphytica* 96, 31–39.

Gianfranceschi, L., Seglias, N., Tarchini, R., Komjanc, M. and Gessler, C. (1998) Simple sequence repeats for the genetic analysis of apple. *Theoretical and Applied Genetics* 96, 1069–1076.

Gillies, A.C.M., Cornelius, J.P., Newton, A.C., Navarro, C., Hernandez, M. and Wilson, J. (1997) Genetic variation in Costa Rican populations of the tropical timber species *Cedrela odorata* L., assessed using RAPDs. *Molecular Ecology* 6, 1133–1145.

Glaszmann, J.C., Mew, T., Hibino, H., Kim, C.K., Vergel de Dios-Mew, T.I., Vera Cruz, C.M., Notteghem, J.L. and Bonman, J.M. (1996) Molecular variation as a diverse source of disease resistance in cultivated rice. In: Khush, G.S. (ed.) *Rice Genetics III*.

Proceedings of Third International Rice Genetics Symposium. IRRI, Manila, pp. 460–466.

Gonzalez-Candelas, F. and Palacios, C. (1997) Analyzing molecular data for studies of genetic diversity. In: Ayad, W.G., Hodgkin, T., Jaradat, A. and Rao, V.R. (eds) *Molecular Genetic Techniques for Plant Genetic Resources. Report of an IPGRI workshop, 9–11 October 1995, Rome, Italy.* IPGRI, Rome, pp. 55–80.

Harding, K. (1991) Molecular stability of the ribosomal RNA genes in *Solanum tuberosum* plants recovered from slow growth and cryopreservation. *Euphytica* 55, 141–146.

Harding, K. (1994) The methylation status of DNA derived from potato plants recovered from slow growth. *Plant Cell Tissue and Organ Culture* 37, 31–38.

Hayward, M.D. and Sackville-Hamilton, N.R. (1997) Genetic diversity – population structure and conservation. In: Callow, J.A., Ford-Lloyd, B.V. and Newbury, H.J. (eds) *Biotechnology and Plant Genetic Resources.* CAB International, Wallingford, UK, pp. 119–161.

Hokanson, S.C., Szewc-McFadden, A.K., Lamboy, W.F. and McFerson, J.R. (1998) Microsatellite (SSR) markers reveal genetic identities, genetic diversity and relationships in a *Malus* × *domestica* Borkh. core subset collection. *Theoretical and Applied Genetics* 97, 671–683.

Huaman, Z., Ortiz, R., Zhang, D.P. and Rodriguez, F. (2000) Isozyme analysis of entire and core collections of *Solanum tuberosum* subsp *andigena* potato cultivars. *Crop Science* 40, 273–276.

IPGRI/CIAT (1994) Establishment and operation of a pilot *in vitro* active genebank. *Report of a CIAT–IPGRI Collaborative Project Using Cassava (Manihot esculenta Crantz) as a Model.* IPGRI, Rome.

Israeli, Y., Laha, E. and Reuveni, O. (1994) *In vitro* culture of bananas. In: Gowen, S. (ed.) *Bananas and Plantains.* Chapman & Hall, London, pp.147–178.

Jackson, M.T. (1997) Conservation of rice genetic resources: the role of the International Rice Genebank at IRRI. *Plant Molecular Biology* 35, 61–67.

Jackson, M.T., Pham, J.L., Newbury, H.J., Ford-Lloyd, B.V. and Virk, P.S. (1999) A core collection for rice: needs, opportunities, and constraints. In: Johnson, R.C. and Hodgkin, T. (eds) *Core Collections for Today and Tomorrow.* IPGRI, Rome, pp. 18–27.

Karp, A. and Edwards, K.J. (1997) Molecular techniques in the analysis of the extent and distribution of genetic diversity. In: Ayad, W.G., Hodgkin, T., Jaradat, A. and Rao, V.R. (eds) *Molecular Genetic Techniques for Plant Genetic Resources: Report of an IPGRI Workshop, 9–11 October 1995, Rome, Italy.* IPGRI, Rome.

Karp, A., Isaacs, P.G. and Ingram, D.S. (1998) *Molecular Tools for Screening Biodiversity.* Chapman & Hall, London.

Kearsey, M.J. (1997) Genetic resources and plant breeding. In: Callow, J.A., Ford-Lloyd, B.V. and Newbury, H.J. (eds) *Biotechnology and Plant Genetic Resources: Conservation and Use.* CAB International, Wallingford, UK, pp. 175–202.

Lanham, P.G. and Brennan, R.M. (1998) Characterization of the genetic resources of redcurrant (*Ribes rubrum*: subg. *Ribesia*) using anchored microsatellite markers. *Theoretical and Applied Genetics* 96, 917–921.

Lawrence, M.J. and Marshall, D.F. (1997) Plant population genetics. In: Maxted, N., Ford-Lloyd, B.V. and Hawkes, J.G. (eds) *Plant Genetic Conservation: the* In-situ *Approach.* Chapman & Hall, London, pp. 99–113.

Lawrence, M.J., Marshall, D.F. and Davies, P. (1995) Genetics of genetic conservation. 1. Sample size when collecting germplasm. *Euphytica* 84, 89–99.

Lebot, V. (1999) Biomolecular evidence for plant domestication in Sahul. *Genetic Resources and Crop Evolution* 46, 619–628.

Liu, C.J. and Date, R.A. (1997) The use of genetic markers to improve seed and RNB collection and genetic conservation. *Tropical Grasslands* 31, 355–358.

Martin, C., Juliano, A., Newbury, H.J., Lu, B.-R., Jackson, M.T. and Ford-Lloyd, B.V. (1997) The use of RAPD markers to facilitate the identification of *Oryza* species within a germplasm collection. *Genetic Resources and Crop Evolution* 44, 175–183.

Maxted, N., Ford-Lloyd, B.V. and Hawkes, J.G. (eds) (1997) *Plant Genetic Conservation: the* in-situ *Approach*. Chapman & Hall, London.

Michelmore, R.W., Paran, W.I. and Kesseli, R.V. (1991) Identification of markers linked to disease-resistance genes by bulked segregant analysis; a rapid method to detect markers in specific genomic regions by using segregating populations. *Proceedings of the National Academy of Sciences USA* 88, 9828–9832.

Munthali, M.T., Newbury, H.J. and Ford-Lloyd, B.V. (1996) The detection of somaclonal variants of beet using RAPD. *Plant Cell Reports* 15, 474–478.

Nei, M. (1973) Analysis of gene diversity in subdivided populations. *Proceedings of the National Academy of Sciences USA* 70, 3321–3323.

Nevo, E., Baum, B., Beiles, A. and Johnson, D.A. (1998) Ecological correlates of RAPD DNA diversity of wild barley, *Hordeum spontaneum*, in the Fertile Crescent. *Genetic Resources and Crop Evolution* 45, 151–159.

Newbury, H.J. and Ford-Lloyd, B.V. (1997) Estimating genetic diversity. In: Maxted, N., Ford-Lloyd, B.V. and Hawkes, J.G. (eds) *Plant Genetic Conservation: the* in-situ *Approach*. Chapman & Hall, London, pp. 192–206.

Newbury, H.J., Howell, E.C., Crouch, J.H. and Ford-Lloyd, B.V. (2000) Natural and culture-induced genetic variation in plantains (*Musa* spp. AAB group). *Australian Journal of Botany* 48, 493–500.

Parsons, B., Newbury, H.J., Jackson, M.T. and Ford-Lloyd, B.V. (1999) The genetic structure and conservation of aus, aman and boro rices from Bangladesh. *Genetic Resources and Crop Evolution* 46, 587–598.

Parsons, B.J., Newbury, H.J., Jackson, M.T. and Ford-Lloyd, B.V. (1997) Contrasting genetic diversity relationships are revealed in rice (*Oryza sativa* L.) using different marker types. *Molecular Breeding* 3, 115–125.

Perera, L., Russell, J.R., Provan, J., McNicol, J.W. and Powell, W. (1998) Evaluating genetic relationships between indigenous coconut (*Cocos nucifera* L.) accessions from Sri Lanka by means of AFLP profiling. *Theoretical and Applied Genetics* 96, 545–550.

Perera, L., Russell, J.R., Provan, J. and Powell, W. (2000) Use of microsatellite DNA markers to investigate the level of genetic diversity and population genetic structure of coconut (*Cocos nucifera* L.). *Genome* 43, 15–21.

Petit, R.J., El Mousadik, A. and Pons, O. (1998) Identifying populations for conservation on the basis of genetic markers. *Conservation Biology* 12, 844–855.

Pickersgill, B. (1997) Genetic resources and breeding of *Capsicum* spp. *Euphytica* 96, 129–133.

Potokina, E., Tomooka, N., Vaughan, D.A., Alexandrova, T. and Xu, R.Q. (1999) Phylogeny of *Vicia* subgenus *Vicia* (Fabaceae) based on analysis of RAPDs and RFLP of PCR-amplified chloroplast genes. *Genetic Resources and Crop Evolution* 46, 149–161.

Prasada Rao, K.E. and Ramanatha Rao, V. (1995) The use of characterization data in developing a core collection of sorghum. In: Hodgkin, T., Brown, A.H.D., van

Hintum, Th.J.L. and Morales, E.A.V. (eds) *Core Collections of Plant Genetic Resources*. John Wiley & Sons, Chichester, UK, pp. 109–115.

Ribaut, J.M. and Betran, J. (1999) Single large-scale marker-assisted selection (SLS-MAS). *Molecular Breeding* 5, 531–541.

Ribaut, J.M. and Hoisington, D. (1998) Marker-assisted selection: new tools and strategies. *Trends in Plant Science* 3, 236–239.

Roberts, E.H. (1973) Predicting the storage life of seeds. *Seed Science and Technology* 1, 499–514.

Rodriguez, J.M., Berke, T., Engle, L. and Nienhuis, J. (1999) Variation among and within *Capsicum* species revealed by RAPD markers. *Theoretical and Applied Genetics* 99, 147–156.

Sabir, A., Newbury, H.J., Todd, G. and Ford-Lloyd, B.V. (1992) Determination of genetic stability using isozymes and RFLPs in beet plants regenerated *in vitro*. *Theoretical and Applied Genetics* 84, 113–117.

Schoen, D.J., David, J.L. and Bataillon, T.M. (1998) Deleterious mutation accumulation and the regeneration of genetic resources. *Proceedings of the National Academy of Sciences USA* 95, 394–399.

Scowcroft, W.R. (1984) *Genetic Variability in Tissue Culture: Impact on Germplasm Conservation and Utilization*. IBPGR, Rome.

Senior, M.L., Murphy, J.P., Goodman, M.M. and Stuber, C.W. (1998) Utility of SSRs for determining genetic similarities and relationships in maize using an agarose gel system. *Crop Science* 38, 1088–1098.

Shen, Y., Newbury, H.J. and Ford-Lloyd, B.V. (1996) The taxonomic characterization of annual *Beta* germplasm in a genetic resources collection using RAPD markers. *Euphytica* 91, 205–212.

Shen, Y., Ford-Lloyd, B.V. and Newbury, H.J. (1998) Genetic relationships within the genus *Beta* determined using both PCR-based marker and DNA sequencing techniques. *Heredity* 80, 624–632.

Smith, J.S.C., Kresovich, S., Hopkins, M.S., Mitchell, S.E., Dean, R.E., Woodman, W.L., Lee, M. and Porter, K. (2000) Genetic diversity among elite sorghum inbred lines assessed with simple sequence repeats. *Crop Science* 40, 226–232.

Tohme, J., Jones, P., Beebe, S. and Iwanaga, M. (1995) The combined use of agroecological and characterization data to establish the CIAT *Phaseolus vulgaris* core collection. In: Hodgkin, T., Brown, A.H.D., van Hintum, Th.J.L. and Morales, E.A.V. (eds) *Core Collections of Plant Genetic Resources*, John Wiley & Sons, Chichester, UK, pp. 95–107.

Tohme, J., Beebe, S. and Iglesias, C. (1999) Molecular characterization of the CIAT bean and cassava core collections. In: Johnson, R.C. and Hodgkin, T. (eds) *Core Collections for Today and Tomorrow*. IPGRI, Rome, pp. 28–36.

UNCED (1992) *Convention on Biological Diversity*. United Nations Conference on Environment and Development, Geneva.

Virk, P., Newbury, H.J., Jackson, M.T. and Ford-Lloyd, B.V. (1995) The identification of duplicate accessions within a rice germplasm collection using RAPD analysis. *Theoretical and Applied Genetics* 90, 1049–1055.

Virk, P., Ford-Lloyd, B.V., Jackson, M.T. and Newbury, H.J. (1996) Predicting quantitative variation within rice germplasm using molecular markers. *Heredity* 76, 296–304.

Virk, P.S., Zhu, J., Newbury, H.J., Bryan, G.J., Jackson, M.T. and Ford-Lloyd, B.V. (2000a) Effectiveness of different classes of molecular marker for classifying and revealing variation in rice (*Oryza sativa*) germplasm. *Euphytica* 112, 275–284.

Virk, P.S., Newbury, H.J., Jackson, M.T. and Ford-Lloyd, B.V. (2000b) Are mapped markers more useful for assessing genetic diversity? *Theoretical and Applied Genetics* 100, 607–613.

Von Braun, J. and Virchow, D. (1998) Plant genetic resources between supply and demand – development of institutional framework conditions for conservation and use. *Berichte uber Landwirtschaft* 76, 74–86.

Vuylsteke, D. (1989) *Shoot-tip Culture for the Propagation, Conservation and Exchange of Musa Germplasm. Practical Manuals for Handling Crop Germplasm* in vitro 2. IBPGR, Rome.

Wang, Z.Y. and Tanksley, S.D. (1989) Restriction fragment length polymorphism in *Oryza sativa* L. *Genome* 32, 1113–1118.

Westman, A.L. and Kresovich, S. (1997) Use of molecular marker techniques for description of plant genetic variation. In: Callow, J.A., Ford-Lloyd, B.V. and Newbury, H.J. (eds) *Biotechnology and Plant Genetic Resources: Conservation and Use.* CAB International, Wallingford, UK, pp. 9–48.

Westman, A.L. and Kresovich, S. (1999) Simple sequence repeat (SSR)-based marker variation in *Brassica nigra* genebank accessions and weed populations. *Euphytica* 109, 85–92.

Zhu, J., Gale, M.D., Quarrie, S., Jackson, M.T. and Bryan, G.J. (1998) AFLP markers for the study of rice biodiversity. *Theoretical and Applied Genetics* 96, 602–611.

Zimmer, E.A., Jupe, E.R. and Walbot, V. (1988) Ribosomal gene structure, variation, and inheritance in maize and its ancestors. *Genetics* 120, 1125–1136.

Chapter 6

Applications of Molecular Marker Techniques to the Use of International Germplasm Collections

M. WARBURTON AND D. HOISINGTON

Applied Biotechnology Center, International Maize and Wheat Improvement Center, CIMMYT, México

Introduction

Molecular genetic markers have the potential to be a powerful tool in managing plant germplasm collections, both *in situ* and *ex situ*. As molecular techniques advance, the efficiency, amount of information generated and ease of use increase while the cost and time involved decrease. Conversely, costs associated with field space, repository space, and labour associated with curation of a germplasm repository (including seed testing, regeneration and seed dissemination) are increasing. Therefore, every tool available which make the collection less expensive to store and analyse, and more useful to plant improvement, must be exploited to its fullest potential.

The measurement of genetic diversity and the fingerprinting of genotypes using molecular markers are two distinct, but related, techniques. Genetic diversity is a relative measure of the genetic distances between genotypes in a defined set or study using a pre-selected number of markers. These distances depend on the composition of the genotypes in the set, and the markers used for measurement. Fingerprinting is an absolute measure of the genetic makeup of an individual or a line, and must be unique to that individual or line in order to distinguish it from all others (not only in one study, but from all others in existence, except for twins, clones and completely identical inbreds or doubled haploids). Both genetic diversity studies and fingerprinting use the same techniques, and in some instances in this chapter will be used interchangeably.

Genetic markers have been proposed to aid in plant genetic resource management in resource acquisition (sampling strategies and determining which groups may be under-represented in a collection), maintenance (maintaining trueness to type and monitoring changes in allele frequencies in populations), characterization (of newly acquired accessions when few or no data

© CAB *International* 2001. *Plant Genotyping: the DNA Fingerprinting of Plants* (ed. R.J. Henry)

are available) and utilization (aiding in pre-breeding and introgression and marker-assisted selection) (Bretting and Widrlechner, 1995). Specifically, fingerprinting and genetic diversity measures can be of use to germplasm bank curators and breeders in the following capacities:

1. To search for correlations of traits and markers (without mapping) in related individuals using a common database to store multiple types of data.
2. To narrow the search for new alleles in loci of interest.
3. To verify pedigrees and fill in the gaps in incomplete pedigree or selection history.
4. To assign lines and populations to heterotic groups.
5. To choose parents for mapping, marker-assisted selection and backcrossing schemes.
6. To monitor changes in allele frequencies in populations.
7. To study the evolutionary history of wild relatives.

Fingerprinting applications for the maize and wheat collections of CIMMYT

The CIMMYT genebank consists of over 145,000 accessions of maize and wheat and their relatives, and every year CIMMYT breeders release many new breeding lines of both species. The importance of these lines and accessions to international agriculture cannot be underestimated (Mann, 1997). Many of the uses of molecular markers cited above are highlighted in this chapter using ongoing CIMMYT fingerprinting projects of these diverse materials.

The International Crop Information System (ICIS): the power of databases; correlation of traits and markers (without mapping) in related individuals

Currently, if one wishes to find new information on genetic linkages (for example between markers and loci of traits of interest), one must conduct an extensive mapping experiment to identify and confirm that two or more loci are indeed linked. When data on molecular markers generated in a fingerprinting study are included in a database with phenotypic and pedigree data, it becomes possible to find putative linkages without mapping. All traits of interest can be correlated with the marker information for all related individuals for which there are data, and close linkages will show up as a constant association between certain alleles of the marker and certain phenotypes in most or all of the related individuals. Because recombination will break the association of loci that are not found close to each other on the chromosomes, only closely linked markers will be identified in this way. In the past, some associations may have been serendipitous in small diversity studies (Bretting and Wildrlechner, 1995) but large-scale fingerprinting studies and further advances in database capabilities will allow for directed searches of linkages to occur in a more efficient manner. Large-scale fingerprinting studies in wheat (funded by the private Eislen foundation of Germany and the GTZ agency of the German government) and in maize (funded by the BMZ agency of the German government) will provide marker information on thousands of individuals that will be stored in ICIS. Field data, phenotypic characterization, pedigrees and passport data are available for thousands of individuals stretching back several decades. In addition, CIMMYT and the Consultative Group on International Agricultural Research (CGIAR) are developing the ICIS Project, 2000. ICIS is a database of pedigree and origin (passport) data, phenotypic and field trial data, and molecular marker data that will soon allow these different types of data to be stored and searched simultaneously. The more data that are stored and simultaneously accessible on each individual, the more efficient the search and discovery process will become.

Narrowing the search for new alleles in loci of interest using information on pedigree and seeking new alleles of previously cloned genes

Germplasm collections are known to be the repository of many different alleles for agronomically useful loci. The main hindrance in the exploitation of these alleles has been the lack of characterization for most of these loci. Many of the loci have only recently been discovered, and may be well-characterized at the molecular level (with a known sequence or linked molecular markers) or may be known only for the phenotypic effect. Either way, it is a top priority for useful exploitation of genebank resources to search for previously unknown alleles for these loci. These alleles can then be tested for favourable performance and targeted for transfer to new cultivars, via backcross breeding or genetic transformation. However, it would be nearly impossible to check all accessions in the genebank for new alleles, even for the simplest case of a well-characterized single gene trait. An efficient method to maximize the probability of finding the most new alleles possible while testing the fewest accessions possible would be to group accessions based on relationship and check only a few accessions within each group. Using the assumption that closely related lines would more often carry the same allele at all loci than distantly related lines, one could avoid redundant testing of lines which have a low probability of carrying a unique allele at the loci of interest.

CIMMYT has developed a series of approximately 120 wheat sister lines known as the 'Bobwhite' lines. These lines originated from a cross of Aurora//Kalyan/Bluebird/3/Woodpecker (Skovmand *et al.*, 1997). These lines have been selected at different generations of inbreeding in many different environments by different breeders around the world, and the resulting sister lines segregate for many traits. One trait of particular interest is the ability to regenerate in tissue culture, which has been demonstrated to be very easy for some

'Bobwhite' lines (Cheng *et al.*, 1997; Fellers *et al.*, 1997). We are testing which lines regenerate the best, and whether closely related lines (those derived from the same immediate parent, for example) share the same level of regenerability as opposed to more distantly related lines. This will help to clear up confusion in different labs using different 'Bobwhite' lines for transformation and will determine which line or lines are the best for this purpose. At the same time, we will test the use of relationship information to narrow down the search for new alleles.

Further mining of data stored in databases will one day open up a most important function of genebanks: the retrieval of novel alleles of previously cloned loci. Many of these new alleles are currently hidden in undesirable genetic backgrounds (poorly performing landraces, wild or weedy ancestors, etc.) and will not be uncovered by conventional germplasm characterization programmes (Tanksley and McCouch, 1997). These alleles can be discovered by seeking new polymorphisms in the sequence of the loci of interest, or novel expression patterns using DNA chip technology. Individuals containing new polymorphisms or levels in expression would be subjected to rigorous characterization for the trait of interest; alleles that display new expression patterns would be transferred to breeding lines via backcrossing or genetic transformation for further characterization of agronomic performance. If data on polymorphism in the gene of interest are not already in the database, rather than try to sequence this loci for all lines in the genebank, data from markers linked to this region could provide an idea of which individuals in the genebank to sequence. One could choose only those lines containing unique variation in this region of the genome, as indicated by unique alleles of markers linked to this region. Conversely, this method can be used to choose the most diverse (and potentially novel) individuals for DNA expression characterization. In this way, data mining will lead to allele mining and the under-exploited variation in the genebanks will be targeted for use in the most efficient manner possible. The wheat characterization project funded by Eislen and GTZ and the maize characterization project funded by BMZ will provide much of the genetic diversity data in the database.

Verification of pedigrees and filling in the gaps in incomplete pedigree or selection history

Databases which store both pedigree and marker data can also be used to verify that reported pedigrees are indeed correct. In the process of developing new breeding lines or cultivars, repeated crossing, selection, and growing out and harvesting of progeny seed provide many possibilities for mislabelling or accidental mixing or exchanging of seeds. Furthermore, in the past, many records were kept by hand and re-copied multiple times until they were eventually (in most cases) computerized. All these opportunities for mistakes could build up until a reported pedigree is not actually correct. In a database, for any given individual, reported pedigrees can be confirmed using markers for which data are reported in the individual and its ancestors. Alleles of the markers should be inherited in a Mendelian fashion; if more than one or two markers are found which do not agree with predicted segregation (one or two could be attributed to errors in the fingerprinting data) one should assume that the reported pedigree is incorrect. In addition, the database and marker data could be used to assign a putative parent to individuals with a pedigree that has been demonstrated to be incorrect.

The 'Bobwhite' study highlights this use of marker data and another important use in collections: the ability to fill in gaps in pedigree or selection history. Some of the 'Bobwhite' sister lines were selected by breeders outside of CIMMYT and seed returned to the genebank; however, selection history for these lines was frequently missing. Cluster analysis of all the 'Bobwhite' sister lines using molecular data allowed us to estimate which of the sister lines these unknown lines were most closely related to, thus allowing us to fill in the most likely selection history for these lines. Once molecular data are routinely generated for all new breeding lines and cultivars, these data can be stored in a database and checked for Mendelian inheritance against the parents reported for these lines. This can quickly uncover any discrepancies, which would indicate an incorrect pedigree, and allow us to determine a more probable pedigree for the incorrectly labelled line. This could greatly improve not only record keeping but, more importantly, gain from selection, as mistakes in pedigree cause different results in crosses from those predicted from the pedigree of the lines being crossed. This may also be one cause of the low correlation between coefficient of parentage and relationships calculated based on molecular genetic data (Plaschke *et al.*, 1995; Warburton and Bliss, 1996).

Assigning lines and populations to heterotic groups

CIMMYT maize breeders and, more recently, wheat breeders are working to improve the performance of hybrids by refining the heterotic groups to which lines, pools and populations are assigned (Hallauer and Miranda, 1981). New germplasm is also being generated, which must first be placed in the proper heterotic group for use in heterosis breeding. Placing a line or population in its correct heterotic group can be time- and labour-intensive if one uses crosses to define this. Decreasing the number of crosses by using testers can cause mistakes in assigning lines and populations to a heterotic group if a sufficient number of testers are not used, or if the testers chosen are not representative of the heterotic group from which they were chosen. The use of molecular markers has been suggested as an alternative to assign lines and populations to heterotic groups. Although it is not currently possible to predict the exact level of heterosis expected for a given cross based on genetic distance estimates (Lee *et al.*, 1989; Bernardo, 1998) success has been reported for assigning the lines to the proper heterotic group (Smith *et al.*, 1990; Melchinger, 1993).

CIMMYT is currently conducting several studies related to this topic which we hope will aid in hybrid breeding of maize and wheat. In wheat, a project will look at assigning wheat lines to heterotic groups in an attempt to break the 'yield barrier' experienced by many wheat breeders in the world. Wheat hybrid breeding is a fairly new effort, and little is known about establishing heterotic groups. This will be an ideal use of molecular markers and fingerprinting for increased breeding efficiency. In maize, the Asian Maize Biotechnology Network has worked to assign tropical inbred lines, many of which have no information available on heterotic groups, to existing or new heterotic groups (Fig. 6.1). These assignments will have to be verified using some crosses and measurement of F_1 progeny performance, but the number of crosses needed has been dramatically reduced using molecular fingerprinting. Finally, in maize breeding, a project is aimed at assigning inbred lines, pools and populations to heterotic groups; a task which has very rarely been worked on in the past due to the complexity of fingerprinting heterogeneous populations. However, fingerprinting represents the best hope for measuring the complex diversity associated with populations, which cannot reasonably be undertaken in the field for large numbers of populations.

Choosing parents for mapping, marker-assisted selection and backcrossing schemes

Molecular markers have been used very successfully in CIMMYT's Applied Biotechnology Center laboratories to map genes involved in the expression of both qualitative and quantitative (complex) traits and to use markers closely linked to these traits as an aid to the breeders via marker-assisted selection (Ribaut *et al.*, 1997; Khairallah *et al.*, 1998; Ribaut and Hoisington, 1998). When choosing parents of the mapping population, it is desirable to choose parents that have different levels of expression of the trait of interest and also to be as different in the background genome as possible. This will increase the probability of polymorphisms at as many marker loci as possible, which are essential when determining linkage distances between loci. When choosing parents for a backcrossing scheme (either using markers or not) it is important to choose a donor line which differs as little as possible from the recurrent parent in all but the gene(s) to be transferred. The more similar the rest of the genome is between the two lines, the more quickly a return to the recurrent genotype (except for the trait of interest) can be achieved. Fingerprinting studies can provide the data necessary so that the most efficient possible parents for mapping or backcrossing projects are chosen.

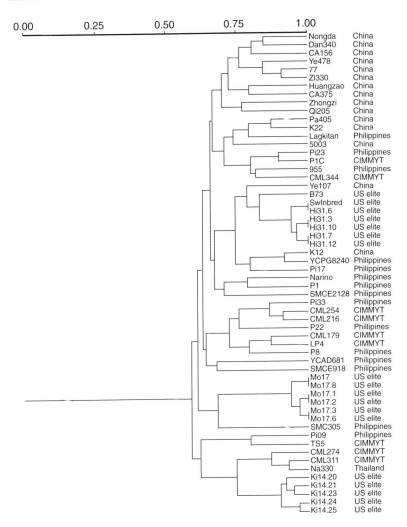

Fig. 6.1. Dendrogram resulting from simple matching coefficients of amplified fragment length polymorphism and simple sequence repeat markers measured in 40 tropical inbred lines from China, the Philippines, Thailand and CIMMYT. Eighteen commercial temperate inbreds belonging to known heterotic groups were included as a reference.

Monitoring changes in allele frequencies in populations

Allele frequencies can change in populations over time due to selection, mutation, genetic drift and migration (Falconer, 1989). These changes can be monitored using molecular markers. Changes in allele frequencies due to selection can be desirable, as in the case of breeding, or undesirable, as in the case of genotypes being maintained in germplasm collections. Mutation can only be monitored for known genes, but changes due to mutation are so rare that they can be ignored in genebanks and selection programmes. Genetic drift and migration can cause unwanted changes in allele frequencies, particularly in *in situ* or *ex situ* germplasm collections.

The CIMMYT wheat programme has, in recent years, begun an effort to increase the genetic diversity present in released breeding lines and cultivars. It is hoped that an increase in genetic diversity will lead to an increase in the yield stability of CIMMYT wheat cultivars in different environments; a decrease in the vulnerability to new pests or pathogens due to genetic uniformity; and to greater gains via selection using CIMMYT breeding lines. In a study of 32 CIMMYT wheat cultivars released over three decades, the number of polymorphic bands for both amplified fragment length polymorphism (AFLP) and simple sequence repeat (SSR) markers increased significantly (Almanza, 2000). This suggests that the genetic diversity *per se* of these lines has also increased. The CIMMYT maize programme has begun to investigate the suitability of farmer participatory breeding and maintenance of *in situ* collections of maize. In Oaxaca, Mexico, maize seeds are being collected from populations of maize grown by local farmers and compared with varieties of the same names collected in the past. We hope to monitor changes in allele frequencies of the populations which would be due to selection, migration or genetic drift, and determine the suitability of these conditions to either maintain germplasm in an unchanging form, or select new populations adapted to the specific environments of the farmers' fields.

Studying the evolutionary history of wild relatives of maize and wheat

In order to expand our access to secondary and tertiary genepools and the traits they may contain, a better understanding of relationships among and within the ancestral races and species of maize and wheat is necessary. Wide crosses allow access to genes from closely related species, but the organization of the diversity contained in these species is poorly understood. Such wide crossing strategies can be more efficient for species and for individuals that have a genomic makeup more similar to the domesticated species than to those that are more distantly related. Much has been learned about how each of the wild species is related to maize and wheat, but little is known about the variation within each of these species. It would be greatly advantageous, therefore, to study the genetic diversity within ancestral species; to group the ancestral species into distinct clusters based on this genetic diversity; and to determine from which cluster it is most likely that the cultivated species originated. This would provide information on which individuals from the ancestral species should cross most easily with the domesticated species.

CIMMYT has a potentially very valuable resource in several series of synthetic wheats constructed by crossing tetraploid durum wheats with 'D' genome diploid species. The most extensive of these are the *Aegelops tauchii*-derived synthetic series, which has several different *A. tauchii* parents and multiple durum parents. The synthetics, advanced backcross lines to domesticated hexaploid bread wheat, and their bread wheat, durum and *A. tauchii* parents, will be fingerprinted using genome-specific SSR markers. This will allow the determination of how the *A. tauchii* parents are related to each other; how the synthetics and the advanced backcross lines differ from their durum and bread wheat parents; and for which chromosomal segments the offspring differ. CIMMYT has also begun the fingerprinting of various tropical maize races and landraces, which contain a bewildering array of genetic variation that has yet to be unravelled. In addition, a proposed project to study the genetic diversity in the wild ancestral species *Teosinte* and the related species *Tripsacum* may provide information on the patterns of diversity within these species, and between these species and domesticated maize. Full exploitation of many potentially useful traits may await a more extensive characterization of these three species.

Considerations in large-scale fingerprinting of plant collections

Past fingerprinting projects at CIMMYT, and at most public institutions, have generally been quite small (containing up to a few hundred individuals). However, with 17,000 collections of maize and teosinte, and 130,000 wheat and related *Triticeae* accessions in the CIMMYT gene bank facilities, future fingerprinting projects must include thousands of individuals in order to begin to understand the patterns of diversity these collections contain. New molecular marker technologies will make it possible for very large-scale fingerprinting projects to be undertaken. New databasing capabilities will allow the storage and analysis of the data these projects will produce. A fully integrated approach to data generation, storage, analysis and dissemination will be necessary to allow the most efficient use of the data with the fewest inputs of time, reagents and money.

In changing the scales of fingerprinting to larger projects and more efficient procedures, considerations must include very rapid and reliable DNA extraction methods; high throughput marker analyses; the use of automatic sequencers for fragment analysis and data input; and the use of markers (AFLPs and/or SSRs) which generate a great deal of information per reaction. SSRs in particular are amenable to high throughput analysis because they can be automated, can be multiplexed for further efficiency, and are usually mapped. Some are even linked to, or form part of the sequence of, genes of interest. It is also possible to amplify SSRs using smaller amounts of DNA, and of a lower purity, than with AFLPs. Finally, a database capable of storing, querying and making reports of marker data, in conjunction with pedigree and phenotypic data for large data sets, is a priority.

The ABC laboratories are currently equipped with an ABI prism 377XL automatic DNA sequencer (Perkin Elmer). SSR markers for both maize and wheat have been optimized for use on the sequencer in multiplexed reactions; however, more will be tested so that up to 100 markers per species are available for use, if necessary. Several methods of DNA extraction are being tested, including the use of a Sap Extractor (Wenig and Koch, Hanover, Germany), tissue grinding with metal beads (Sigma Engineering, White Plains, New York) and the Matrix Mill DNA extraction system (Harvester Technology Inc., Lansing, New York). The quality of DNA from each method must also be tested. The International Crop Information System, ICIS, is currently under development by CIMMYT, the International Rice Research Institute and the University of Queensland. A working prototype is currently available for storage of various data types, and work continues to improve the database for use in all crops and cropping systems (ICIS Project 2000; http://www.cgiar.org/icis/homepage.htm).

Conclusions

Molecular markers are proving useful in the maintenance of and access to the diversity present in germplasm collections. In order to screen even a fraction of the total number of lines in a large collection, high throughput marker techniques and efficient data management are essential components of a system that must be optimized before undertaking such a monumental task. The importance of data storage and analysis cannot be understated; access to marker data in an integrated database will allow future analysis and experiments *in silico* which were not imagined at the time the lines were fingerprinted but which will be able to take advantage of the existing data without the need for further lab work.

In very large germplasm collections such as CIMMYT's maize and wheat collections, it is not possible – or even desirable – to adapt all accessions for use by breeders. However, using markers, we can quickly screen accessions for the presence of novel alleles of useful genes. In the future, as the genomics revolution provides sequences of genes of interest to breeders, markers in the gene itself will allow us to maximize the true potential of germplasm collections by screening for new alleles of these loci which can be transferred by introgression or transformation to elite germplasm. This represents a more directed search for new alleles of known loci, which may provide increased levels of adaptation to poor environments, resistance to pests, yield and new metabolites than currently available to breeders and geneticists.

References

Almanza, M.I. (2000) Estudio comparativo de diversidad genetica en trigos harineros de primavera (*Triticum aestivum* L. em. Thell) utilizando coeficientes de parentesco y marcadores moleculares. MSc thesis, Colegio de Post-graduados, Mexico.

Bernardo, R. (1998) Predicting the performance of untested single crosses: trait and marker data. In: Lamkey, K.R. and Staub, J.E. (eds) *Concepts and Breeding of Heterosis* in Crop Plants. CSSA Special Publication 25. CSSA, Madison, Wisconsin, pp. 117–127.

Bretting, P.K. and Widrlechner, M.P. (1995) Genetic markers and plant genetic resource management. In: Jules Janick (ed.) *Plant Breeding Reviews*, vol. 13. John Wiley & Sons, New York, pp. 11–86.

Cheng, M., Fry, J.E., Pang, S., Zhou, H., Hironaka, C.M., Duncan, D.R., Conner, T.W. and Wan, Y. (1997) Genetic transformation of wheat mediated by *Agrobacterium tumefaciens*. *Plant Physiology* 115, 971–980.

Falconer, D.S. (1989) *Introduction to Quantitative Genetics*. Longman Scientific and Technical, New York.

Fellers, J.P., Guenzi, A.C. and Porter, D.P. (1997) Marker proteins associated with somatic embryogenesis of wheat callus cultures. *Journal of Plant Physiology* 151, 201–208.

Hallauer, A.R. and Miranda, J.B. (1981) *Quantitative Genetics in Maize Breeding*. Iowa State University Press, Ames, Iowa.

ICIS Project (2000) *International Crop Information System CD* V1.0. CIMMYT and IRRI, Mexico.

Khairallah, M.M., Bohn, M., Jiang, C., Deutsch, J.A., Jewell, D.C., Mihm, J.A., Melchinger, A.E., González de León, D. and Hoisington, D.A. (1998) Molecular mapping of QTL for southwestern corn borer resistance, plant height and flowering in tropical maize. *Plant Breeding* 117, 309–318.

Lee, M., Godshalk, E.B., Lamkey, K.R. and Woodman, W.W. (1989) Association of restriction fragment length polymorphisms among maize inbreds with agronomic performance of their crosses. *Crop Science* 29, 1067–1071.

Mann, C. (1997) Reseeding the Green Revolution. *Science* 277, 1038–1043.

Melchinger, A.E. (1993) Use of RFLP markers for analysis of genetic relationships among breeding materials and prediction of hybrid performance. In: Buxton, D.R., Shibles, R., Forsberg, R.A., Blad, B.L., Asay, K.H., Paulson, G.M. and Wilson, R.F. (eds) *International Crop Science I*. CSSA, Madison, Wisconsin, pp. 621–628.

Plaschke, J., Ganal, M.W. and Roder, M.S. (1995) Detection of genetic diversity in closely related bread wheat using microsatellite markers. *Theoretical and Applied Genetics* 91, 1001–1007.

Ribaut, J.M. and Hoisington, D.A. (1998) Marker-assisted selection: new tools and strategies. *Trends in Plant Science* 3, 236–239.

Ribaut, J.M., Jiang, C., González de León, D., Edmeades, G.O. and Hoisington, D.A. (1997) Identification of quantitative trait loci under drought conditions in tropical maize 2: yield components and marker-assisted selection strategies. *Theoretical and Applied Genetics* 94, 887–896.

Skovmand, B., Villareal, R., van Ginkel, M., Rajaram, S. and Ortiz-Ferrara, G. (1997) *Semi-dwarf Bread Wheats: Names, Pedigrees, and Origins*. CIMMYT, Mexico DF.

Smith, O.S., Smith, J.S.C., Bowen, S.L., Tenborg, R.A. and Wall, S.J. (1990) Similarities among a group of elite maize inbreds as measured by pedigree, F_1 grain yield, heterosis, and RFLPs. *Theoretical and Applied Genetics* 80, 833–840.

Tanksley, S.D. and McCouch, S.R. (1997) Seed banks and molecular maps: unlocking genetic potential from the wild. *Science* 277, 1063–1066.

Warburton, M.L. and Bliss, F.A. (1996) Genetic diversity in peach (*Prunus persica* L. Batch) revealed by RAPD markers and compared to inbreeding coefficients. *Journal of the American Society for Horticultural Science* 121, 1012–1019.

Chapter 7

Molecular Analysis of Wild Plant Germplasm: the Case of Tea Tree (*Melaleuca alternifolia*)

L.S. LEE, M. ROSSETTO, L. HOMER AND R.J. HENRY

Centre for Plant Conservation Genetics, Southern Cross University, Lismore, Australia

Introduction

Molecular markers have been widely used in studies of the genetics of wild plant populations. The technique enables researchers to answer questions about speciation, genotypic diversity within and between populations, gene flow characteristics and levels of outcrossing. Co-dominant markers such as isozymes, restriction fragment length polymorphisms and microsatellites are particularly useful in wild populations where identification of parentage is usually a problem. The work reported here is an example of how microsatellite markers were utilized, in association with phenotypic analysis, to reveal novel information about the genetic structure of the wild population of the Australian native species *Melaleuca alternifolia* (tea tree).

Tea tree oil

Tea tree is an Australian native plant species of the *Myrtaceae* family. The leaf oil has a long history of use in pharmaceutical and cosmetic applications because of its antiseptic properties (Belaiche, 1985; Altman, 1989). It is also used as an insect repellent and antipruritic. The oil is extracted commercially from harvested leaf material by steam distillation.

During the last two decades a significant tea tree plantation industry has developed in Australia, principally in the region to which the species is endemic, in the north-eastern corner of New South Wales. In recent times, tea tree production has been established in north Queensland and in several other countries. As a result, the traditional bush-harvesting industry is disappearing. Hand cutting of leaf material from wild stands of tea tree and oil extraction in crude stills is being replaced by sophisticated mechanical farming operations and quality-controlled distillation facilities (Newman, 1992). These changes raise issues of germplasm exploitation, both from a biological resource conservation perspective, and from a need for efficient resource utilization.

Current practice in plantation production of tea tree entails the high density row cropping of seedling plants which are mechanically harvested and allowed to coppice for subsequent leaf production each year. A plantation will continue to produce for many years. Seeds for new plantings are collected from mature trees growing in wild stands. Variability in oil yield and oil quality between trees has been observed from the earliest times of the industry, long before plantation production began (Newman, 1992). Accordingly, to produce nursery plants for plantation establishment, parent trees for seed collection have usually been selected for their superior oil quality. However, because biomass yield cannot be reliably estimated from mature trees in the wild, particularly as it would pertain to coppiced plantation production, little if any selection has been undertaken for yield maximization.

Varietal improvement

Selection of parent trees based upon superior oil characteristics has presumably resulted in some enhancement in oil yield and quality compared with average performance of wild trees. However, tea tree is a largely outcrossing species and the use of wild germplasm results in high levels of phenotypic variability in plantations. Breeding and selection for high biomass yield, high oil yield and superior oil quality would result in significant improvements to plantation productivity and product quality (Butcher *et al.*, 1996).

Crop improvement programmes must be based on comprehensive understanding of the genetic characteristics of the species in question. The emergence of the tea tree plantation industry led to the availability of limited resources to conduct genetic research (Butcher *et al.*, 1992; Butcher, 1994). Recently, a major investment was made to conduct genotypic and phenotypic characterization of *M. alternifolia* wild populations from throughout the known range of the species. This research (Rossetto *et al.*, 1999a,b; Homer *et al.*, 2000), described below, provides a sound foundation for designing tea tree breeding programmes (Lee *et al.*, 1999) which have excellent prospects for developing significantly improved varieties for commercial production. Strategies such as vegetative propagation and seed production by controlled pollination have been developed to minimize progeny variability.

Conservation

The information derived from this study provides valuable insights into the conservation requirements of wild stands of the species. Continuing land clearing and modification of *M. alternifolia* habitat has the potential to affect the species. Assessment of the significance of this impact must be founded upon an understanding of the genetic diversity and structure of the wild population. Traditional bush-cutting from wild stands of tea tree was a very damaging practice leading to prevention of flowering and sometimes to the death of large stands of trees. Although the replacement of bush-cutting by plantation production has largely overcome this problem, it creates the possible new problem of 'genetic pollution' by the introduction of foreign genotypes with the potential to cross-pollinate with local wild plants. This is not a problem while plantations remain in production, whereby they are regularly harvested prior to flowering, but it is conceivably a problem when plantations are abandoned. The reality of this problem can only be assessed through an understanding of the genetic structure of the species population.

This chapter reports on the use of microsatellite markers specifically characterized for *M. alternifolia* applied to the wild population of the species. A comparison of the genotypic distribution with that of the leaf oil chemistry phenotypes (chemotype) across the known range of the species is also discussed, as are the germplasm conservation issues raised by this study.

Tea tree genotyping

Plant sampling

Samples of tea tree leaf material were collected from 15 individual trees no less than 100 m apart in wild stands at each of 40 locations throughout the known geographic range of the species (Rossetto *et al.*, 1999b). The aim was to select a comprehensive representation of the entire species' population. The samples were air dried before being used for DNA extraction and for oil analysis as described below.

The sample sites, shown in Fig. 7.1, covered three river catchments: the Severn in southern Queensland, the Richmond in far northern New South Wales, and the Clarence to the west and south of the Richmond. The latter two river catchments were divided into several subcatchments as shown in Fig. 7.1, and the locations of the individual sampling sites are also indicated.

Microsatellite analysis

DNA simple sequence repeats (microsatellites) were the molecular marker of choice in this project. This is a powerful tool for genetic analysis and the technology is discussed in detail elsewhere in this volume. The development and application of the tea tree microsatellites is described fully in Rossetto *et al.* (1999a,b).

An enriched DNA library of simple sequence repeats (SSRs) was produced

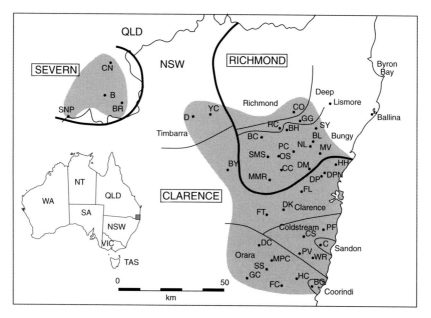

Fig. 7.1. Sample sites, catchments and subcatchments (reproduced from Rossetto *et al.* (1999b), with permission from Blackwell Science Ltd).

based on the method of Edwards *et al.* (1996). Two dinucleotide and 11 trinucleotide repeat sequences were targeted and 831 clones were sequenced. Of these, 86% contained microsatellite sequences. Primers were designed from the flanking regions of 139 of the most suitable SSRs (Rossetto *et al.*, 1999a) and 93 of these exhibited microsatellite polymorphism among as few as five individual test plants.

Five microsatellite markers (Table 7.1) were selected to genotype the tea tree population represented in the samples as described above. Of the 600 potentially available samples, 484 individuals were analysed. Only those individuals which exhibited amplification products for all five loci were employed in the genotyping.

Genetic analysis

The outcomes of the genetic analysis of the sample population are described in detail in Rossetto *et al.* (1999b). Briefly, observed heterozygosity was high (H_o = 0.724) and the greatest genetic variation occurred within single sites rather than between them. Table 7.2 shows that when catchments and subcatchments were treated as discrete groups, variation between the groups and between the individual sites within each group was always low, whereas over 90% ($P <$ 0.001) of the genetic variation occurred between individuals within sites. The mean selfing rate was only 14%.

Table 7.1. Microsatellites used in genotyping the tea tree population (adapted from Rossetto et al. (1999b), with permission from Blackwell Science Ltd).

Microsatellite locus	Repeat unit	No. of alleles[a]
scu008TT	$(AG)_n$	28
scu013TT	$(AG)_n$	31
scu044TT	$(AG)_n$	25
scu023TT	$(GCC)_n$	7
scu031TT	$(GCC)_n ACC (GCC)_n$	7

[a] Number of alleles in the sample population.

Table 7.2. Percentage molecular genetic variation between groups of sampling sites (adapted from Rossetto et al. (1999b), with permission from Blackwell Science Ltd).

Groups (of sites)	No. of groups	Between groups %	P	Between sites within groups %	P	Within sites %	P
States (2)	2	11.02	< 0.001	4.34	< 0.001	84.64	< 0.001
Catchments (3)	3	4.75	< 0.001	4.23	< 0.001	91.02	< 0.001
Subcatchments (10)	10	3.86	< 0.001	3.88	< 0.001	92.26	< 0.001
NSW subcatchments (9)	9	0.97	< 0.001	4.07	< 0.001	94.96	< 0.001

NSW, New South Wales.

Gene flow between sites was high as shown by an Nm of 1.7 for $R_{ST} = 0.128$. There was, however, significant allelic difference between sites. There was an inverse correlation ($R^2 = 0.436$) between genetic distance and geographic distance. The greatest genetic divergence was between the more distant subcatchments. On the other hand, most alleles were represented to some degree in each subcatchment. The apparent contradiction between this genetic structuring on the one hand and the predominance of genetic diversity at the within-site level on the other, arises largely due to genetic distinction between the Severn catchment subpopulation (in the State of Queensland) and the remaining sites (all in the State of New South Wales). This is illustrated in Table 7.2. When the data are segregated between states, a much higher level of genetic variance is attributable to those groupings than when the data are segregated by catchments or by subcatchments. This conclusion is reinforced by the observation that when the New South Wales subcatchments alone are analysed, less than 1% of the genetic variance can be attributed to differences between these groups.

We believe that the slight but significant genetic distinction between the Severn subpopulations and the remainder of the species is due to a founder effect stemming from the relatively recent origin of the Severn sites. The Severn subpopulations grow in a cool, highland environment, very different from the lowland, warm, coastal environment of the Richmond and Clarence catchment subpopulations. Furthermore, the Severn catchment is isolated by distance and by topography from the Richmond and Clarence catchments which adjoin each other on the coastal lowlands. Gene flow between the Severn and the other subpopulations was much lower ($Nm < 1.0$) than between the Clarence and Richmond ($Nm = 4.75$).

Tea tree oil phenotyping

Oil components

Tea tree oil is a complex mixture of approximately 100 different compounds and is comprised primarily of terpenes and sesquiterpenes. The relative concentrations of terpinen-4-ol, 1,8-cineole and terpinolene are the most significant determinants of commercial oil quality. For commercial purposes, high oil yield is important but secondary to good oil quality. Oil constituents of all 600 samples were analysed by head-space gas chromatography as detailed in Homer *et al.* (2000). The data were subjected to analysis of variance of oil yield and chemotype.

Six discrete oil chemotypes were identified among the 600 individual trees analysed based on the relative concentrations of terpinen-4-ol, 1,8-cineole and terpinolene (Fig. 7.2). One of these had not been previously identified (Homer *et al.*, 2000). The six chemotypes fell into three main groups, one characterized by high terpinen-4-ol (chemotype 1), another by high terpinolene (chemotype 2), while the remaining four displayed high 1,8-cineole with various combinations of terpinen-4-ol and terpinolene concentrations (chemotypes 3–6).

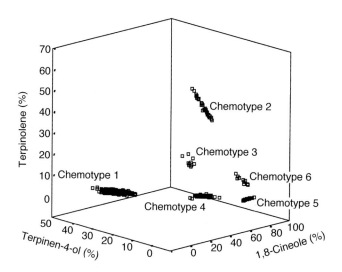

Fig. 7.2. Relative concentrations of terpinen-4-ol, 1,8-cineole and terpinolene in the six distinct chemotypes observed in the sample population (reprinted from *Biochemical Systematics and Ecology*, 28, Homer *et al.*, 367–382. Copyright (2000), with permission from Elsevier Science Ltd).

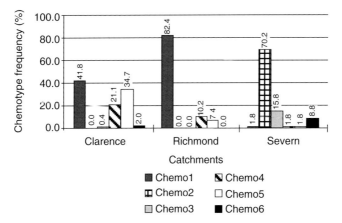

Fig. 7.3. Relative frequencies of the six chemotypes within each of the three catchments.

Geographic distribution

The geographic distribution of the chemotypes among catchments and sub-catchments was investigated. The relative frequencies of the six chemotypes (nominally 1–6) within each of the three catchments are shown in Fig. 7.3. Clearly, chemotypic distribution differed between catchments. Clarence catchment had a preponderance of chemotypes 1, 4 and 5, whereas the Richmond was dominated by chemotype 1 and had much lower incidence of chemotypes 4 and 5. The other three chemotypes (2, 3 and 6) were very rare or absent in these catchments. Conversely, the Severn catchment showed quite a different chemotypic composition. Chemotype 1, which dominated in the other catchments, was very rare, whereas chemotype 2 made up over 70% of the individuals in this catchment. Unlike the Clarence and Richmond catchments, chemotypes 3 and 6 were well represented in the Severn catchment. The three catchments are therefore differentiated on the basis that the Clarence is unique in the occurrence of high representation of 1,8-cineole chemotypes (4 and 5), the Richmond catchment is unique in its predominance of the high terpinen-4-ol chemotype (1) and the Severn is unique in its predominance of the terpinolene chemotype (2), and is the only location where this chemotype occurred (Homer *et al.*, 2000).

Figure 7.4 shows the relative frequencies between catchments for each chemotype. There is a clear distinction between the Severn catchment and the other two in chemotypes 1–5. The Clarence and Richmond catchments are very distinct with regard to their relative occurrences of chemotypes 4, 5 and 6. Again, the distinction between the Clarence, Richmond and Severn catchments in terms of the predominance of 1,8-cineole, terpinen-4-ol and terpinolene chemotypes, respectively, can be seen.

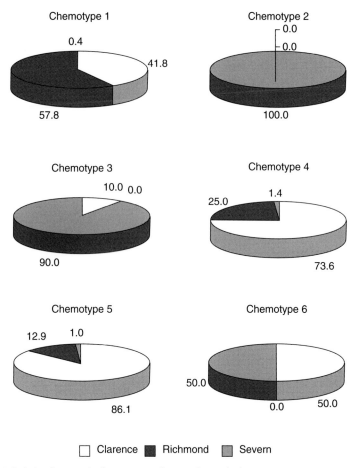

Fig. 7.4. Relative frequencies between catchments for each chemotype.

Relationships between genotype, chemotype and geographic distribution

The relationship between genotypes and chemotypes was assessed by analysis of molecular variance incorporating 1000 permutations using the genetic structure analysis software ARLEQUIN, version 2.000 (Schneider et al., 2000). The analysis was conducted based on groups defined by river catchment areas (see Fig. 7.1 and Table 7.2). Grouping by states differentiated between the coastal lowland catchments of New South Wales (Richmond and Clarence) on the one hand, and the highland catchment in Queensland (Severn) on the other. Grouping by catchments differentiated between the three separate river catchments. Grouping by subcatchments differentiated between the distinct tributary systems within the catchments; and grouping by New South Wales subcatchments allowed differentiation between the lowland subcatchments alone.

The level of genetic variation due to chemotypes was low (4.34%) but significant ($P < 0.001$). This observation closely reflects the situation with the amount of genetic variation due to different catchments (see Table 7.2). The level of genetic diversity within individual sites was so high that despite there being a significant relationship between location and chemotype, each chemotype showed a broad representation of genotypes. Accordingly, the predominant factor in genotypic variability is its predominance within sites rather than between geographic locations. Similarly, the relationship between genotype and chemotype is largely overshadowed by the high within-sites genetic variation.

None the less, this high level of genetic variation within sites does not completely mask the distinction, in terms of genotype, between catchments. Of particular note is the fact that the geographic delineation of genotypes approximates that of the geographic delineation of chemotypes, and that the boundaries of these units coincide with discrete river catchments. Geographically, the Severn catchment is very different from the other two. Elevation is much higher, rainfall is lower, temperatures are lower and have a greater range, and geology and soil type are different (Table 7.3).

Table 7.3. Geographical characteristics of the river catchments.

	Clarence	Richmond	Severn	Reference
Geology	Palaeozoic to Mesozoic	Palaeozoic to Mesozoic	Mesozoic	ASGO NDGD
Origin	Non-metamorphic	Non-metamorphic	Granitic	ASGO NDGD
Soil texture	Clay loam to sandy clay loam	Clay loam to sandy clay loam	Coarse sand to gravelly	CMA
Sedimentary basin	Clarence–Moreton Basin	Clarence–Moreton Basin	New England Fold Belt	ASGO NDGD
Site elevations (m)	< 100[a]	< 100	> 750	CMA
Mean annual rainfall (mm)	1075	1107	784	BOM
Temperature (°C)	[Grafton]	[Casino]	[Applethorpe]	BOM
Mean daily max (range)	25.6 (20.3–30.0)	26.8 (21.2–31.4)	20.4 (13.7–26.1)	
Mean daily min (range)	13.4 (6.0–19.6)	13.1 (6.6–18.8)	8.9 (1.9–15.2)	
Mean annual humidity (%) (3:00 pm)	51	51	52	BOM

AGSO, Australian Geological Survey Organisation; NDGD, National Digital Geoscience Datasets; BOM, Australian Bureau of Meteorology; CMA, Central Mapping Authority of NSW.
[a]One upstream site ~150 m, another ~480 m.

Implications

Because *M. alternifolia* is restricted to swampy and riparian environments, and because it produces minute seeds in small woody capsules, dissemination is presumed to be primarily due to water flow but with some recruitment via attachment to birds and other animals. The genotype and chemotype distribution characteristics raise questions about the evolution of the species and its interaction with the environments in which it grows. Issues about the conservation of the species and strategies for breeding commercially superior varieties can also be addressed in terms of the genetic data.

Conservation

M. alternifolia occurs in a region of intensive farming development, particularly grazing. Significant clearing for pastures has been carried out over the last century. Although this species' preference for swampy locations not well suited to grazing has afforded it some refuge, it has been subject to considerable reduction by land clearing and also by bush-cutting for tea tree oil harvesting in the past. The present study provides valuable information pertaining to the conservation of the species.

Although conservation of natural habitat is a priority for the 21st century, the high degree of genetic diversity within discrete sites, the low level of genetic differentiation between sites and the high gene flow suggest that the genetic diversity of *M. alternifolia* is under little threat of adverse impact. Furthermore, because there was found to be a relatively high degree of genotypic homogeneity between sites and considerable gene flow, the prospect of 'genetic pollution' of wild stands from commercial plantations would not be a problem.

Evolution

While *Melaleuca* is an old genus, the evolution of its species has been driven by cyclical climatic changes in the Australian environment. The contraction of populations during prolonged arid periods has led to the divergence of species (Barlow, 1988). Being confined largely to moist environments, *M. alternifolia* may conform well to this model. The greatest genotypic partitioning was found between the Queensland (Severn) and New South Wales subpopulations. Furthermore, there were no private alleles revealed in the Queensland population (Rossetto *et al.*, 1999b). These findings are consistent with the allozyme studies of Butcher *et al.* (1992). Whereas gene flow was shown to be high between the subpopulations, it was lowest between the Severn and the other two catchments. This is presumably due to the separation of the Severn from the others by mountainous topography. It appears that the population in the Severn catchment originated with a narrow genetic base from the Clarence catchment subpopulation (Rossetto *et al.* 1999b).

The occurrence of a range of chemotypes in each catchment is consistent with the genotypic data. This, however, does not adequately explain the prominent chemotypic divergence between the three catchments. It would be reasonable to hypothesize that differentiation of chemotypes between the Queensland and New South Wales catchments could be in response to distinct environmental conditions. The chemotypic structure of the Severn subpopulation is very different (Fig. 7.3) from the other two populations. Whether this effect is the result of a genetic distinction, or a physiological response triggered by environmental conditions, cannot be determined with the existing data. None the less, this observation is in accordance with the genotypic data. It is more difficult, however, to attribute the chemotypic difference between the Clarence and Richmond subpopulations to environmental factors, without considerable ecological research.

Agricultural considerations

The molecular genetics data raise a number of issues pertaining to the exploitation of tea tree for commercial oil production. Coupled with the chemotype observations, significant improvements to agricultural practices are possible.

Current practice involves growing plantations from seed collected from wild trees known to yield high quality leaf oil. Considerable variability among individual plantation trees is apparent and clearly results from the high outcrossing rates in combination with the high degree of genetic variability within sites, as demonstrated in this study. Strategies to control genetic variability among plants grown for plantation production would result in improved yield and oil quality. Vegetative propagation, controlled pollination or use of isolated elite parent tree seed plots would be useful techniques.

The genetic data are also valuable in designing tea tree breeding programmes (Lee *et al.*, 1999). Maintaining heterozygosity is considered vital in genetic improvement of tree crops (Wright, 1976) and this is achieved by conducting wide crosses. The molecular data from this study serve to identify individual trees which are genetically distinct and would therefore make the most suitable parents, after consideration of superiority of oil traits. An extension of the current work for the purposes of tea tree genetic improvement would be the application of marker-assisted selection. It is possible that certain molecular marker profiles could be associated with desirable phenotypic traits: not only oil quality attributes but also characteristics such as pest and disease tolerance, superior growth rates and favourable coppicing. The extensive suite of microsatellite markers now available would provide a sound foundation for such work.

Conclusion

The application of molecular analysis to study the genetics of wild populations of *M. alternifolia* provides a model for other species. This study of genetic diversity, as revealed by microsatellite markers and oil phenotype, enabled researchers to address issues of the species' conservation status, hypothesize about its evolution and develop a structured strategy for its genetic improvement for agricultural production purposes. It is a good example of the utility of molecular markers for addressing a wide range of questions pertaining to wild plant populations.

References

Altman, P. (1989) Australian tea tree oil – a natural antiseptic. *Australian Journal of Biotechnology* 3, 247–248.

Barlow, B.A. (1988) Patterns of differentiation in tropical species of *Melaleuca* L. (Myrtaceae). *Proceedings of the Ecological Society of Australia* 15, 239–247.

Belaiche, P. (1985) Treatment of skin infections with the essential oil of *M. alternifolia* Cheel. *Phytotherapy* 15, 15–17.

Butcher, P.A. (1994) Genetic diversity in *Melaleuca alternifolia*: implications for breeding to improve production of Australian tea tree oil. PhD thesis, Department of Forestry, Australian National University, Canberra.

Butcher, P.A., Bell, C.J. and Moran, G.F. (1992) Patterns of genetic diversity and nature of breeding system in *Melaleuca alternifolia* (Myrtaceae). *Australian Journal of Botany* 40, 365–375.

Butcher, P.A., Matheson, A.C. and Slee, M.U. (1996) Potential for genetic improvement of oil production in *Melaleuca alternifolia* and *M. linariifolia*. *New Forests* 11, 31–51.

Edwards, K.J., Barker, J.H.A., Daly, A., Jones, C. and Karp, A. (1996) Microsatellite libraries enriched for several microsatellite sequences in plants. *BioTechniques* 20, 759–760.

Homer, L., Leach, D., Lea, D., Lee, L.S., Henry, R. and Baverstock, P. (2000) Natural variation in essential oil content of *Melaleuca alternifolia* Cheel (Myrtaceae). *Biochemical Systematics and Ecology* 28, 367–382.

Lee, L.S., Rossetto, M., Homer, L.E., Henry, R.J. and Leach, D.N. (1999) A tea tree breeding strategy based on molecular genetics. In: Langridge, P., Barr, A., Auricht, G., Collins, G., Granger, A., Handford, D. and Paull, J. (eds) *Proceedings of the 11th Australian Plant Breeding Conference*, Vol 2. Adelaide, April 1999, University of Adelaide, Australia, pp. 92–93.

Newman, M. (1992) *Australia's Own – Tea-Tree Oil: a Local History of the Industry*. Mid-Richmond Historical Society, Coraki, New South Wales.

Rossetto, M., McLauchlan, A., Harriss, F., Henry, R., Baverstock, P., Lee, L.S., Maguire, T. and Edwards, K. (1999a) Abundance and polymorphism of microsatellite markers in tea tree (*Melaleuca alternifolia* – Myrtaceae). *Theoretical and Applied Genetics* 98, 1091–1098.

Rossetto, M., Slade, R., Baverstock, P., Henry, R. and Lee, L.S. (1999b) Microsatellite variation and assessment of genetic structure in tea tree (*Melaleuca alternifolia* – Myrtaceae). *Molecular Ecology* 8, 633–643.

Schneider, S., Roessli, D. and Excoffier, L. (2000) Arlequin: a software for population genetics data analysis. Ver 2.000. Genetics and Biometry Laboratory, Department of Anthropology, University of Geneva.

Wright, J.W. (1976) *Introduction to Forest Genetics.* Academic Press, New York.

Chapter 8
Genotyping Pacific Island Taro (*Colocasia esculenta* (L.) Schott) Germplasm
I.D. GODWIN, E.S. MACE AND NURZUHAIRAWATY
School of Land and Food Sciences, The University of Queensland, Brisbane, Queensland, Australia

Introduction

Taro (*Colocasia esculenta* (L.) Schott) is an important root crop throughout the tropics, particularly in west Africa, east and Southeast Asia and the Pacific Islands. It is the most important edible member of the aroid family (*Araceae*); other edible family members include *Xanthosoma*, *Alocasia* and *Cyrtosperma*. The *Araceae* also include many familiar ornamental genera such as *Caladium*, *Dieffenbachia*, *Philodendron* and *Anthurium*. *Colocasia* is believed to have originated in the Indo-Malay region (Plucknett *et al.*, 1970) or possibly in India/Bangladesh (Kuruvialla and Singh, 1981; Lebot and Aradhya, 1991).

The species is well adapted to the wet tropics, and is often grown in swampy or flooded regions, although it can grow under drier conditions. Taro cultivation is often found in irrigated lowland systems in Asia, often in the same areas as rice is cultivated. Although the species flowers and sets seed in the tropics, it is almost exclusively propagated vegetatively by small cormels or cuttings. Cultivated types are mostly diploid ($2n = 2x = 28$), although some triploids are also found ($2n = 3x = 42$). There are two major taxonomic varieties: the dasheen type (*C. esculenta* var. *esculenta*), with a large central corm, with suckers and stolons; and the eddoe type (*C. esculenta* var. *antiquorom*), with a small central corm and a large number of smaller cormels (Purseglove, 1972). In addition to the corm, the leaf is an important green vegetable.

Taro forms an important part of both the subsistence and cash crop systems of the islands of Melanesia, including Papua New Guinea, Micronesia and Polynesia. The Pacific taros are of the dasheen type, almost exclusively, and play a central role in many gardens in subsistence-based agriculture in Melanesia, and parts of Polynesia. Until 1993–1994, with the outbreak of taro leaf blight (*Phytophthera colocasiae*) in Samoa and American Samoa, taro was the basis of cash cropping, and an important export commodity. Other major constraints to taro productivity are drought and viruses such as dasheen mosaic virus and the alomae–bobone virus complex, caused by a large and a small taro bacilliform virus (Gollifer and Brown, 1972; Zettler *et al.*, 1978; Pearson, 1981).

Overall, the germplasm of the Pacific taros is not well characterized or documented. Even though it is assumed that the clonally propagated taros are highly heterozygous, there is an extreme narrowing of diversity among the Polynesian taros, suggesting a genetic bottleneck in the Solomon Islands. This can, and indeed has, with the spread of *Phytophthora* leaf blight into the Pacific, led to genetic vulnerability. While this has to some extent been recognized in the past, there is nevertheless a lack of a well-maintained germplasm collection among the South Pacific nations. There is no readily accessible literature from which to draw information on genetic aspects of the species, its phylogeny, geographic distribution and breeding. In fact, there are no truly well-documented publications where the genetic analysis of the inheritance for any trait (agronomic, stress response or quality) has been elucidated, nor any measure of the level of heterozygosity within taro cultivars.

The germplasm of the South Pacific taros does not appear to represent a very diverse sample of the species. Indeed, the cultivated Polynesian taros are not very distinct at the genotypic level. A study of isozyme variation among taros from Asia and Oceania revealed that there is virtually no genetic diversity among Polynesian cultivated taros, whereas there is substantial variation among Melanesian taros. Among 193 cultivated taros from Polynesia, all gave the same isozyme fingerprint for six enzymes, and variation for one enzyme was found in only three cultivars from French Polynesia. Hence, the cultivated taros in Samoa, Hawaii, Easter Island, The Cook Islands, Niue, Tonga and New Zealand exhibit an extremely narrow genetic diversity. It appears that the distinct morphotypes among these taros are, in all probability, somatic mutations of the same original clonal propagule. In contrast, there is a wealth of diversity among the Melanesian and Indonesian taros as revealed by isozymes (Lebot and Aradhya, 1991) and random amplifieid polymorphic DNA (RAPD) data (Irwin *et al.*, 1998).

Molecular marker data and DNA fingerprinting have the potential to play an important role in collection, characterization and management of the Pacific taro germplasm. It should be acknowledged that 'molecular data as a measure of diversity are a substitute for direct analysis of traits' (Marshall, 1997), hence molecular markers are only cost effective if they are a useful predictor of diversity and provide an advantage in terms of cost and time. Avise (1994) stated that molecular markers are used most intelligently when they address areas of contention or are used to solve particular problems which are otherwise intractable using traditional techniques.

The impetus for the application of DNA fingerprinting to Pacific Island taros stems from the outbreak of taro leaf blight in Samoa, and the recognition that the apparent lack of genetic diversity within the Polynesian cultivars created great vulnerability to disease epidemics, as foreseen by Lebot (1992). National and regional collections have been made possible with funding from the Australian and European governments. DNA fingerprinting will be applied to assess diversity, help in identifying representative samples, screen duplicates and mislabellings, and to assess the genetic integrity of accessions stored *in vitro*.

Considerations for marker systems for Pacific Island countries

There are numerous types of molecular markers available, and an even greater number of variants upon these with increasingly complex acronyms to describe them. All methods have characteristic advantages and disadvantages, and there is a need to consider these carefully before applying them to taro. Added to this is a major consideration that if such techniques are to be applied in South Pacific countries, the markers system(s) should:

1. Deliver maximum informativeness in a cost-effective manner;
2. Be achievable on basic equipment, and not require complex scanners or automation;
3. Not be reliant on very expensive, perishable consumables; and
4. Be robust with good lab-to-lab transferability.

We will not, therefore, include discussion of expensive automation or technically and equipment intensive methods such as DNA sequencing.

Isozyme markers

Isozymes may be allelic (allozymes) or at independent loci, and can be separated on starch or polyacrylamide gels on the basis of size and charge at a set pH. This is one of the cheaper forms of molecular marker, and large populations can be screened relatively quickly (Lebot and Aradhya, 1991). However, there are serious limitations with the use of isozymes. The number of polymorphic loci is very limited within a gene pool, and polymorphism may be very low, as seen with the Polynesian taro, where there was little or no informative polymorphism. For this reason alone, the use of isozymes would not be recommended in taro. Isozymes are also phenotypic markers, in that they can be affected by the tissue, growth stage and conditions of plant growth. Tissues need also to be fresh or properly treated before protein extraction, or erroneous results may be generated. Hence DNA markers are favoured for most purposes.

Restriction fragment length polymorphism (RFLP) markers

Numerous studies have demonstrated the utility of RFLP analysis for DNA fingerprinting and genetic diversity analysis within species, such as rice (Wang and Tanksley, 1989) and sorghum (Tao *et al.*, 1993). RFLP markers are based on Southern hybridization of restricted DNA with genomic or cDNA clones. Due to probe sequence homology, such markers are ideally suited to phylogenetic analysis between related species, such as within the genus *Musa* (Bhat *et al.*, 1997). RFLPs are a very robust methodology, with generally good transferability between labs. They are co-dominant, hence can be good estimators of heterozygosity. Once a library of probes is generated, RFLPs can be generated without any sequence information. For a permanent germplasm collection, RFLP data are very useful.

However, there are considerable disadvantages with the technology, not least of which is the low levels of polymorphism seen within some species such as groundnut (Kochert *et al.*, 1991). For a previously unexplored species, such as taro, it is necessary to develop a suitable probe library. The generation of RFLP data is time-consuming, particularly with single copy probes, and the assay is one of the most costly to perform as many steps are involved and radioactivity is required. Large quantities of DNA are also required, generally 5–10 µg per digest and, as a result, whole plants would be needed for DNA extractions. Probes also need to be distributed to collaborating labs and, overall, the generation of RFLPs is moderately technically demanding.

PCR-based markers

PCR-based markers share a number of general advantages over RFLP technology. The major advantages are the speed with which results are generated, low amounts of genomic DNA required for PCR (5–25 ng, which is up to 1000-fold less than RFLP analysis) and the ability to share information on primer sequences without the need to exchange DNA. We will consider three major types of markers: RAPD, microsatellites or simple sequence repeats (SSR) and interSSR (ISSR) markers.

RAPD markers

RAPDs were the first arbitrarily primed PCR markers to be developed (Williams *et al.*, 1990), and there have been a number of modifications made to the technique since, predominantly in primer length and detection methodology. The advantages of RAPDs are that they are easy to generate, rapid, multilocus and do not require radioactivity. Hence, they have many suitable qualities for use in a lab with little equipment except for a PCR thermal cycler, gel electrophoresis and photographic equipment. However, there are some reliability problems and most of the markers generated are dominant. There is also a lack of cross-transferability, and it must be acknowledged that in some cases, fragments which are the same length may not necessarily be the same sequence. Some of these problems can be overcome by cloning and partially sequencing the fragments, to turn these into sequence characterized amplified regions (Paran and

Michelmore, 1993), which are usually more robust than RAPDs. This does, however, lose the attraction of multilocus markers for diversity analysis, and does not always overcome the problem of dominance.

SSR markers

SSRs are also known as microsatellite or variable number of tandem repeat markers. These are short, 2–8 nucleotide repeats such as CA or AGC, which are repeated in tandem up to hundreds of times at many independent loci, and are ubiquitous in eukaryote genomes (Lagercrantz et al., 1993). They are generally very highly polymorphic, mainly based on the number of tandem repeat units. SSR markers can be assayed in a similar manner to RFLPs using a short synthetic oligonucleotide probe such as $(CA)_{20}$ to hybridize to blots. However, to speed up and simplify the process, SSR markers may be sequence tagged (Morgante and Oliveiri, 1993). This requires sequencing the flanking regions of a specific SSR locus and designing primers which will amplify the SSR. Polymorphism will be based on either the flanking sequence or, more usefully, the number of repeats.

SSRs are very powerful markers in that they are single locus, co-dominant and multi-allelic. They do not require radioactivity for detection, although this is sometimes used on polyacrylamide gels to detect accurately alleles which differ by one repeat unit (as little as 2 bp). They are extremely robust and easily exchanged between labs, and multiplex reactions can be run to speed up the assay, where the products have non-overlapping size ranges.

The greatest disadvantage is the initial cost in finding and sequencing loci, because although SSRs are ubiquitous, there needs to be considerable effort put into their isolation, hence they have a higher cost of establishment than other systems. SSRs also have limited use for phylogenetic analysis because of their high mutation rate. It is considered by some that due to the limited cross-transferability and long start-up time, where speed is essential, SSRs are not the best choice (Karp and Edwards, 1997). However, once they are established, SSR markers are not only a permanent, highly informative resource for germplasm fingerprinting and management, but they will be useful for mapping, as has been demonstrated in many plant species including eucalyptus (Brondani et al., 1998), soybean (Akkaya et al., 1995), wheat (Korzun et al., 1999), rice (Chen et al., 1997; Temnykh et al., 2000) maize (Taramino and Tingey, 1996) and triticum (Salina et al., 2000).

ISSR markers

InterSSR fingerprinting was developed such that no sequence knowledge was required. Primers based on a repeat sequence, such as $(CA)_n$, can be made with a degenerate 3′-anchor, such as $(CA)_8RG$ or $(AGC)_6TY$. The resultant PCR amplifies the sequence between two SSRs, yielding a multilocus marker system useful for fingerprinting, diversity analysis and genome mapping (Zietkiewicz et al., 1994). PCR products may be radiolabelled with ^{32}P or ^{33}P via end labelling or PCR-incorporation, and separated on a polyacrylamide sequencing gel prior to autoradiographic visualization. A typical reaction yields 20–100 bands per lane depending on the species and primer (Godwin et al., 1997a). Markers may also be separated on agarose gels with ethidium bromide visualization, or polyacrylamide gels with silver staining techniques.

ISSR markers have compared favourably with RFLP and RAPD markers with applications to DNA fingerprinting and diversity analysis in sorghum (Yang et al., 1996), finger millet (Salimath et al., 1995), maize (Kantety et al., 1995) and sweet potato (Huang and Sun, 2000). By virtue of greater numbers of bands amplified and detected per primer (over 100 in some cases), ISSR analysis is quicker to apply than other methodologies with the possible exception of amplified fragment length polymorphism analysis.

The advantage of greater band amplification is augmented; ISSR markers reportedly reveal substantially higher levels of polymorphism than RFLP markers in maize (Kantety et al., 1995) and finger millet (Salimath et al., 1995) and RAPD markers in finger millet (Salimath et al., 1995) and sorghum (Yang et al., 1996). Both ISSR and RFLP analyses were more effective than RAPD analysis in genotypic discrimination in sorghum; four genotype pairs gave identical RAPD fingerprints, but were easily distinguishable with RFLP or ISSR fingerprinting. Yang et al. (1996) found that 10% of RAPD bands in pairwise sorghum genotype comparisons were unreliable, whereas the error rate for ISSR bands was less than half that.

Materials and methods

Plant material and genomic DNA isolation

A core set of *C. olocasia esculenta* var. *esculenta* (dasheen type) accessions was selected based on country of origin, passport data and, where available, isozyme electrophoretic patterns (TANSAO, 1999), to be representative of the genetic diversity in the Pacific Island region (Table 8.1). An accession of *C. esculenta* var. *antiquorum* (eddoe type) and a *Xanthosoma* species, both from Fiji, were also included for comparative purposes.

Single-plant samples of leaf tissue (0.1 g) were ground and genomic DNA was extracted and purified using a Nucleon PhytoPure Plant DNA Extraction Kit (Amersham). DNA concentration was measured both on a fluorometer (Hoefer TKP 100) following the manufacturer's instructions, and by agarose gel (0.8%) electrophoresis.

Construction of an enriched microsatellite library and sequencing of clones

The microsatellite enrichment procedure was based on that described by Edwards *et al.* (1996). The following synthetic oligonucleotide fragments have

Table 8.1. List of *C. esculenta* var. *esculenta* accessions assayed.

Accession number	Origin	Variety name	Source
E122	Vanuatu	Akastem	SPC, Fiji
E271	Vanuatu	Navenahihirig	SPC, Fiji
E791	Hawaii	Niocoa	SPC, Fiji
E782	Hawaii	Pialli	SPC, Fiji
E872	Samoa	85045-9	SPC, Fiji
E424	FSM	Pastora	SPC., Fiji
E698	Fiji	106/5	SPC, Fiji
E21	Fiji	Toakula	SPC, Fiji
E720	Fiji	123/102	SPC, Fiji
E368	Niue	Maga Tea	SPC, Fiji
BC826	PNG (MP, Lae)	Numkowec koko	Bubia, PNG
BC772	PNG (MP, Siboma)	Tatoma	Bubia, PNG
BC880	PNG (MP, Finschafen)	Kipora+A1	Bubia, PNG
BC803	PNG (MP, Sisi)	Modam-2	Bubia, PNG
BC776	PNG (MP, Salamaua)	Mosine	Bubia, PNG
BC853	PNG (MP, Nawae)	Gopap	Bubia, PNG

FSM, Federated States of Micronesia; PNG, Papua New Guinea.

been used to enrich for microsatellite containing DNA fragments: $[CT]_{15}$, $[CA]_{20}$, $[GA]_{15}$, $[ACC]_{10}$, $[GT]_{15}$, $[GAC]_{10}$, $[CAT]_{10}$, $[AGC]_{10}$. Approximately 200 ng of genomic DNA was digested with the restriction enzyme *Rsa*I. One µg of an *Mlu*I adaptor was then ligated to the digested DNA, which served as a priming site for the subsequent PCR. The resulting amplification products were then hybridized to a membrane to which SSR-containing probes had been covalently attached. Non-hybridizing genomic DNA was washed away in a series of high stringency washes; five times in $2 \times$ SSC/0.01% SDS (5 min per wash) followed by three times in $0.5 \times$ SSC/0.01% SDS (5 min per wash). The remaining fragments, which were enriched for microsatellites, were then eluted from the membrane and amplified in a second PCR. Through the PCR amplification, not only were second strands of the fragments formed, but also the amount of DNA available for cloning was increased. The amplification products were digested with *Mlu*I and, following the removal of the adaptors with QIAquick PCR purification column (QIAGEN), were ligated into 10 ng of pJV1 (a modified pUC19 vector). Transformed cells were plated on to LB-agar plates containing 100 µg ml^{-1} ampicillin and 50 µg ml^{-1} X-galactosidase. The putatively positive clones were identified by plating out the genomic library, making colony lifts with nylon membranes, positively charged (Roche Diagnostics) following Boehringer Mannheim's protocol (Boehringer Mannheim, 1995) and hybridizing the membranes with digoxigenin-labelled probes containing the SSR motifs being searched. The putatively positive clones were then grown overnight in 5 ml of LB broth with 100 µg ml^{-1} ampicillin. Plasmid DNA was extracted using Promega's Wizard® *Plus* SV Minipreps DNA Purification System. Sequencing was performed by the dideoxynucleotide chain termination method using an M13 forward 24-mer. DNA sequencing was carried out on an ABI377 sequencer at a commercial sequencing centre.

SSR primer design and PCR analysis

Primers flanking the repeat motifs for trinucleotide repeats greater than four and dinucleotide repeats greater than seven were designed using dedicated primer design software (PRIME; GCG). Primer selection criteria were based on size, GC content, melting temperature curve and lack of secondary structure. Primers were screened against the core set of *C. olocasia esculenta* accessions described, including the original genotype from which the library was constructed (as a positive size control). PCR was carried out in a total volume of 25 μl under the following conditions: 10 ng of template DNA, 50 ng of each primer, 0.2 mM of each dNTP, 2.5 mM $MgCl_2$, 1 unit of *Taq* DNA polymerase (Promega). The conditions of amplification were as follows:

1. Initial denaturation was at 94°C for 5 min;
2. 35 cycles of:
   - denaturing at 94°C for 30 s;
   - optimal annealing temperature (ranging from 50°C to 55.5°C, see Table 8.2) for 1 min;
   - primer extension at 72°C for 2 min;
3. A single extension at 72°C for 10 min.

The amplification products were resolved using two electrophoretic methods. Polyacrylamide gels (10%) were run at 300 V for 35 min in 1 × TBE and visualized by silver staining. High resolution agarose gels (4%) were run at 100 V for 3 h and stained with ethidium bromide. Gels were scored using Kodak Digital

Table 8.2. List of primers developed for *Colocasia* accessions using enrichment methods.

SSR	Primer sequence 5' to 3'	Annealing temperature (°C)	Size (bp)	Polymorphism present	No. of alleles
$(CAC)_4$	Fwd: TGTGGAGCACTGTGAGTAGC Rvs: GGTGGTGGCAATAATGGTGG	53.4	107	Yes	3
$(GT)_6$	Fwd: TGTCCCTTTTGATCTGTACAAG Rvs: CTCAACGGCTCATACACAC	52.5	191	Yes	3
$(TG)_9$	Fwd: ACAGACCCTAGCTGAGTCCTAC Rvs: ACACGTCGCATCACAACAC	52.3	92	No	2
$(CT)_{15}$	Fwd: ATGCCAATGGAGGATGGCAG Rvs: CGTCTAGCTTAGGACAACATGC	54.7	164	Yes	4
$(CA)_8$	Fwd: TCAGAACAACACACACACG Rvs: TCAACCTTCTCCATCAGTCC	52.5	194	No	2
$(CT)_8(CA)_{14}$	Fwd: GCGTGGACTAACAGACAGAAG Rvs: ATTAAGAGAGAGGGGGCCAAG	55.5	151	Yes	5
$(CT)_{18}$	Fwd: CATCGCCCTCAAAAGAAGAC Rvs: GTTGCTTGTTATCTCCCACG	52.4	266	Yes	4
$(CAT)_9$	Fwd: CACACATACCCACATACACG Rvs: CCAGGCTCTAATGATGATGATG	50.0	94	Yes	3

Science™ 1D Image Analysis Software (version 2.0.1), and the sizes of the amplified fragments were determined by comparison with a 100 bp ladder.

ISSR protocol

A total of 33 primers were tested for ISSR amplification in the taro genome. Four primers were selected to continue with based on the clarity of the bands, the polymorphism observed and the repeatability of the banding patterns: $(GA)_9T$, $(GA)_9AC$, $(GA)_9AT$, $(ACC)_6Y$. A single primer was used in each PCR, which was carried out in a total volume of 20 μl containing 10 ng of genomic DNA, 2 mM $MgCl_2$, 0.1 mM of each dNTP, 10 μM of primer, 1 unit of *Taq* DNA polymerase (Promega). Amplification conditions were as follows:

1. Initial denaturation was at 94°C for 2 min;
2. Thirty-five cycles of:
   - denaturation at 94°C for 15 s;
   - annealing at 58°C for 15 s;
   - extension at 72°C for 1 min;
3. A final extension at 72°C for 5 min.

The amplification products were resolved using both polyacrylamide gel electrophoresis (10%), followed by silver staining as above, and agarose gel electrophoresis (1.5%) with detection by staining with ethidium bromide.

Data analysis

Pair-wise comparisons of accessions, based on the presence or absence of unique and shared fragments produced by ISSR and SSR amplification, both separately and combined, were used to generate similarity matrices based on three different measures:

1. Nei and Li's (1979) definition of similarity: $Sij = 2a/(2a + b + c)$, where Sij is the similarity between two individuals, i and j, a is the number of bands present in both i and j, b is the number of bands present in i and absent in j, and c is the number of bands present in j and absent in i; this is also known as the Dice coefficient (1945).
2. Jaccard's coefficient (Jaccard, 1908): $Sij = a/a + b + c$.
3. The simple matching (SM) coefficient (Sokal and Michener, 1958): $Sij = a + d/a + b + c + d$, where d is the number of bands absent in both i and j.

The matrices of similarity were then analysed using various clustering methods:

1. UPGMA (unweighted pairgroup method; Sokal and Michener, 1958).
2. WPGMA (weighted pairgroup method; Sneath and Sokal, 1973).

3. Complete linkage (Lance and Williams, 1967).
4. Single linkage (Lance and Williams, 1967).

Analyses were performed using the software NTSYS-pc, version 1.80 (Rohlf, 1998). The dendrograms were created with the TREE program of NTSYS, and the goodness of fit of the clustering to the data was calculated using the COPH and MXCOMP programs (Rohlf, 1998). In addition, a principal coordinate analysis (PCO) was carried out in NTSYS-pc using the DCENTER and EIGEN procedures.

Results

SSR library enrichment

Of the 98 clones sequenced to date, 100% contained inserts of which 21 (21.4%) contained microsatellite repeat motifs. The average insert size was 336 bp, but sizes ranged from 114 to 669 bp. The majority of the repeat motifs were either dinucleotide or trinucleotide repeats, of which 40% were compound perfect repeats and the remaining 60% were compound imperfect repeats. One heptamer was also recorded with two repeats. The dinucleotide repeats had, on average, higher numbers of repeats (nine) than the trinucleotide repeats (seven). The most common motif found in taro was GT/CA.

To date, eight microsatellites have been used for primer design as they contained either dinucleotide repeats with more than seven repeat motifs or trinucleotides with more than four repeat motifs, in addition to having sufficient flanking sequence to permit the design of a PCR primer pair. They all produced a fragment of the expected size, with six (75%) showing polymorphism (Table 8.2 and Fig. 8.1). The maximum number of bands detected across all microsatellite loci was five, with an average of 3.25. The occurrence of non-specific banding patterns did not affect the interpretation of results as they fell outside the expected size range.

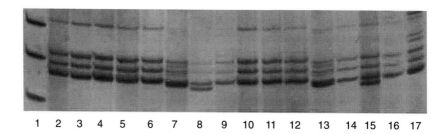

Fig. 8.1. Polymorphism of microsatellite $[CT]_{15}$ in clonal genotypes of taro. Primer pair for the $(CT)_{15}$ repeat on a 10% polyacrylamide gel, visualized by silver staining. Lanes: 1, 100 bp ladder; 2 and 3, Vanuatu taro cvs; 4 and 5, Hawaiian cvs; 6, Samoan cv; 7, Federated States of Micronesia cv; 8–10, Fijian cvs; 11, Niuen cv; 12–17, Papua New Guinea cvs.

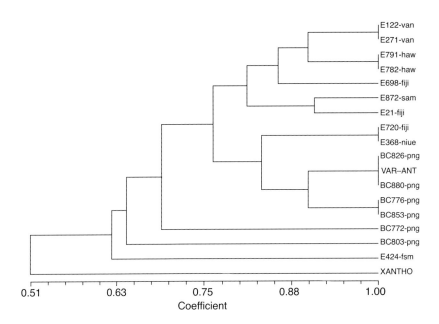

Fig. 8.2. A dendogram based on Jaccard's similarity coefficient and UPGMA clustering methodology on 17 *Colocasia* accessions and one *Xanthosoma* accession for the SSR data set.

Distribution of variation between and within Pacific Island countries

The dendrograms constructed, based on the preliminary data sets available to date, and using the three different similarity matrices (Dice, Jaccard's and SM) and various different clustering methods (UPGMA, WPGMA, complete linkage and single linkage) were examined and the cophenetic correlation values produced by each coefficient compared. The cophenetic correlation values were very similar between the different combinations, and in addition the clusterings produced were almost identical. Figure 8.2 shows the dendrogram based on Jaccard's similarity coefficient and the UPGMA clustering method.

The dendrogram indicates that the majority of accessions cluster together at a level of 75% similarity. There appears to be no variation between the accessions from either Vanuatu or Hawaii, in contrast to the variation observed between the accessions from both Fiji and Papua New Guinea, the latter showing a very high level of within-country variation. However, due to the very limited size of the data sets obtained to date, care should be taken in interpreting these results. The accession from the Federated States of Micronesia is also shown to be quite distinct from the other Pacific Island accessions, only grouping with them at a level of 63% similarity. As expected, the *Xanthosoma* accession is the most distantly related accession, grouping at a level of just 50% similarity. Interestingly, though, the *C. esculentum* var. *antiquorum* accession is

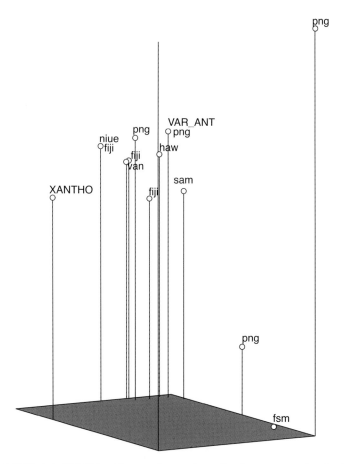

Fig. 8.3. A PCO plot of 17 *Colocasia* accessions and one *Xanthosoma* accession for the SSR data set. Accessions are labelled according to country of origin (haw, Hawaii; van, Vanuatu; sam, Samoa; fiji, Fiji; niue, Niue; fsm, Federated States of Micronesia; png, Papua New Guinea).

not distinguishable from the *C. esculenta* var. *esculenta* accessions, based on the preliminary data sets available to date.

This is also reflected in the PCO plot (Fig. 8.3), where the x axis represents 28% of the total variation, the y axis represents the next 23% and the z axis represents the next 18% of the total variation (69% represented in total). In particular the accessions from Papua New Guinea are shown to be very distinct, indicating the high level of diversity which exists between these accessions.

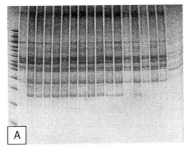 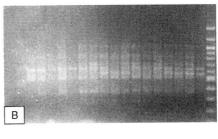

Fig. 8.4. ISSR banding profile for the primer $(GA)_9T$ on silver-stained 10% polyacrylamide gels (a) and 1.5% agarose gels stained with ethidium bromide (b) for 17 *Colocasia* accessions and one *Xanthosoma* accession (for details see Table 8.1).

ISSRs

ISSR bands could be most clearly distinguished on 10% polyacrylamide gels (Fig. 8.4). A total of 34 bands were scored for the four ISSR primers with nine (27%) of them monomorphic. Primer $(GA)_9T$ generated 12 bands of which ten (83%) were variable. Primers $(GA)_9AC$ and $(GA)_9AT$ generated eight bands of which six (75%) and seven (87.5%) were variable, respectively. Primer $(ACC)_6Y$ generated the most monomorphic bands; four from the total of six bands scored (only 33% variable). The size of the bands amplified by the ISSR primers ranged from 100 bp to 1500 bp with a peak concentration around 700 bp, although the specific pattern varied depending on the primer.

The dendrograms constructed using the three different similarity matrices (Dice, Jaccard's and SM) and various different clustering methods (UPGMA, WPGMA, complete linkage and single linkage) were examined, as for the SSR data set, and the cophenetic correlation values produced by each coefficient compared. The cophenetic correlation values were all very high (Table 8.3), where $r > 0.9$ indicates a very good fit; $r = 0.9–0.8$ indicates a good fit; and $r < 0.8$ indicates a poor fit, and in addition the clusterings produced were almost identical between the different construction methodologies. Figure 8.5 shows the dendrogram based on the Dice similarity coefficient and the UPGMA clustering method. The dendrogram indicates that the majority of accessions cluster together at a slightly higher level of similarity (82%) than shown by the dendrogram constructed using the SSR data set (75%). Once again, the accessions from Vanuatu and Hawaii cluster together (at a level of 90% similarity), and the Fiji accessions group together more distantly (at only 82% similarity). However, in contrast to the SSR data set, Fig. 8.5 indicates that the accessions from Papua New Guinea cluster together at a high level of similarity (approximately 90%). The accession from the Federated States of Micronesia groups with the accessions from the other Pacific Island countries at a much higher level of similarity (82%) than indicated by the dendrogram constructed from the SSR data set (63%). Overall, the level of variation observed is much lower

Table 8.3. A comparison of the cophenetic correlation values obtained from the various similarity coefficients and clustering techniques employed for the *Colocasia* accessions for the ISSR data set.

	[Dice]	[Jaccard's]	[SM]
UPGMA	0.984	0.978	0.972
WPGMA	0.983	0.975	0.966
Complete linkage	0.940	0.909	0.906
Single linkage	0.981	0.971	0.966

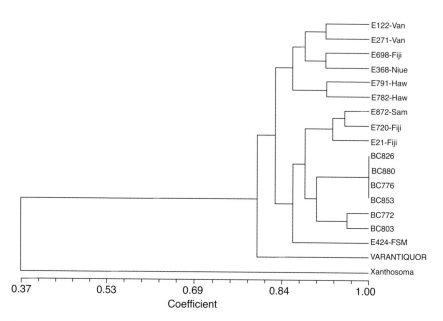

Fig. 8.5. A dendogram based on the Dice similarity coefficient and UPGMA clustering for 17 *Colocasia* accessions and one *Xanthosoma* accession for the SSR data set (based on four primers).

between the Pacific Island country accessions for the dendrogram constructed from the ISSR data set than with the SSR data set. The PCO plot (Fig. 8.6), where the x axis represents 54% of the total variation, the y axis represents the next 11% and the z axis represents the next 8% of the total variation (73% represented in total), also indicated that the *Colocasia* accessions group together very closely, and are distinct from the *Xanthosoma* accession included for comparative purposes.

Combined SSR and ISSR data analysis

Cluster analysis was carried out on combined ISSR and SSR data sets. As before, dendrograms were constructed using different similarity coefficients and

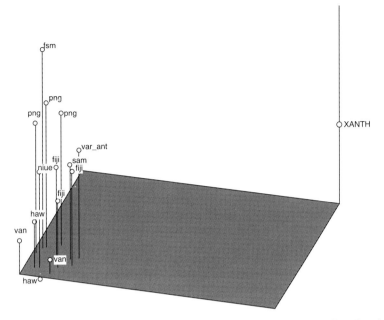

Fig. 8.6. A PCO plot of 17 *Colocasia* accessions and one *Xanthosoma* accession based on the ISSR data set. Accessions are labelled according to country of origin (haw, Hawaii; van, Vanuatu; sam, Samoa; fiji, Fiji; niue, Niue; fsm, Federated States of Micronesia; png, Papua New Guinea).

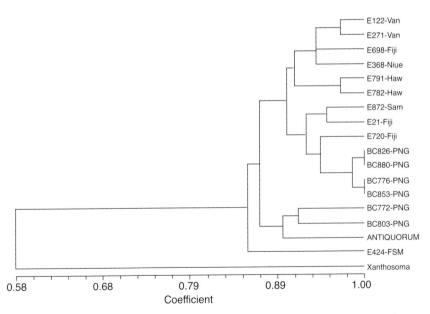

Fig. 8.7. A dendogram based on the Dice similarity coefficient and WPGMA clustering for 17 *Colocasia* accessions and one *Xanthosoma* accession for the combined ISSR and SSR data set.

clustering methods, and cophenetic correlation values produced by each coefficient compared. Figure 8.7 shows the dendrogram based on the Dice similarity coefficient and the WPGMA clustering method, which produced the highest cophenetic correlation value ($r = 0.97$). It shows that the majority of accessions screened (14 from a total of 18) do have unique fingerprints, and indicates the power of combining data sets for the unique identification of individuals (78% uniquely identified with the combined data sets, as opposed to just 39% with the SSR data set alone and 68% with the ISSR data set alone).

Conclusions

DNA fingerprinting for identification

The major objective of DNA fingerprinting is identification of accessions. A clear DNA identity tag can then be recorded for each accession, which allows for resolution of mislabelling and duplication, particularly that which may arise from the use of local names for widely adapted germplasm. These preliminary results demonstrate the power of both SSR and ISSR markers in fingerprinting Pacific Island taros. The combination of the two different marker types was able to provide fingerprints for all but four of the accessions, which were all from Morobe province in Papua New Guinea. The results highlight that there are insufficient markers in this data set, and further work is in progress to develop a larger set of SSR markers for taro.

Applications of DNA markers to taro germplasm management

DNA markers have a range of applications for the management and use of germplasm, and these can be viewed as the specific applications of the DNA fingerprint alone as an identifier for an accession, and the wider ranging genetic implications of these markers in the development of a knowledge base of the collection and its subsequent use. Analysis of DNA fingerprinting data is a valuable resource for germplasm management and utilization. Genetic diversity can be assessed, both within localized gene pools, such as for Polynesia or Melanesia, and for the diversity of the gene pool across the species' geographical distribution. This technique has already been used for many species, including rice (Ford-Lloyd *et al.*, 1997) and sweet potato (Connolly *et al.*, 1994).

The use of DNA data has also been proposed for the development of core collections (Mackay, 1995), particularly in aiding decision making for setting the collection size. For a clonally reproduced crop such as taro, information could be generated on the level of heterozygosity among cultivated and wild accessions. It may be that the level of heterozygosity among the Polynesian taros is lower than that found in Melanesia and Asia. Co-dominant markers, such as SSRs, are required for the determination of heterozygosity. Where tissue culture has been used, the genetic integrity of regenerants can be assessed, as has been performed for banana (Damasco *et al.*, 1996) and rice (Godwin *et al.*, 1997b). As it is planned that a representative sample of up to 500 accessions will be held *in vitro* at the SPC Germplasm Facility in Suva, Fiji, these markers will prove invaluable for determining the genetic integrity of the long-term *in vitro* collection.

By including wild species and genera, phylogenetic information can be gathered, as has been achieved within the *Eleusine* genus (Salimath *et al.*, 1995). Further molecular marker data can be gathered to look at issues such as genome organization and the synteny of genomes as first achieved between tomato and potato by Bonierbale *et al.* (1988). Once markers linked to traits of interest in mapping and breeding programmes have been developed, they may be used to screen germplasm for identification of accessions with desirable traits. The quantitative trait loci approach has also been proposed, whereby germplasm for rice has been screened for desirable traits (Virk *et al.*, 1996).

Molecular germplasm databases

The taro DNA fingerprint data will be stored in a database management system (ICIS, the International Crop Information System; http://www.cgiar.org/icis/homepage.html) in order to facilitate the management of the germplasm within and between Pacific nations. The storage of molecular weight data, together with the scanned image of gels or autoradiographs, offers an efficient and convenient way for different nations to compare their collections, to identify duplicates and to assess the geographic distribution of genetic variation. In the database, the DNA fingerprint data can be integrated with other data types, such as passport data, morphological and agronomic trait data, which allows for more efficient management and utilization of taro germplasm.

References

Akkaya, M.S., Shoemaker, R.C., Specht, J.E., Bhagwat, A.A. and Cregan, P.B. (1995) Integration of simple sequence repeat DNA markers into a soybean linkage map. *Crop Science* 35, 1439–1445.

Avise, J.C. (1994) *Molecular Markers, Natural History and Evolution.* Chapman & Hall, New York.

Bhat, K.V., Lakhanpaul, S., Chandel, K.P.S. and Jarret, R.L. (1997) In: Ayad, W.G., Hodgkin, T., Jaradat, A. and Rao, V.R. (eds) *Molecular Genetic Techniques for Plant Genetic Resources.* IPGRI, Rome, pp. 107–117.

Boehringer Mannheim (1995) *The DIG System User's Guide for Filter Hybridisation.* Boehringer Mannheim GmbH, Biochemica, Mannheim, Germany.

Bonierbale, M.W., Plaisted, R.L. and Tanksley, S.D. (1988) RFLP maps based on a common set of clones reveal modes of chromosomal evolution in potato and tomato. *Genetics* 120, 1095–1103.

Brondani, R.P.V., Brondani, C., Tarchini, R. and Grattapaglia, D. (1998) Development, characterisation and mapping of microsatellite markers in *Eucalyptus grandis* and *E. urophylla*. *Theoretical and Applied Genetics* 97, 816–827.

Chen, X., Temnykh, S., Xu, Y., Cho, Y.G. and McCouch, S.R. (1997) Development of a microsatellite framework map providing genome-wide coverage in rice (*Oryza sativa* L.). *Theoretical and Applied Genetics* 95, 553–567.

Connolly, A.G., Godwin, I.D., Cooper, M. and DeLacy, I.H. (1994) Interpretation of randomly amplified polymorphic DNA data for fingerprinting sweet potato. *Theoretical and Applied Genetics* 88, 332–336.

Damasco, O.P., Graham, G.C., Henry, R.J., Adkins, S.W., Smith, M.K. and Godwin, I.D. (1996) Random amplified polymorphic DNA (RAPD) detection of dwarf off-types in micropropagated Cavendish bananas. *Plant Cell Reports* 16, 118–123.

Dice, L.R. (1945) Measures of the amount of ecologic association between species. *Ecology* 26, 297–302.

Edwards, K.J., Barker, J.H.A., Daly, A., Jones, C. and Karp, A. (1996) Microsatellite libraries enriched for several microsatellite sequences in plants. *BioTechniques* 20, 758–760.

Ford-Lloyd, B.V., Jackson, M.T. and Newbury, H.J. (1997) Molecular markers and the management of genetic resources in seed genebanks: a case study of rice. In: Callow, J.A., Ford-Lloyd, B.V. and Newbury, H.J. (eds) *Biotechnology and Plant Genetic Resources: Conservation and Use*. CAB International, Wallingford, UK, pp. 103–115.

Godwin, I.D., Aitken, E.A.B. and Smith, L.W. (1997a) Applications of inter simple sequence repeat (ISSR-PCR) markers to plant genetics. *Electrophoresis* 18, 1524–1528.

Godwin, I.D., Sangduen, N., Kunanuvatchaidach, K., Piperidis, G. and Adkins, S.W. (1997b) RAPD polymorphisms among variant and phenotypically normal rice (*Oryza sativa* var. *indica*) somaclonal progenies. *Plant Cell Reports* 16, 320–324.

Gollifer, D.E. and Brown, J.F. (1972) Virus diseases of *Colocasia esculenta* in the British Solomon Islands. *Plant Disease Reporter* 56, 597–599.

Huang, J.C. and Sun, M. (2000) Genetic diversity and relationships of sweet potato and its wild relatives in *Ipomoea* series *Batatas* (Convolvulaceae) as revealed by inter-simple sequence repeat (ISSR) and restriction analysis of chloroplast DNA. *Theoretical and Applied Genetics* 100, 1050–1060.

Irwin, S.V., Kaufusi, P., Banks, K., de la Peña, R. and Cho, J.J. (1998) Molecular characterisation of taro (*Colocasia esculenta*) using RAPD markers. *Euphytica* 99, 183–189.

Jaccard, P. (1908) Nouvelles recherches sur la distribution florale. *Bulletin de la Societé Vaudoise des Sciences Naturelles* 44, 223–270.

Kantety, R.V., Zeng, X.P., Bennetzen, J.L. and Zehr, B.E. (1995) Assessment of genetic diversity in dent and popcorn (*Zea mays* L.) inbred lines using inter-simple sequence repeats. *Molecular Breeding* 1, 365–373.

Karp, A. and Edwards, K.J. (1997) Molecular techniques in the analysis of the extent and distribution of genetic diversity. In: Ayad, W.G., Hodgkin, T., Jaradat, A. and Rao, V.R. (eds) *Molecular Genetic Techniques for Plant Genetic Resources*. IPGRI, Rome, pp. 11–22.

Kochert, G., Halward, T., Branch, W.D. and Simpson, C.E. (1991) RFLP variability in peanut (*Arachis hypogaea* L.) cultivars and wild species. *Theoretical and Applied Genetics* 81, 565–570.

Korzun, V., Röder, M.S., Wendehake, K., Pasqulone, A., Lotti, C., Ganal, M.W. and Blanco, A. (1999) Integration of dinucleotide microsatellites from hexaploid bread wheat

into a genetic linkage map of durum wheat. *Theoretical and Applied Genetics* 98, 1202–1207.
Kuruvialla, K.M. and Singh, A. (1981) Karyotypic and electrophoretic studies on taro and its origins. *Euphytica* 30, 405–412.
Lagercrantz, U., Ellegren, H. and Andersson, L. (1993) The abundance of various polymorphic microsatellite motifs differs between plants and vertebrates. *Nucleic Acids Research* 21, 1111–1115.
Lance, G.N. and Williams, W.T. (1967) A general theory of classificatory sorting strategies 1. Hierarchical systems. *Computer Journal* 9, 373–380.
Lebot, V. (1992) Genetic vulnerability of Oceania's traditional crops. *Experimental Agriculture* 29, 309–323.
Lebot, V. and Aradhya, K.M. (1991) Isozyme variation in taro (*Colocasia esculenta* (L.) Schott.) in Asia and Oceania. *Euphytica* 56, 55–66.
Mackay, M.C. (1995) One core collection or many? In: Hodgkin, T., Brown, A.H.D., van Hintum, T.J.L. and Morales, E.A.V. (eds) *Core Collections of Plant Genetic Resources.* John Wiley and Sons, Chichester, UK, pp. 199–210.
Marshall, D.F. (1997) Meeting training needs in developing countries. In: Ayad, W.G., Hodgkin, T., Jaradat, A. and Rao, V.R. (eds) *Molecular Genetic Techniques for Plant Genetic Resources.* IPGRI, Rome, pp. 128–132.
Morgante, M. and Olivieri, A.M. (1993) PCR-amplified microsatellites as markers in plant genetics. *Plant Journal* 3, 175–182.
Nei, M. and Li, W.-H. (1979) Mathematical model for studying genetic variation in terms of restriction endonucleases. *Proceedings of the National Academy of Sciences USA* 76, 5269–5273.
Paran, I. and Michelmore, R.W. (1993) Development of reliable PCR-based markers linked to downy mildew resistance genes in lettuce. *Theoretical and Applied Genetics* 85, 985–993.
Pearson, M.N. (1981) Virus diseases of taro. *Harvest* 7, 136–138.
Plucknett, D.L., de la Peña, R.S. and Obrero, F. (1970) Taro (*Colocasia esculenta*). *Field Crop Abstracts* 23, 413–426.
Purseglove, J.K. (1972) *Tropical Crops. Monocotyledons.* Longman, London.
Rohlf, F.J. (1998) NTSYS-pc: Numerical Taxonomy and Multivariate Analysis System. Version 2.02i, Exeter Software, New York.
Salimath, S.S., de Oliveira, A.C., Godwin, I.D. and Bennetzen, J.L. (1995) Assessment of genome origins and genetic diversity in the genus *Eleusine* with DNA markers. *Genome* 38, 757–763.
Salina, E., Börner, A., Leonova, I., Korzun, V., Laikova, L., Maystrenko, O. and Röder, M.S. (2000) Microsatellite mapping of the induced sphaerococcoid mutation genes in *Triticum aestivum. Theoretical and Applied Genetics* 100, 686–689.
Sneath, P.H.A. and Sokal, R.R. (1973) *Numerical Taxonomy: the Principles and Practice of Numerical Classification.* Freeman, San Francisco.
Sokal, R.R. and Michener, C.D. (1958) A statistical method for evaluating systematic relationships. *University of Kansas Scientific Bulletin* 38, 1409–1438.
TANSAO (1999) Taro Network for South East Asia and Oceania: evaluation and breeding for rainfed cropping systems in South East Asia and Oceania. Annual Report 1999. Port Vila, Vanuatu.
Tao, Y., Manners, J.M., Ludlow, M.M. and Henzell, R.G. (1993) DNA polymorphisms in grain sorghum (*Sorghum bicolor* (L.) Moench). *Theoretical and Applied Genetics* 86, 679–688.

Taramino, G. and Tingey, S. (1996) Simple sequence repeats for germplasm analysis and mapping in maize. *Genome* 39, 277–287.

Temnykh, S., Park, W.D., Ayres, N., Cartinhour, S., Hauck, N., Lipovich, L., Cho, Y.G., Ishii, T. and McCouch, S.R. (2000) Mapping and genome organization of microsatellite sequences in rice (*Oryza sativa* L.). *Theoretical and Applied Genetics* 100, 697–712.

Virk, P.S., Ford-Lloyd, B.V., Jackson, M.T., Pooni, H.S., Clemeno, T.P. and Newbury, H.J. (1996) Predicting quantitative variation within rice germplasm using molecular markers. *Heredity* 76, 296–304.

Wang, Z.Y. and Tanksley, S.D. (1989) Restriction fragment length polymorphism in *Oryza sativa* L. *Genome* 32, 1113–1118.

Williams, J.G.K., Kubelik, A.R., Livak, K.J., Rafalski, J.A. and Tingey, S.V. (1990) DNA polymorphisms amplified by arbitrary primers are useful as genetic markers. *Nucleic Acids Research* 18, 6531–6535.

Yang, W., de Oliveira, A.C., Godwin, I., Schertz, K. and Bennetzen, J.L. (1996) Comparison of DNA marker technologies in characterizing plant genome diversity: high levels of variability observed in Chinese sorghum. *Crop Science* 36, 1669–1676.

Zettler, F.W., Abo El-Nil, M.M. and Hartman, R.D. (1978) Dasheen mosaic virus. *CMI/AAB Description of Plant Viruses*. No. 191. Kew, UK.

Zietkiewicz, E., Rafalski, A. and Labuda, E. (1994) Genome fingerprinting by simple sequence repeats (SSR) anchored PCR amplification. *Genomics* 20, 176–183.

Chapter 9
Molecular Marker Systems for Sugarcane Germplasm Analysis

G.M. CORDEIRO

Centre for Plant Conservation Genetics, Southern Cross University, Lismore, Australia

Introduction

Varietal identification is essential in any plant improvement programme and associated research. Identification of newly introduced clones is important to verify their role in the parentage of progeny where legal protection is required for improved varieties. A means of clonal identification can also restore cases of suspected clonal mislabelling. Ideally, the method for identification would be rapid, simple and inexpensive. In addition, obtaining information on genetic diversity is important for the selection of parental clones and in the prediction of hybrid performance. Predicting heterosis before carrying out crosses would allow a considerable reduction in the number of crosses to be performed and the number of resulting progeny screened, thereby presenting a saving in time and resources.

This chapter assesses the need for markers in sugarcane germplasm analysis and summarizes the marker techniques that have been applied for this species.

Sugarcane systematics and genetics

As a staple source of sweetening agents in the world, sugarcane (*Saccharum* sp.) is an important crop. Although sugarcane is grown mainly for sugar production in over 127 countries, in Brazil it has also been used for ethanol as a primary automotive fuel (Rash, 1995).

This large perennial grass belongs to the *Saccharum* genus of the Andropogoneae tribe from the Poacea (grass) family and is related to the genera *Zea* and *Sorghum*. The *Saccharum* genus comprises six recognized species and is characterized by aneuploidy and by having a high ploidy level (Daniels and Roach, 1987). Two of these species are considered wild and four domesticated.

Saccharum officinarum is the sugar-producing species often referred to as the 'noble' cane with $2n = 80$ and a proposed basic chromosome number of $x = 10$ (Bremer, 1961a; D'Hont *et al.*, 1998). This species has the commercial qualities of high sugar content, purity and low fibre and starch (Sreenivasan *et al.*, 1987). Hybridization between *S. officinarum* and *Saccharum spontaneum* has provided most of the cultivars of modern sugarcane, together with some contribution from *Saccharum sinense* ($2n = 111–124$) and *Saccharum barberi* ($2n = 81–120$), two groups thought to be derived from natural hybridization between *S. officinarum* and *S. spontaneum* (Price, 1968). Both *S. officinarum* and *S. spontaneum* are thought to have an autopolyploid origin, although no close diploid relatives are known. It has been proposed they both derive from the wild species, *Saccharum robustum* ($2n = 80$), found in Papua New Guinea (Grassl, 1946).

S. spontaneum is a wild species distributed from Japan and Papua New Guinea to the Mediterranean and Africa. This species carries the desirable agronomic traits of disease resistance and a high tolerance to abiotic stress, making it highly adaptable to varied habitats. It comprises very diverse euploid and aneuploid members ($2n = 40–128$) and has a proposed base chromosome number of $x = 8$ (Sreenivasan *et al.*, 1987). *S. spontaneum* has contributed to the origin of current sugarcane cultivars particularly through crossing with *S. officinarum*.

Saccharum edule, thought to be an intergeneric hybrid of *Miscanthus floridus* and *S. robustum*, is a sterile species cultivated in Melanesia for its edible flowers. *S. sinense* has been used for sugar production in areas of India, China, Japan, the Philippines and Hawaii, while *S. barberi* is a native of India and has been used for sugar production.

In addition to the six species, other allied genera including *Erianthus* sect. *Ripidium*, *Miscanthus* sect. *Diandra* Keng, *Narenga* Bor and *Sclerostachya* (Hack.) A. Camus constitute a closely related interbreeding group suggested to be involved in the origin of sugarcane and thus termed the '*Saccharum* complex' (Mukherjee, 1957; Daniels and Roach, 1987). The mutual relationships and actual contribution of these different genera, however, remain unclear due to their high and variable ploidy levels (D'Hont *et al.*, 1996).

Modern sugarcane is derived from a series of interspecific crosses made early in the 20th century (Berding and Roach, 1987) by Dutch sugarcane breeders in Indonesia. Hybridization of *S. officinarum*, notably with *S. spontaneum* and *S. barberi*, has been necessary to increase disease resistance, adaptability and tolerance to stress conditions. Transmission of the somatic chromosomes ($2n = 80$) from *S. officinarum* in early generation hybrids has resulted in modern varieties having a chromosome number usually greater than 100, with over 80 contributed by *S. officinarum* and the remainder from *S. spontaneum* (Bremer, 1961a,b,c,d, 1962, 1963). Despite this, modern cultivars still display bivalents with few uni- or multivalents during meiosis (Price, 1963). Because of the small number of clones of these species used in the primary crosses, the genetic base of modern hybrid varieties is narrow. This is a principal cause of the slow rate of sugarcane breeding progress (Berding and Roach, 1987). Understanding and management of the natural variation present within the domestic cultivars and wild relatives of this species is important in the establishment of an efficient programme aimed at crop improvement. Due to the predominance of *S. officinarum* in the genome of cultivars, and the importance of the agronomic characters inherent in this species, it is particularly important to evaluate the diversity existing within this species and the proportion of this diversity present in the cultivars.

Sugarcane is vegetatively propagated. Physically, sugarcane is a large and vigorous plant, with many tillers often growing in excess of 4 m. Accurate propagation of specific clones relies on identification of morphological traits specific to a clone and on accurate labelling and sampling. Numerous errors have been detected even in international germplasm collections. Until recently, there has been no reliable independent method of identifying clones of sugarcane. Management could be improved by grouping similar clones and eliminating duplicates. Documenting information on the occurrence of potentially useful genes and gene complexes within the collection could increase utility.

The International Society of Sugarcane Technologists has identified fingerprinting of germplasm as an important issue. In particular the questions under consideration are:

1. What is the appropriate method for fingerprinting?
2. Where would the fingerprinting be performed?
3. What type of material is to be fingerprinted?

In addition, a need was identified to create an inventory of *Saccharum* complex germplasm worldwide and to rationalize and coordinate world collections through the use of standardized molecular markers (Egan, 1996).

Molecular markers need to be applied to the sugarcane industry in the following areas:

- protection of plant breeders' rights;
- management of breeding programmes through marker-assisted selection;
- ensuring field grown cane is true to type;
- confirmation that the identity of parents selected in breeding programmes is true to type;
- determination of genetic diversity of parents selected in breeding programmes (heterosis);
- determination of genetic diversity in commercial sugarcane cultivars.

Molecular markers used in sugarcane

Molecular biology has begun to play an important role in agriculture due to its ability to modify microorganisms, plants, animals and agricultural processes. Particularly, it can aid conventional plant breeding programmes using molecular markers. Several types of DNA markers and molecular breeding strategies are now available to plant breeders and geneticists to help overcome many problems faced in conventional breeding (Caetano-Anollés *et al.*, 1991; Weising *et al.*, 1998).

Sequences (and restriction site analyses) are the only molecular markers that contain a comprehensible record of their own history. Hence, appropriate analysis based on sequence data (or restriction site data) can provide hypotheses on the relationships between the different genotypic categories (or species) that they class together (Karp *et al.*, 1996).

DNA markers have the potential to enable individual clones to be differentiated reliably and unambiguously; such markers are widely used in human forensics and paternity testing, in animal paternity testing and in fingerprinting a wide range of plants, animals and microorganisms. To assess germplasm diversity within the genus *Saccharum*, a diversity of methods have been employed. One of the initial molecular methods for varietal identification of sugarcane cultivars has been in the development of isozymes and restriction fragment length polymorphism (RFLP). The development of PCR-based molecular markers such as random amplified polymorphic DNA (RAPD), amplified fragment length polymorphism (AFLP) and, more recently, microsatellites or simple sequence repeats (SSRs) has allowed for fine-scale genetic characterization of germplasm collections previously thought impossible. Of these techniques, SSRs appear promising in relation to reproducibility and interchangeability of results, which is required for cultivar identification.

Other methods trialled for determining general relationships between the different *Saccharum* species include low copy nuclear sequences (Lu *et al.*, 1994a,b), polymorphisms of nuclear ribosomal DNA (Glaszmann *et al.*, 1990) and polymorphisms of cytoplasmic DNA (D'Hont *et al.*, 1994). Studies relating to the characterization of the double genome structure and the genome organization of modern sugarcane cultivars (Irvine, 1999) using genomic *in situ* hybridization have also been published (D'Hont *et al.*, 1994, 1996; Grivet *et al.*, 1996).

Isozymes

Early research on sugarcane genotyping had indicated that isozymes could be used to differentiate sugarcane clones (Heinz, 1969; Thom and Maretzki, 1970; Waldron and Glasziou, 1971; Fautret and Glaszmann, 1988). This, however, had been limited to detecting the identity of varietal material and the verification of hybrid origins of offspring after a cross. Glaszmann *et al.* (1989) took this a step further and used isozymes to differentiate between wild and noble canes. Their research showed strong differentiation between *S. spontaneum* and *S. officinarum*, revealing that most of the diversity among sugarcane varieties is related to the presence or absence of *S. spontaneum* genes. Their research also indicated that multiple bands of unequal intensities were often present and this was consistent with the high polyploidy present in the various groups of sugarcane. This brings about practical difficulties in characterizing clones due to the high number of bands that may migrate at similar distances.

Gallacher *et al.* (1995) found that some isozyme markers are not polymorphic, may be weak and unreliable, and may produce different results in different laboratories (Ortiz, 1983; Glaszmann *et al.*, 1989). Genetic interpretation of isozyme data in the highly polyploid *Saccharum* spp. is often complex. The inability to interpret much of the observed variation limits the practical application of isozyme visualization (Gallacher *et al.*, 1995) and no set of isozymes can differentiate among all clones. They also give insufficient information to indicate confidently genetic similarity between clones. In addition, due to the requirement that samples used are freshly collected, their wider application is limited by the requirement for local cultivation of the plants for varietal identification (Eksomtramage *et al.*, 1992).

RFLPs

RFLPs (Burnquist *et al.*, 1992; Lu *et al.*, 1994a,b; D'Hont *et al.*, 1995; Jannoo *et al.*, 1999) are the original plant DNA marker. They are robust, reliable, dominant and are simply inherited, naturally occurring Mendelian characters. These markers, first used in the construction of genetic maps (Botstein *et al.*, 1980), have now been successfully used in the assessment of genetic variability and in the elucidation of phylogenetic relationships among plant populations (Song *et al.*, 1988, 1990; Debener *et al.*, 1990; Miller and Tanksley, 1990; Wang *et al.*, 1992).

Several research groups (Burnquist *et al.*, 1992; Lu *et al.*, 1994a,b; Grivet *et al.*, 1996) have tested this marker technique in sugarcane. It has been used to show a strong molecular differentiation between *S. officinarum* and *S. spontaneum* and that the major part of the diversity among sugarcane cultivars is due to the *S. spontaneum* fraction of the genome. However, only a limited variability within *S. officinarum* was identified, although only a small number of clones have been assessed. A more recent and larger study (Jannoo *et al.*, 1999) used RFLP probes on 53 clones of *S. officinarum* and 109 cultivars from different sugarcane-growing regions of the world. It revealed a larger degree of RFLP polymorphism with the surveyed material, thereby allowing an analysis of the organization of genetic diversity within *S. officinarum* and the studied cultivars. Their results contrasted with those of Lu *et al.* (1994a,b) in indicating that a considerable diversity does exist within *S. officinarum*.

Lu *et al.* (1994b) showed RFLP markers to be efficient for research on sugarcane genetic diversity and taxonomy and were able to identify the relationships between the different sugarcane species and use the marker as a powerful tool for variety identification (Lu *et al.*, 1994a). RFLP markers have, in addition, been shown to be significantly associated with expression of agronomically important traits such as Brix (a measure of sugar content) (DaSilva *et al.*, 1993), stalk number and stalk diameter (Sills *et al.*, 1995).

The work carried out to date gives an indication that RFLPs in sugarcane can be useful for elucidating the genomic constitution of modern varieties of interspecific origin. However, despite these positive results, RFLPs do not appear to have been widely adopted for fingerprinting sugarcane. The technique does require large quantities of DNA, is costly, time-consuming and difficult to automate (Beckmann and Soller, 1983; Gale et al., 1990), which may be contributing reasons for the less than widespread use of the technique. RFLPs have, however, been more widely used as a means to map sugarcane genetically (DaSilva et al., 1993, 1995; D'Hont et al., 1994; Grivet et al., 1996).

RAPDs

RAPDs were the original PCR-based DNA marker, are relatively easy to use, rapid, moderately reliable and dominant (Welsh and McClelland, 1990; Williams et al., 1990), and suited for efficient non-radioactive DNA fingerprinting of genotypes (dos Santos et al., 1994; Thormann et al., 1994). In sugarcane, RAPDs have been used to detect polymorphisms in a quick and reproducible manner (Oropeza and Degarcia, 1997); to determine genetic diversity successfully in 20 commercial sugarcane hybrids (Harvey and Botha, 1996) as well as between members of the *Saccharum* complex; and in resolving taxonomical groups in cluster analyses (Harvey and Botha, 1996; Nair et al., 1999). RAPDs have also been shown to be efficient in detecting gross genetic changes in sugarcane cultivars subject to prolonged periods in tissue culture, although sensitivity was lacking when the technique was applied to detecting minor genetic changes such as that which occurs during sugarcane genetic transformation (Taylor et al., 1995).

The use of RAPD markers in determining relationships between distantly related species or genera has, however, been questioned (Pan et al., 1997) because markers are non-locus specific (Kesseli et al., 1994; Karp et al., 1996). Difficulties with reproducibility and amplification (Karp et al., 1997) have also been recorded.

AFLPs

This method rapidly generates hundreds of reproducible markers from DNA of any organism allowing the high-resolution genotyping of fingerprinting quality. The key feature of this technique is its capacity for the simultaneous screening of many different regions distributed randomly throughout the genome (Mueller and Wolfenbarger, 1999). As a means of determining genetic diversity, they have been found to be of value in many crop species (Lu et al., 1996; Sensi et al., 1996; Tohme et al., 1996; Muluvi et al., 1999; Loh et al., 2000) including sugarcane (Besse et al., 1998).

These markers have also been highly successful as a means of producing high density linkage maps (Van Eck *et al.*, 1995; Schondelmaier *et al.*, 1996; Boivin *et al.*, 1999; Jin *et al.*, 2000) or as a means of saturating a map region containing a trait of interest (Thomas *et al.*, 1995; Haanstra *et al.*, 1999; Hartl *et al.*, 1999).

The use of AFLP markers for determining genetic diversity and mapping in sugarcane has been assessed by Besse *et al.* (1998) and Xu *et al.* (1999). Besse *et al.* (1998) used these markers as a means of providing a preliminary characterization of the levels of diversity in wild collections of members of the *Saccharum* complex with the ultimate aim of using the technique to assist in the selection of diverse accessions as possible parents in introgression programmes. AFLP markers were able to reveal the major *Saccharum* complex groups corresponding to the genera under study, and the results were in accordance with data obtained using RFLP markers (Lu *et al.*, 1994b; Besse *et al.*, 1997). These results indicate that these markers have the potential as genetic markers in sugarcane to analyse relationships within and between species within the same genus, with the additional advantage of rapid data generation.

The method has been promoted as being highly reproducible allowing for high-resolution genotyping of fingerprinting quality. However, despite the encouraging results of Besse *et al.* (1998), there are questions regarding their reproducibility in sugarcane (L. McIntyre, CSIRO, personal communication).

SSRs

SSRs are the most frequent DNA marker used for fingerprinting and have been developed for use in sugarcane (Cordeiro *et al.*, 1999, 2000). The principal reason for the increasing success of SSRs as a molecular tool is that they provide a higher incidence of detectable polymorphisms than other techniques such as RFLPs and RAPDs (Powell *et al.*, 1996a).

An example of the success of this technique in fingerprinting a vegetatively propagated crop can be found in grape. With over 6000 varieties identified worldwide based on morphological criteria, there has been a need for molecular markers for fingerprinting. SSRs have now been studied intensively in this crop. The use of these markers for fingerprinting (Cipriani, 1994; Lamboy and Alpha, 1998; Sefc *et al.*, 1998) has led to the successful reconstruction of the complex genetic relationship among a number of European grapevine cultivars (Sefc *et al.*, 1998). SSRs have also been used for germplasm management (Lopes *et al.*, 1999) in the Portuguese grapevine gene pool. Studies in grape varietal identification have shown that SSRs (as compared with RFLPs and AFLPs) are the most promising molecular method for reproducible and interchangeable cultivar identification (Sanchez-Escribano *et al.*, 1999).

The ability of SSRs to reveal high allelic diversity is particularly useful in distinguishing between genotypes. The success of using these markers in other crop species like barley (*Hordeum vulgare*) (Saghai Maroof *et al.*, 1994; Russell *et al.*, 1997), rice (*Oryza sativa*) (Wu and Tanksley, 1993), wheat (*Triticum aestivum*) (Röder *et al.*, 1995), apple (*Malus* × *domestica*) (Szewc-McFadden *et al.*,

1996) and avocado (*Persea americana*) (Lavi *et al.*, 1994) has encouraged the testing of SSRs in sugarcane. Where characterization and identification of germplasm for purposes of research, product development, conservation, measurement and monitoring of genetic diversity in agriculture and for support of intellectual property is concerned, microsatellite repeats exceed the capabilities of RFLPs (Smith *et al.*, 1997).

Information content and ease of genotyping are two important criteria in the selection choice of an assay (Rafalski and Tingey, 1993). Polymorphisms based on SSRs are the prime method in mammalian genome research and are already having an impact on plant genetics. This is because SSRs provide a co-dominant PCR-based assay which is compatible with the requirements of plant breeding and population genetics programmes (Powell *et al.*, 1996a). A system based on SSR markers would provide unique DNA profiles of sugarcane clones, thereby satisfying all the requirements necessary for fingerprinting sugarcane germplasm, thus resolving issues such as those discussed earlier.

Much of the early characterization of SSRs has relied on database searches of published sequences or on the construction of genomic libraries. The recent development of new microsatellite enrichment techniques (Edwards *et al.*, 1996; Cordeiro *et al.*, 1999) has, however, increased the efficiency of microsatellite characterization in species for which little or no previous sequence knowledge is available. However, in contrast to methods such as RFLPs that do not require previous sequence knowledge, the development of SSRs requires an initial high cost and labour-intensive development.

Because of the high cost of SSR development, a consortium of sugarcane biotechnologists developed a collaborative effort to isolate approximately 200 microsatellite markers through the creation of an enriched SSR DNA library (Edwards *et al.*, 1996; Cordeiro *et al.*, 1999). Testing of these markers on a small sample population of five sugarcane genotypes revealed each marker to have between three and 12 alleles, with an average of eight. Markers showing polymorphisms had a polymorphism information content (PIC) (Weir, 1996) value of between 0.48 and 0.8, with a mean value of 0.72 (Cordeiro *et al.*, 2000). This PIC value, a determination of the value of a marker in detecting polymorphism, indicates that sugarcane microsatellite markers will be suitable for use in genotypic identification.

A semi-automated genotyping system is currently being developed for *Saccharum* spp. using microsatellite markers and a capillary electrophoresis system (ABI 310). This system will allow for large-scale fingerprinting of sugarcane germplasm, genotyping of progenitor species and breeding populations of *Saccharum*; and protection of plant breeders' rights. PCR products from a subset of five fluorescently labelled primer sets tested on 20 sugarcane genotypes of Australian origin were run on an ABI Prism 310 genetic analyser (Applied Biosystems). A sample output of the electropherograms (Fig. 9.1) produces clear peaks (bands), which can then be transferred into a tabulated format (Table 9.1) for an accurate assessment of the data obtained. The high ploidy number in sugarcane is an advantage in this situation, allowing unique fingerprints to be cre-

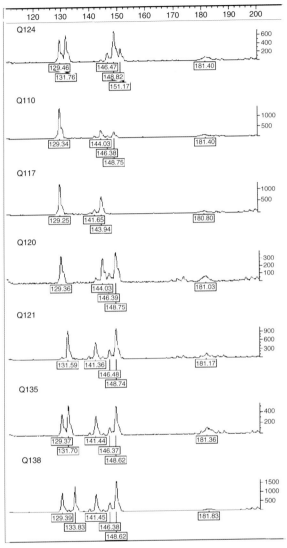

Fig. 9.1. Electropherogram of a sample of commercial sugarcane genotypes amplified with primer pairs of sugarcane microsatellite marker SMC371CG. The number of loci per genotype is greater than in most crop plants; however, the method is sufficiently sensitive to provide clear scoring. Peaks in the electropherogram correspond to bands that would normally be visualized on a gel.

ated for the sample population represented using only a single marker. An unweighted pair group method using arithmetic averages clustering analysis of the same 20 cultivars was performed. Relationships closely resembling pedigree information were established with data from seven microsatellite markers (G.M. Cordeiro, 2000, unpublished results). Prudence is, however, always necessary

Table 9.1. Tabulation of a sample of cultivar allele sizes from the electropherogram of Fig. 9.1. Primer pairs of marker SMC371CG were used. In this instance, between four and six alleles are present in each genotype tested, with a total of nine alleles in the sample population. The high ploidy number in sugarcane is an advantage in this situation, allowing unique fingerprints to be created for this sample population using only a single marker.

Cultivar	Allele 1	Allele 2	Allele 3	Allele 4	Allele 5	Allele 6	Allele 7	Allele 8	Allele 9
Q110	199	204	209	211			224	227	
Q117	199		209	211	213			227	
Q120			209	211	213		224	227	
Q121	199		209	211	213		224		238
Q124	199	204	209						238
Q135			209	211	213		224		
Q136		204	209	211	213			227	
Q137	199	204	209	211					
Q138	199		209	211	213	219			

when using a small number of primers to identify a set of varieties, as the same set may not be sufficient to distinguish individuals from a larger sample group. A good knowledge of the varietal diversity of the species is also essential before choosing the primer combination for a small set of varieties (Tessier *et al.*, 1999).

A separate study on 20 SSR markers derived from sugarcane expressed sequence tag libraries (see Scott, Chapter 15 this volume) has also identified at least three markers specific to either *S. spontaneum* or *S. officinarum* (G.M. Cordeiro, 2000, unpublished results). These markers will be particularly useful for studies into the improvement of commercial varieties through introgression.

The assessment of genetic diversity in sugarcane germplasm is currently based on pedigree records and phenotypic traits. Screening and evaluating the available genetic diversity in sugarcane with SSR repeats could both optimize and facilitate the breeding process. It will at the same time assign unique genetic fingerprints to varieties and provide information on molecular-based genetic relationships. Whenever SSRs have been compared with other marker systems, they have generally revealed a high level of polymorphism (Tautz, 1989; Morgante and Olivieri, 1993; Koreth *et al.*, 1996). In addition, their ease of use and reproducibility make them a prime candidate in sugarcane germplasm analysis.

Summary

Few direct comparisons have been made between different molecular marker techniques for sugarcane germplasm analysis, except that between RAPDs and SSRs made by Harvey and Botha (1996). The benefits and drawbacks of systems so far applied to sugarcane are summarized in Table 9.2. Direct comparative and reproducibility tests using different marker systems have, however, been

Table 9.2. Summary of the different molecular marker techniques used in sugarcane.

Technique	Benefit	Drawback
Isozymes	Able to identify and differentiate between closely and distantly related canes	Unequal banding intensities; difficult to distinguish between large number of bands migrating at close distances; poor exchange of data between laboratories; difficulty with interpretation of observed variation
RFLPs	Able to show strong molecular differentiation between closely related *Saccharum* species; reproducible; reveals relationships	Require large quantities of DNA; time-consuming and costly
RAPDs	Easy to perform; quick and reproducible; suited for fingerprinting	Moderately reliable; not suited for determination of genetic diversity – results differ with other techniques; not as polymorphic
AFLPs	Able to reveal major groups in *Saccharum* complex; dominant; easy to perform; can automate; reveals relationships	Questions regarding its reproducibility in sugarcane; poor exchange of data between laboratories
SSRs	Easy to use; able to distinguish between clones; able to determine relationships; highly polymorphic; effective for fingerprinting; can automate; EST-derived markers are transferable to closely related genera; ease of exchange between laboratories	High cost of development

RFLP, restriction fragment length polymorphism; RAPD, random amplified polymorphic DNA; AFLP, amplified fragment length polymorphism; SSR, simple sequence repeat; EST, expressed sequence tag.

carried out on crops such as barley (Powell *et al.*, 1996b; Russell *et al.*, 1997), poplar (Jones *et al.*, 1997) and maize (Pejic *et al.*, 1998). The comparisons have broadly indicated that each technique has merits, and the choice of a marker system is dependent on the application.

RFLPs and SSRs require an initial investment in terms of probe or sequence information. SSRs in particular are very costly and time-consuming to develop owing to this requirement. However, both these techniques allow probes of primer pairs that are non-polymorphic in the germplasm set or population to be excluded from future consideration. AFLPs and RAPDs, however, do not require any prior sequence information, but do allow empirical determination of which primer pairs and restriction enzymes are best for a given germplasm (Powell *et al.*, 1996b). Their dominant nature of inheritance, however, results in a lack of comparative information at each assayed locus thus precluding an accurate assessment of true genetic relationships (Russell *et al.*, 1997). SSRs and AFLPs are, however, much simpler to apply, more sensitive and have the advantage that they can be automated (Pejic *et al.*, 1998). SSRs, due to their multiallelism and co-dominance, appear suited for the analysis of outcrossing heterozygous individuals, which makes them ideal for sugarcane. AFLPs, unfortunately, have problems with reproducibility when used in sugarcane.

Hence, the level of polymorphism, dominant or co-dominant inheritance of the marker, convenience, technical difficulty, availability of species-specific probes/primers, reproducibility, quantity of DNA required and the ease of exchange of data between laboratories are all factors contributing to the choice of marker. It would be difficult to find a marker that meets all criteria, but a marker system can be identified that would fulfil a number of the desired qualities.

References

Beckmann, J.S. and Soller, M. (1983) Restriction length polymorphisms in general improvement: methodologies, mapping and cost. *Theoretical and Applied Genetics* 67, 35–43.

Berding, N. and Roach, B.T. (1987) Germplasm collection, maintenance, and use. In: Heinz D.J. (ed.) *Sugarcane Improvement through Breeding*. Elsevier, Amsterdam, pp. 143–210.

Besse, P., McIntyre, C.L. and Berding, N. (1997) Characterisation of *Erianthus* sect. *Ripidium* and *Saccharum* germplasm (Andropogoneae-Saccharinae) using RFLP markers. *Euphytica* 93, 283–292.

Besse, P., Taylor, G., Carroll, B., Berding, N., Burner, D. and McIntyre, C.L. (1998) Assessing genetic diversity in a sugarcane germplasm collection using an automated AFLP analysis. *Genetica* 104, 143–153.

Boivin, K., Deu, M., Rami, J.F., Trouche, G. and Hamon, P. (1999) Towards a saturated sorghum map using RFLP and AFLP markers. *Theoretical and Applied Genetics* 98, 320–328.

Botstein, D., White, R.L., Skolnick, M.H. and Davis, R.W. (1980) Construction of a genetic map in man using restriction fragment length polymorphisms. *American Journal of Human Genetics* 32, 314–331.

Bremer, G. (1961a) Problems in breeding and cytology of sugarcane. I. A short history of sugarcane breeding – the original forms of *Saccharum*. *Euphytica* 10, 59–78.

Bremer, G. (1961b) Problems in breeding and cytology of sugarcane. I. A short history of sugarcane breeding – the original forms of *Saccharum*. *Euphytica* 10, 121–133.

Bremer, G. (1961c) Problems in breeding and cytology of sugarcane. I. A short history of sugarcane breeding – the original forms of *Saccharum*. *Euphytica* 10, 229–243.

Bremer, G. (1961d) Problems in breeding and cytology of sugarcane. I. A short history of sugarcane breeding – the original forms of *Saccharum*. *Euphytica* 10, 325–342.

Bremer, G. (1962) Problems in breeding and cytology of sugarcane. I. A short history of sugarcane breeding – the original forms of *Saccharum*. *Euphytica* 11, 65–80.

Bremer, G. (1963) Problems in breeding and cytology of sugarcane. I. A short history of sugarcane breeding – the original forms of *Saccharum*. *Euphytica* 12, 178–188.

Burnquist, W.L., Sorrells, M.E. and Tanksley, S. (1992) Characterization of genetic variability in *Saccharum* germplasm by means of restriction fragment length polymorphism (RFLP) analysis. *XXI Proceedings of the International Society of Sugarcane Technologists* 2, 355–365.

Caetano-Anollés, G., Bassam, B.J. and Gresshoff, P.M. (1991) DNA-fingerprinting: a strategy for genome analysis. *Plant Molecular Biology Reporter* 9, 294–307.

Cipriani, G., Frazza, G., Peterlunger, E. and Testolin, R. (1994) Grapevine fingerprinting using microsatellite repeats. *Vitis* 33, 211–215.

Cordeiro, G.M., Maguire, T.L., Edwards, K.J. and Henry, R.J. (1999) Optimisation of a microsatellite enrichment technique in *Saccharum* spp. *Plant Molecular Biology Reporter* 17, 225–229.

Cordeiro, G.M., Taylor, G.O. and Henry, R.J. (2000) Characterisation of microsatellite markers from sugarcane (*Saccharum* sp.), a highly polyploid species. *Plant Science* 155, 161–168.

Daniels, J. and Roach, B.T. (1987) Taxonomy and evolution. In: Heinz, D.J. (ed.) *Sugarcane Improvement through Breeding.* Elsevier, Amsterdam, pp. 7–84.

DaSilva, J., Burnquist, W.L. and Tanksley, S.D. (1993) RFLP linkage map and genome analysis of *Saccharum spontaneum*. *Genome* 36, 782–791.

Debener, T., Salamini, F. and Gebhardt, C. (1990) Phylogeny of wild and cultivated *Solanum* species based on nuclear restriction fragment length polymorphisms (RFLPs). *Theoretical and Applied Genetics* 79, 360–368.

D'Hont, A., Lu, Y.H., Gonzalez de Leon, D., Grivet, L., Feldmann, P., Lanaud, C. and Glaszmann, J.C. (1994) A molecular approach to unravelling the genetics of sugarcane, a complex polyploid of the Andropogoneae tribe. *Genome* 37, 222–230.

D'Hont, A., Rao, S., Feldmann, P., Grivet, P., Islam-Faridi, L., Berding, N. and Glaszmann, J.C. (1995) Identification and characterisation of intergeneric hybrids, *Saccharum officinarum* × *Erianthus arundinaceus*, with molecular markers and *in situ* hybridization. *Theoretical and Applied Genetics* 91, 329–326.

D'Hont, A., Grivet, L., Feldmann, P., Rao, S., Berding, N. and Glaszmann, J.C. (1996) Characterisation of the double genome structure of modern sugarcane cultivars (*Saccharum* spp.) by molecular cytogenetics. *Molecular and General Genetics* 250, 405–413.

D'Hont, A., Ison, D., Alix, K., Roux, C. and Glaszmann, J.C. (1998) Determination of basic chromosome numbers in the genus *Saccharum* by physical mapping of ribosomal RNA genes. *Genome* 41, 221–225.

dos Santos, J.B., Nienhuis, J., Skroch, P., Tivang, J. and Slocum, M.K. (1994) Comparison of RAPD and RFLP genetic markers in determining genetic similarity among *Brassica oleracea* L. genotypes. *Theoretical and Applied Genetics* 87, 909–915.

Edwards, K.J., Barker, J.H.A., Daly, A., Jones, C. and Karp, A. (1996) Microsatellite libraries enriched for several microsatellite sequences in plants. *Biotechniques* 20, 759–760.

Egan, B.T. (1996) Role of ISSCT in promoting sugarcane germplasm collections and their maintenance. In: Croft, B.J., Piggin, C.M., Wallis, E.S. and Hogarth, D.M. (eds) *Sugarcane Germplasm Collections and their Maintenance,* ACIAR Proceedings No. 67. Paragon Printers, Fyshwick, Australia, pp. 12–13.

Eksomtramage, T.F., Laulet, F., Noyer, J.L., Feldmann, P. and Glaszmann, J.C. (1992) Utility of isozymes in sugarcane breeding. *Sugar Cane* 3, 14–21.

Fautret, A. and Glaszmann, J.C. (1988) Isozyme electrophoresis for sugarcane clonal identification at IRAT CIRAD. International Society of Sugarcane Technologists Cane Disease Committee, Pathology Workshop.

Gale, M.D., Chao, S. and Sharp, P.J. (1990) RFLP mapping in wheat progress and problems. In: Gustafson, J.P. (ed.) *Gene Manipulation and Plant Improvement II*. Plenum Press, New York, pp. 353–364.

Gallacher, D.J., Lee, D.J. and Berding, N. (1995) Use of isozyme phenotypes for rapid discrimination among sugarcane clones. *Australian Journal of Agricultural Research* 46, 601–609.

Glaszmann, J.C., Fautret, A., Noyer, J.L., Feldmann, P. and Lanaud, C. (1989) Biochemical genetic markers in sugarcane. *Theoretical and Applied Genetics* 78, 537–543.

Glaszmann, J.C., Lu, Y.H. and Lanaud, C. (1990) Variation of nuclear ribosomal DNA in sugarcane. *Journal of Genetics and Breeding* 44, 191–198.

Grassl, C.O. (1946) *Saccharum robustum* and other wild relatives of noble sugarcanes. *Journal of the Arnold Arboretum Harvard University* 27, 234–252.

Grivet, L., D'Hont, A., Roques, D., Feldmann, P., Lanaud, C. and Glaszmann, J.C. (1996) RFLP mapping in cultivated sugarcane (*Saccharum* spp) – genome organization in a highly polyploid and aneuploid interspecific hybrid. *Genetics* 142, 987–1000.

Haanstra, J.P.W., Wye, C., Verbakel, H., Meijer-Dekens, F., van den Berg, P., Odinot, P., van Heusden, A.W., Tanksley, S., Lindhout, P. and Peleman, J. (1999) An integrated high density RFLP-AFLP map of tomato based on two *Lycopersicon esculentum* $\times$ *L. pennelii* F-2 populations. *Theoretical and Applied Genetics* 99, 254–271.

Hartl, L., Mohler, V., Zeller, F.J., Hsam, S.L.K. and Schweizer, G. (1999) Identification of AFLP markers closely linked to the powdery mildew resistance genes Pm1c and Pm4a in common wheat (*Triticum aestivum* L.). *Genome* 42, 322–329.

Harvey, M. and Botha, F.C. (1996) Use of PCR-based methodologies for the determination of DNA diversity between *Saccharum* varieties. *Euphytica* 89, 257–265.

Heinz, D.J. (1969) Isozyme prints for variety identification. *International Society of Sugarcane Technologists Cane Breeders' Newsletter* 24, 8.

Irvine, J.E. (1999) *Saccharum* species as horticultural classes. *Theoretical and Applied Genetics* 98, 186–194.

Jannoo, N., Grivet, L., Seguin, M., Paulet, F., Domaingue, R., Rao, P.S., Dookun, A., D'Hont, A. and Glaszmann, J.C. (1999) Molecular investigation of the genetic base of sugarcane cultivars. *Theoretical and Applied Genetics* 99, 171–184.

Jin, H., Domier, L.L., Shen, X.J. and Kolb, F.L. (2000) Combined AFLP and RFLP mapping in two hexaploid oat recombinant inbred populations. *Genome* 43, 94–101.

Jones, C.J., Edwards, K.J., Castaglione, S., Winfield, M.O., Sala, F., Vandewiel, C., Bredemeijer, G., Vosman, B., Matthes, M., Daly, A., Brettschneider, R., Bettini, P., Buiatti, M., Maestri, E., Malcevschi, A., Marmiroli, N., Aert, R., Volckaert, G., Rueda, J., Linacero, R., Vazquez, A. and Karp, A. (1997) Reproducibility testing of RAPD, AFLP and SSR markers in plants by a network of European laboratories. *Molecular Breeding* 3, 381–390.

Karp, A., Seberg, O. and Buiatti, M. (1996) Molecular techniques in the assessment of botanical diversity. *Annals of Botany* 78, 143–149.

Karp, A., Edwards, K.J., Bruford, M., Funk, S., Vosman, B., Morgante, M., Seberg, O., Kremer, A., Boursot, P., Arctander, P., Tautz, D. and Hewitt, G.M. (1997) Molecular technologies for biodiversity evaluation – opportunities and challenges. *Nature Biotechnology* 15, 625–628.

Kesseli, R.V., Paran, H. and Michelmore, R.W. (1994) Analysis of a detailed genetic linkage map of *Lactuca sativa* (Lettuce) constructed from RFLP and RAPD markers. *Genetics* 136, 1435–1446.

Koreth, J., O'Leary, J.J. and McGee, J.O. (1996) Microsatellites and PCR genome analysis. *Journal of Pathology* 178, 239–248.

Lamboy, W.F. and Alpha, C.G. (1998) Using simple sequence repeats (SSRs) for DNA fingerprinting germplasm accessions of grape (*Vitis* L.) species. *Journal of the American Society of Horticultural Science* 123, 182–188.

Lavi, U., Akkaya, M., Bhagwat, A., Lahav, E. and Cregan, P.B. (1994) Methodology of generation and characteristics of simple sequence repeat DNA markers in avocado (*Persea americana* M). *Euphytica* 80, 171–177.

Loh, J.P., Kiew, R., Hay, A., Ke, A., Gan, L.H. and Gan, Y.Y. (2000) Intergeneric and interspecific relationships in Araceae tribe Caladieae and development of molecular markers using amplified fragment length polymorphism (AFLP). *Annals of Botany* 85, 371–378.

Lopes, M.S., Sefc, K.M., Eiras Dias, E., Steinkellner, H., Laimer da Câmara Machado, M. and da Câmara Machado, A. (1999) The use of microsatellites for germplasm management in a Portuguese grapevine collection. *Theoretical and Applied Genetics* 99, 733–739.

Lu, J., Knox, M.R., Ambrose, M.J., Brown, J.K.M. and Ellis, T.H.N. (1996) Comparative analysis of genetic diversity in pea assessed by RFLP- and PCR-based methods. *Theoretical and Applied Genetics* 93, 1103–1111.

Lu, Y.H., D'Hont, A., Paulet, F., Grivet, L., Arnaud, M. and Glaszmann, J.C. (1994a) Molecular diversity and genome structure in modern sugarcane varieties. *Euphytica* 78, 217–226.

Lu, Y.H., D'Hont, A., Walker, D.I.T., Rao, P.S., Feldmann, P. and Glaszmann, J.C. (1994b) Relationships among ancestral species of sugarcane revealed with RFLP using single copy maize nuclear probes. *Euphytica* 78, 7–18.

Miller, J.C. and Tanksley, S.D. (1990) RFLP analysis of phylogenetic relationships and genetic variation in the genus *Lycopersicon*. *Theoretical and Applied Genetics* 80, 437–448.

Morgante, M. and Olivieri, A.M. (1993) PCR-amplified microsatellites as markers in plant genetics. *Plant Journal* 3, 175–182.

Mueller, U.G. and Wolfenbarger, L.L. (1999) AFLP genotyping and fingerprinting. *Trends in Ecology and Evolution* 14, 389–394.

Mukherjee, S.K. (1957) Origin and distribution of *Saccharum*. *Botany Gazette* 119, 55–61.

Muluvi, G.M., Sprent, J.I., Soranzo, N., Provan, J., Odee, D., Folkard, G., McNicol, J.W. and Powell, W. (1999) Amplified fragment length polymorphism (AFLP) analysis of genetic variation in *Moringa oleifera* Lam. *Molecular Ecology* 8, 463–470.

Nair, N.V., Nair, S., Sreenivasaan, T.V. and Mohan, M. (1999) Analysis of genetic diversity and phylogeny in *Saccharum* and related genera using RAPD markers. *Genetic Resources and Crop Evolution* 46, 73–79.

Oropeza, M. and Degarcia, E. (1997) Use of molecular markers for identification of sugar cane varieties (*Saccharum* sp). *Phyton-International Journal of Experimental Biology* 61, 81–85.

Ortiz, Q.R. (1983) Culture de tissus somatiques chez *Saccharum*. Analyse de la variabilité enzymatique des plantes régénérées. These de Docteur-Ingénieur, Développment et Amelioration des Végétauz, Université Paris-Sud, Orsay, France.

Pan, Y.B., Burner, D.M., Ehrlich, K.C., Grisham, M.P. and Wei, Q. (1997) Analysis of primer-derived, non-specific amplification products in RAPD-PCR. *Biotechniques* 22, 1071–1077.

Pejic, L., Ajmone-Marsan, P., Morgante, M., Kozumplick, V., Castiglioni, P., Taramino, G. and Motto, M. (1998) Comparative analysis of genetic similarity among maize inbred lines detected by RFLPs, RAPDs, SSRs, and AFLPs. *Theoretical and Applied Genetics* 97, 1248–1255.

Powell, W., Machray, G.C. and Provan, J. (1996a) Polymorphism revealed by simple sequence repeats. *Trends in Plant Science* 7, 215–222.

Powell, W., Morgante, M., Andre, C., Hanafey, M., Vogel, J., Tingey, S. and Rafalski, A. (1996b) The comparison of RFLP, RAPD, AFLP and SSR (microsatellite) markers for germplasm analysis. *Molecular Breeding* 2, 225–238.

Price, S. (1963) Cytogenetics of modern sugar canes. *Economic Botany* 17, 97–105.

Price, S. (1968) Cytology of Chinese and North Indian sugarcanes. *Economic Botany* 22, 155–164.

Rafalski, J.A. and Tingey, S.V. (1993) Genetic diagnostics in plant breeding – RAPDs, microsatellites and machines. *Trends in Genetics* 9, 275–280.

Rash, K. (1995) The structure of technology in Brazilian sugarcane production, 1975–1987 – an application of modified symmetric generalized McFadden cost function. *Journal of Applied Econometrics* 10, 221–232.

Röder, M.S., Plaschke, J., König, S.U., Börner, A., Sorrells, M.E., Tanksley, S.D. and Ganal, M.W. (1995) Abundance, variability and chromosomal location of microsatellites in wheat. *Molecular and General Genetics* 246, 327–333.

Russell, J.R., Fuller, J.D., Macaulay, M., Hatz, B.G., Jahoor, A., Powell, W. and Waugh, R. (1997) Direct comparison of levels of genetic variation among barley accessions detected by RFLPs, AFLPs, SSRs and RAPDs. *Theoretical and Applied Genetics* 95, 714–722.

Saghai Maroof, M.A., Biyashev, R.M., Yang, G.P., Zhang, Q. and Allard, R.W. (1994) Extraordinarily polymorphic microsatellite DNA in barley: species diversity, chromosomal locations, and population dynamics. *Proceedings of the National Academy of Sciences USA* 91, 466–5470.

Sanchez-Escribano, E.M., Martin, J.R., Carreno, J. and Cenis, J.L. (1999) Use of sequence-tagged microsatellite site markers for characterizing table grape cultivars. *Genome* 42, 87–93.

Schondelmaier, J., Steinrucken, G. and Jung, C. (1996) Integration of AFLP markers into a linkage map of sugar beet (*Beta vulgaris* L.). *Plant Breeding* 115, 231–237.

Sefc, K.M., Steinkellner, H., Glössl, J., Kampfer, S. and Regner, F. (1998) Reconstruction of a grapevine pedigree by microsatellite analysis. *Theoretical and Applied Genetics* 97, 227–231.

Sensi, E., Vignani, R., Rhode, W. and Biricolti, S. (1996) Characterization of genetic biodiversity with *Vitis vinifera* L. sangiovese and colorino genotypes by AFLP and ISTR DNA marker technology. *Vitis* 35, 183–188.

Sills, G.R., Bridges, W., Al-Janabi, S.M. and Sobral, B. (1995) Genetic analysis of agronomic traits in a cross between sugarcane (*Saccharum officinarum* L.) and its presumed progentior (*S. robustum* Brandes and Jesw. ex Grassl). *Molecular Breeding* 1, 355–363.

Smith, J.S.C., Chin, E.C.L., Shu, H., Smith, O.S., Wall, S.J., Senior, M.L., Mitchell, S.E., Kresovich, S. and Ziegle, J. (1997) An evaluation of the utility of SSR loci as molecular markers in maize (*Zea mays* L.): comparisons with data from RFLPs and pedigree. *Theoretical and Applied Genetics* 95, 163–173.

Song, K.M., Osborn, T.C. and Williams, P.H. (1988) *Brassica* taxonomy based on nuclear restriction fragment length polymorphisms (RFLPs). 2. Preliminary analysis of subspecies within *B. rapa* (syn. campestris) and *B. oleracea*. *Theoretical and Applied Genetics* 75, 593–600.

Song, K.M., Osborn, T.C. and Williams, P.H. (1990) *Brassica* taxonomy based on nuclear restriction fragment length polymorphisms (RFLPs). 3. Genome relationships in *Brassica* and related genera and the origin of *B. oleracea* and *B. rapa* (syn. *campestris*). *Theoretical and Applied Genetics* 79, 497–506.

Sreenivasan, T.V., Ahloowalia, B.S. and Heinz, D.J. (1987) Cytogenetics. In: Heinz, D.J. (ed.) *Sugarcane Improvement through Breeding.* Elsevier, Amsterdam, pp. 211–254.

Szewc-McFadden, A.K., Lamboy, W.F., Hokanson, S.C. and McFerson, J.R. (1996) Utilization of identified simple sequence repeats (SSRs) in *Malus* × *domestica* (apple) for germplasm characterization. *Horticultural Science* 31, 619.

Tautz, D. (1989) Hypervariability of simple sequences as a general source for polymorphic DNA. *Nucleic Acids Research* 23, 4407–4414.

Taylor, R.W.J., Geijskes, J.R., Ko, H.L., Fraser, T.A., Henry, R.J. and Birch, R.G. (1995) Sensitivity of random amplified polymorphic DNA analysis to detect genetic change in sugarcane during tissue culture. *Theoretical and Applied Genetics* 90, 1169–1173.

Tessier, C., David, J., This, P., Boursiquot, J.M. and Charrier, A. (1999) Optimization of the choice of molecular markers for varietal identification in *Vitis vinifera* L. *Theoretical and Applied Genetics* 98, 171–177.

Thom, M. and Maretzki, A. (1970) Peroxidase and esterase isozymes in Hawaiian sugarcane. *Hawaiian Planters' Record* 58, 81–94.

Thomas, C.M., Vos, P., Zabeau, M., Jones, D.A., Norcott, K.A., Chadwick, B.P. and Jones, J.D.G. (1995) Identification of amplified restriction fragment polymorphism (AFLP) markers tightly linked to the tomato Cf-9 gene for resistance to *Cladosporium fulvum*. *Plant Journal* 8, 785–794.

Thormann, C.E., Ferreira, M.E., Camargo, L.E.A., Tivang, J.G. and Osborn, T.C. (1994) Comparison of RFLP and RAPD markers to estimating genetic relationships within and among cruciferous species. *Theoretical and Applied Genetics* 88, 973–980.

Tohme, J., Gonzalez, D.O., Beebe, S. and Duque, M.C. (1996) AFLP analysis of gene pools of a wild bean core collection. *Crop Science* 36, 1375–1384.

Van Eck, H.J., van der Voort, J.R., Draaistra, J., van Zandvoort, P., van Enckevort, E., Segers, B., Peleman, J., Jacobsen, E., Helder, J. and Bakker, J. (1995) The inheritance and chromosomal localisation of AFLP markers in a non-inbred potato offspring. *Molecular Breeding* 1, 397–410.

Waldron, J.C. and Glasziou, K.T. (1971) Isozymes as a method of varietal identification in sugarcane. *Proceedings of the International Society of Sugarcane Technologists* 14, 249–256.

Wang, Z.Y., Second, G. and Tanksley, S.D. (1992) Polymorphism and phylogenetic relationships among species in the genus *Oryza* as determined by analysis of nuclear RFLPs. *Theoretical and Applied Genetics* 83, 565–581.

Weir, B.S. (1996) *Genetic-Data Analysis II: Methods for Discrete Population Genetic Data.* Sinauer Assoc. Inc., Sunderland, Massachusetts.

Weising, K., Winter, P., Huttel, B. and Kshl, G. (1998) Microsatellite markers for molecular breeding. *Journal of Crop Production* 1, 113–142.

Welsh, J. and McClelland, M. (1990) Fingerprinting genomes using PCR with arbitrary primers. *Nucleic Acids Research* 18, 7213–7218.

Williams, J.G.K., Kubelik, A.R., Rafalski, J.A. and Tingey, S.V. (1990) DNA polymorphisms amplified by arbitrary primers are useful as genetic markers. *Nucleic Acids Research* 18, 6531–6535.

Wu, K.S. and Tanksley, S.D. (1993) Abundance, polymorphism and genetic mapping of microsatellites in rice. *Molecular and General Genetics* 241, 225–235.

Xu, M.L., Melchinger, A.E., Xia, X.C. and Lubberstedt, T. (1999) High-resolution mapping of loci conferring resistance to sugarcane mosaic virus in maize using RFLP, SSR, and AFLP markers. *Molecular and General Genetics* 261, 574–581.

Chapter 10
Microsatellite Analysis in Cultivated Hexaploid Wheat and Wild Wheat Relatives

A. McLAUCHLAN[1], R.J. HENRY[1], P.G. ISAAC[2] AND K.J. EDWARDS[3]

[1]Centre for Plant Conservation Genetics, Southern Cross University, Lismore, Australia; [2]Agrogene, Moissy-Cramayel, France; [3]IACR-Long Ashton Research Station, Department of Agricultural Sciences, University of Bristol, Bristol, UK

Introduction

Hexaploid bread wheat contains three genomes, A, B and D with the formula AABBDD, amounting to a very large and complex total genome. Although bread wheat, *Triticum aestivum* ($2n = 6x = 42$), is one of the most extensively studied polyploid crops, the probable evolutionary history has only recently been established by studying the various wild relatives. Studies of meiosis and the behaviour of chromosomes in interspecific hybrids during pairing revealed the evolutionary lineages of polyploid wheat species. The A genome is present in both diploid *Triticum monococcum* (AA) and tetraploid *Triticum turgidum* (AABB) wheat. Both are morphologically similar and their hybrids exhibit chromosome pairing. As *T. turgidum* (AABB) and *T. aestivum* (AABBDD) comprise one evolutionary lineage, it was subsequently concluded that *T. monococcum* was the A genome progenitor of polyploid wheat (Kimber and Sears, 1987). It became apparent that einkorn diploid wheat comprised two species, *T. monococcum* L. and *Triticum urartu* Thum., the latter existing only in wild form (Dvorak *et al.*, 1993). A re-examination of the sources of the A genome of *T. turgidum* and *T. aestivum* found it to be contributed by *T. urartu* (Dvorak, 1988) while the A genome of the other evolutionary lineage, *Triticum timopheevii* (AAGG) and *Triticum zhukovskyi* (AAAAGG), was contributed by *T. monococcum* (Dvorak *et al.*, 1993).

Synthesizing a hybrid between tetraploid wheat *T. turgidum* and *Triticum tauschii* (AA) (Coss) (= *Aegilops tauschii*) revealed the source of the D genome,

Schmalh. (also *Aegilops squarrosa* or 'goat grass'). This cross then resulted in plants with 21 pairs of chromosomes and proved that *T. tauschii* (Kihara, 1944; McFadden and Sears, 1946) contributed the D genome. It was thought that the wheat D genome originated by multiple hybridization events with several *T. tauschii* parents involved in the evolution of *T. aestivum*. However, through investigating several accessions of *T. aestivum*, all appear to share the same gene pool having little genetic differentiation among the D genomes (Dvorak *et al.*, 1998). Early attempts to resolve the origin of the B genome in polyploid wheat proved unsuccessful (Dvorak, 1988) and, until recently, this was in dispute. A species of the S genome, *Triticum speltoides*, was proposed as the B genome donor; however, the initial evidence was not unequivocal (Gupta, 1991). Very recently, results have supported the hypothesis that the B genome is derived from *T. speltoides* (Maestra and Naranjo, 1998).

Microsatellites are tandemly repeated DNA sequence units of 2–6 bp. They have abundant and random distribution throughout eukaryotic genomes; their variability in length can be demonstrated by PCR with primers designed from the conserved flanking sequence and visualized as discrete and co-dominant alleles through electrophoretic analysis. Recently, microsatellites have assumed a major role in genetic analysis, initially in studies of the human genome but more recently as valuable genetic markers in a number of plant species. As well as finding application for linkage analysis and genetic mapping, representing sequence-tagged sites and genome fingerprinting, microsatellites have already been demonstrated to be a powerful tool in diversity studies of crop species (Yang *et al.*, 1994; Xiao *et al.*, 1996). Suitable genetic marker systems in wheat have been difficult to establish due to the low genetic variability present in this self-pollinating crop (Röder *et al.*, 1995). In view of the highly informative nature of microsatellite genetic markers, they have been chosen for use in this study to estimate the extent of genetic diversity and relatedness within and between the wild diploid relatives (represented by accessions of each *T. monococcum*, *T. speltoides* and *T. tauschii*) and cultivated hexaploid wheat and therefore to help in defining the phylogenetic origin of the hexaploid genome.

Materials and methods

Genetic stocks/wheat varieties

The wheat genotypes used in the study are drawn from a broad range of the habitats in which wild diploid and cultivated wheat are found worldwide (Tables 10.1 and 10.2). These included 20 cultivated varieties of *T. aestivum* ssp. *vulgare*, as well as four of its subspecies and nine accessions of each of *T. monococcum* (AA), *T. speltoides* (BB) and *T. tauschii* (DD). All seeds were provided by the Australian Winter Cereals Collection, Tamworth, New South Wales. The aneuploid stocks used for determination of chromosomal origins of microsatellite loci were the 18 nullisomic-tetrasomic lines of the wheat variety 'Chinese Spring' kindly supplied by Dr Ken Shepherd (Waite Institute, Adelaide, Australia).

Table 10.1. Hexaploid wheat cultivars and lines.

Aus Number	Name	Species	Ploidy type	Origin
21	Australian Hybrid	*T. aestivum* ssp. *vulgare*	AABBDD	Australia
80	Bunyip	*T. aestivum* ssp. *vulgare*	AABBDD	Australia
85	Cailloux	*T. aestivum* ssp. *vulgare*	AABBDD	France
218	Federation	*T. aestivum* ssp. *vulgare*	AABBDD	Australia
246	Gabo	*T. aestivum* ssp. *vulgare*	AABBDD	Australia
381	Kendee	*T. aestivum* ssp. *vulgare*	AABBDD	Australia
409	King's Red	*T. aestivum* ssp. *vulgare*	AABBDD	Australia
473	Lerma Rojo	*T. aestivum* ssp. *vulgare*	AABBDD	Mexico
687	Napo 63	*T. aestivum* ssp. *vulgare*	AABBDD	Columbo
774	Penkop	*T. aestivum* ssp. *vulgare*	AABBDD	S. Africa
1214	Siete Cerros	*T. aestivum* ssp. *vulgare*	AABBDD	Mexico
1630	Ward's Prolific	*T. aestivum* ssp. *vulgare*	AABBDD	Australia
3520	Webster	*T. aestivum* ssp. *vulgare*	AABBDD	USR
3662	Dirk	*T. aestivum* ssp. *vulgare*	AABBDD	Australia
6027	Kenya 321	*T. aestivum* ssp. *vulgare*	AABBDD	Kenya
14366	Era	*T. aestivum* ssp. *vulgare*	AABBDD	USA
10624	Suwon 92	*T. aestivum* ssp. *vulgare*	AABBDD	Korea
15566	Glenlea	*T. aestivum* ssp. *vulgare*	AABBDD	Canada
19047	Colonias	*T. aestivum* ssp. *vulgare*	AABBDD	Brazil
19361	*T. aestivum*	*T. aestivum* ssp. *spelt*	AABBDD	Unknown
3848	*T. vavilovii*	*T. aestivum* ssp. *vavilovi*	AABBDD	Unknown
17978	*T. compactum*	*T. aestivum* ssp. *compactum*	AABBDD	USR
22497	*T. macha*	*T. aestivum* ssp. *macha*	AABBDD	Unknown
18691	*T. Zhukovskyi*	*T. zhukovski*	AAAABB	USR

Library construction, isolation of microsatellites and primer design

An enriched DNA microsatellite library of 'Chinese Spring' was prepared as part of an international wheat microsatellite consortium. This library was screened for a variety of microsatellites of di-, tri- and tetranucleotide repeats (Edwards *et al.*, 1996). Bacterial stabs of 48 clones were obtained from Agrogene (France). Plasmid DNA was extracted and purified using the method of Sambrook *et al.* (1989). Sizes of all clone inserts were estimated by restriction digests with *Bam*HI and *Eco*RI. All 48 clone inserts were sequenced using the 373A Automated DNA sequencer (ABI Applied Biosystems). Primers were designed from the microsatellite flanking sequences (Agrogene, France). Oligonucleotides were synthesized by the Centre for Molecular and Cellular Biology, Lismore.

DNA extraction

Wheat seeds were germinated on petri dishes and leaf tissue was used for DNA extraction according to the method of Weining and Henry (1995).

Table 10.2. Wild wheat accessions, their species, ploidy and origin (if known).

Aus Number	Name	Species	Ploidy type	Origin
3707	T. monococcum var. flavescens	T. monococcum L.	AA	Unknown
15823	Boeoticum boiss	T. monococcum L.	AA	Unknown
16291	T. monococcum	T. monococcum	AA	Unknown
17977	T. monococcum	T. monococcum	AA	Unknown
19380	T. monococcum–site 236	T. monococcum L.	AA	Unknown
19842	T. monococcum sofianum	T. monococcum L.	AA	Greece
19843	T. monococcum laetissimum	T. monococcum L.	AA	USR
19848	T monococcum vulgare	T. monococcum L.	AA	USR
22274	T. monococcum var. eredvisanur	T. monococcum L.	AA	Unknown
90394	T. thaoudar – 3	T. monococcum L.	AA	UTK
19607	A. speltoides	A. speltoides	BB	Israel
21643	T. speltoides ssp. lingustica	A. speltoides	BB	Turkey
21644	T. speltoides ssp. lingustica	A. speltoides	BB	Turkey
21645	T. speltoides ssp. lingustica	A. speltoides	BB	Turkey
21646	T. speltoides ssp. speltoides	A. speltoides	BB	Turkey
21647	T. speltoides ssp. speltoides	A. speltoides	BB	Syria
21648	T. speltoides ssp. speltoides	A. speltoides	BB	Iraq
21924	T. speltoides	A. speltoides	BB	Turkey
22933	A speltoides	A. speltoides	BB	Australia
23950	A. squarrosa	A. squarrosa	DD	Unknown
24025	A. squarrosa var. typica	A. squarrosa	DD	Unknown
24027	A. squarrosa var. strangulata	A. squarrosa	DD	Unknown
24053	A. squarrosa var. anathera	A. squarrosa	DD	Unknown
24058	A. squarrosa var. meyeri	A. squarrosa	DD	Iran
24139	A. squarrosa var. strangulata	A. squarrosa	DD	Iran
24150	A. squarrosa var. meyeri	A. squarrosa	DD	Iran
24160	A. squarrosa var. typica	A. squarrosa	DD	Iran
24226	A. squarrosa var. anathera	A. squarrosa	DD	Afghanistan

PCR amplification

Each PCR was carried out in a 25 μl reaction mix containing 25–50 ng of template DNA, 200 μM of each dNTP, 0.25 μM of each primer, 0.5 units of *Taq* DNA polymerase (Boeringer-Mannheim) and 1× reaction buffer (Boehringer-Mannheim) containing 1.5 mM $MgCl_2$. The forward primer of each pair was 5′ end-labelled with a fluorescent dye. The temperature cycling was carried out in 0.2 ml tubes on a thermocycler (Perkin Elmer, 9600). The cycling conditions were: initial denaturation at 95°C for 2 min, followed by 30 cycles of 95°C for 20 s, 20 s at 54–60°C and 1 min at 72°C, with a final hold for 2 min at 72°C. Less stringent conditions were adopted for the diploid progenitor accessions. The annealing temperature was reduced to 50°C and Ampli*Taq* Gold (Perkin Elmer) was used at 0.6 units per reaction to maintain a high degree of specificity.

Screening

Initial screening involved visualization of an aliquot of the non-labelled PCR product on 3.5% agarose gels to ensure successful amplification. The forward primer was fluorescently labelled in markers where successful amplification in the expected size range had occurred. The precise size of alleles generated as fluorescently labelled PCR products was established using an ABI Prism 310 Genetic Analyser (Applied Biosystems). This system performed capillary electrophoresis using a replaceable polymer matrix. An aliquot (2.0 µl) of labelled product in 24 µl of formamide along with 0.6 µl of GeneScan-500 TAMRA size standard were transferred to a sample tube for Prism 310. The conditions for capillary electrophoresis were based on the manufacturer's instructions. The effective length (length to detector) was 30 cm. 'POP 4' for performance optimized polymer (Applied Biosystems), a liquid polymer which contained urea, was used for denaturing conditions. The samples were heated to 95°C for 5 min, and electrophoretically loaded to the capillary for 10 s at 7 kV. Alleles were accurately sized using a software program (ABI PRISM Genescan Analysis Version 2.0.2.).

Results

Analysis of the microsatellite library by restriction digestion indicated that the sizes of clone inserts ranged from 130 to 1000 base pairs, with an average insert size of 415 bp (Fig. 10.1). Of the 48 sequences analysed, dinucleotides were the most common microsatellite repeat type, with (CA) then (GT) occurring most frequently. Compound repeats consisting of two or three adjacent blocks of different dinucleotide repeats made up 40% of the dinucleotide repeats. Eleven microsatellite clones of the 48 clones (23%) sequenced were found to be suitable for primer design. Forty per cent of inserts which contained microsatellite repeats were unusable because the repeats occurred too near the insert ends, thus leaving insufficient flanking sequence for primer design.

Preliminary screening, which involved testing unlabelled primers on six

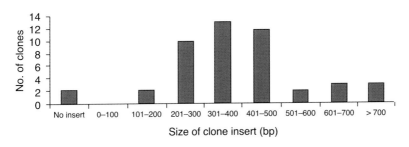

Fig. 10.1. Sizes of clone inserts estimated by restriction digestion using *Bam*HI and *Eco*RI. The vertical scale represents number of clones versus size in base pairs.

Table 10.3. Characteristics of microsatellite primers used in this study.

Marker	Expected size in CS[a]	Primer sequence	Repeat type and length
WMC144	143	F- GGACACCAATCCAACATGAACA R- AAGGATAGTTGGGTGGTGCTGA	$(CA)_{14}$
WMC141	99	F- TGCAAGGGGTTGAGTTCTTAGG R- TTGATAGAGCAAGACGAAGGGC	$(CA)_{15}$
WMC136	198	F- GTATGATGACACATGGTGCTGC R- GACCTGATTCTGCTGTGCTCTG	$(AGGCGG)_{4}$
WMC139	66	F- TGTAACTGAGGGCCATGAAT R- CATCGACTCACAACTAGGGT	$(CA)_{12}$
WMC140	100	F- CCCTCTCCTCGGGTGTTGCTTG R- CCCAGGAGCCCTCATGCATACG	$(GT)_{12}$
WMC146	199	F- CCAAGTGGTTCTTTCCATCATC R- GTGGGTAAAGTGTGCAACCTCT	$(CA)_{31}$
WMC143	212	F- CTTGACCCAAGTAGTTCTTTCC R- GTGGGTAAAGTGTGCAACCTCT	$(CA)_{9}(CA)_{19}$
WMC145	207	F- GGCGGTGGGTTCAAGTCGTCTG R- GGACGAGTCGCTGTCCTCCTGG	$GTGCCG(CGG)TGG(CGG)_{2}$
WMC142	143	F- AACCGCAACTCTTGTGACCGTG R- GCTAGACCTGCCTGTCGGTATG	$(CG)_{2}(ACTC)_{2}(AC)_{2}(GC)_{6}$
WMC138	318	F- AATCCCGGTCAAATGAGAAT R- GAGACATGCATGAATCAGAA	$(CA)14$ 172–199, $(GA)_{8}$ 200–215
WMC137	190	F- CGAGAAGTCTACATATCGAGGG R- CAACAATGACAACAGAAGGGTG	$(GT)_{16}$

[a]CS refers to 'Chinese Spring'.

hexaploid wheat varieties including 'Chinese Spring' (the reference genotype, as a control to ensure products were generated), showed that six of the 11 primers gave bands of the expected size on 3.5% agarose gels. Table 10.3 shows the characteristics of these markers, including primer sequence, size of PCR product in reference variety ('Chinese Spring'), the number of repeat units and a description of the repeat motif. All marker loci are designated 'WMC' for wheat microsatellite consortium. Of the six, one was eliminated (primer pair WMC139), as it was designed to amplify a PCR product of 66 bp in 'Chinese Spring'. Screening of hexaploid genotypes with this primer pair was difficult with both agarose gel electrophoresis used in the preliminary testing or using the ABI Prism 310 Genetic Analyser because the amplicons were close in size to amplification artefacts. This was not a problem with primer pairs designed to give larger products, such as WMC141 and WMC139 at 99 and 100 base pairs, respectively. Marker WMC146 was difficult to analyse because a large number of products occurred in the size range, as reported in another recent microsatellite study in wheat (Bryan *et al.*, 1997). Of the six primers fluorescently labelled, five primers, WMC144, WMC141, WMC136, WMC145 and WMC146, were used for subsequent microsatellite analysis.

Table 10.4. Polymorphisms of microsatellite loci.

Marker	Expected size in CS[a]	Alleles in hexaploid cultivars (bp)	Alleles in diploid accessions (bp)[b]	Location	Additional loci
WMC144	143	135, 137, 139	135, 137 (A, B and D)	2D	
WMC141	99	78, 80, 84, 94, 96, 98	78, 80, 94, 96, 98, 102(A), 80, 92(B), 80, 82, 88, 102(D)	2A, 4A or 3B	
WMC136	198	183, 189, 195	183(A), 178, 183, 190(B), 188, 204, 208, 210(D)	5B, 5D and ?4A	(195) (189) (183)
WMC146	199	191, 195, 199, 203, 207, 209	191, 199, 205(A) 205 (B) 195 (D)	?	
WMC145	207	205	Nil	?	
WMC139	66	Smear	Smear		
WMC140	100	Smear			
WMC143	212	180 (smear)			
WMC138	318	No product			
WMC137	190	120 and 170			

[a]CS refers to 'Chinese Spring'.
[b]A, B and D refer to the accessions of each of hexaploid diploid progenitors, *T. monococcum*, *T. speltoides* and *T. tauschii*, respectively.

It was necessary to use less stringent conditions for amplification of microsatellites from the diploid progenitors. A similar approach was used in a study of related species from the *Poaceae* (Röder *et al*, 1995). Amplification was achieved at 50°C initially using *Taq* DNA polymerase (Boehringer Mannheim); however, the presence of multiple bands (presumably non-specific) made the scoring of alleles difficult. This was overcome by using Ampli*Taq* Gold DNA polymerase (Perkin Elmer) at 50°C, which eliminated these unwanted bands.

Three out of five microsatellites (markers WMC144, WMC141, WMC146) were polymorphic across the hexaploid wheat varieties. Marker WMC144 had a low level of polymorphism with three alleles, WMC141 had six alleles, while WMC136 and WMC145 were monomorphic for the 24 varieties screened (see Table 10.4 for alleles in hexaploid cultivars). WMC146 was screened against 12 varieties and had six alleles. Polymorphisms between four varieties of hexaploid wheat using primer pair WMC141 are shown in Fig. 10.2.

The wild diploid accessions generally showed a high level of heterozygosity with marker WMC144 and marker WMC141, ranging from 60 to 100% in the nine accessions of each *T. monococcum*, *T. speltoides* and *T. tauschii*. Primer pair WMC144 amplified two alleles across all of the diploid accessions at 137 and 139 bp (Table 10.4), which were common to the hexaploid varieties. Primer

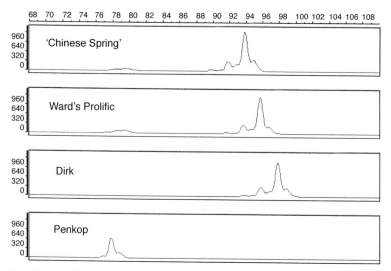

Fig. 10.2. Microsatellite polymorphisms detected by primer pair WMC141 across four hexaploid wheat varieties. PCR-amplified fragments were sized (base pairs) using capillary electrophoresis.

pair WMC141 amplified a wide range of allele sizes in both hexaploid and diploid genotypes. It is possible that this marker was amplifying more than one locus as the sizes of alleles were split into two size ranges: 78–82 bp and 92–102 bp (see Table 10.4 for all allele sizes). Additionally, the number of alleles observed with primer pair WMC141 was not consistent with the primer amplifying a single locus; that is, more than two products were amplified in diploids. Examples of this include *T. monococcum* (accession number 19848) which had an allele at 80, as well as showing heterozygotic alleles of 97 and 101 bp, and *T. tauschii* (accession number 24226) having alleles at 80 as well as 91 and 93 bp.

Some of the fragments amplified by PCR using primer pair WMC136 on wild wheat progenitor and cultivated hexaploid wheat are shown in an electropherogram (Fig. 10.3). With the exception of *Triticum compactum* (a subspecies of *T. aestivum*) which lacked the allele at 189 bp, the pattern was the same throughout the 24 hexaploid varieties (monomorphic), a multilocus pattern of three alleles at 183, 189 and 195 bp. These three alleles are represented in the top window of Fig. 10.3. The accessions of *T. monococcum* (AA), *T. speltoides* (BB) and *T. tauschii* (DD) have either one or two loci. All accessions of *T. monococcum* (AA) are monomorphic with an allele of 183 bp. *T. speltoides* and *T. tauschii*, however, show interspecific polymorphism with marker WMC136. For example, *T. speltoides* (accession number 21644) had an allele at 178 and *T. tauschii* (accession number 23950) was heterozygotic with alleles at 204 and 208 bp, which represent alleles unique to the particular species (see Table 10.3).

Microsatellite loci were assigned to chromosomes corresponding to the aneuploid lines for which either no PCR amplification product was obtained, or

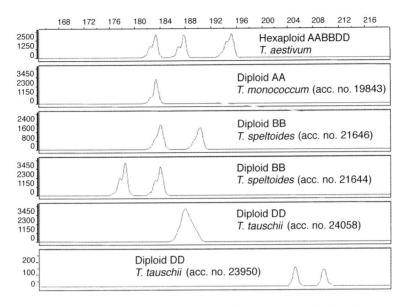

Fig. 10.3. Polymorphisms of PCR-amplified microsatellite DNA marker WMC136 in base pairs across hexaploid and diploid progenitors.

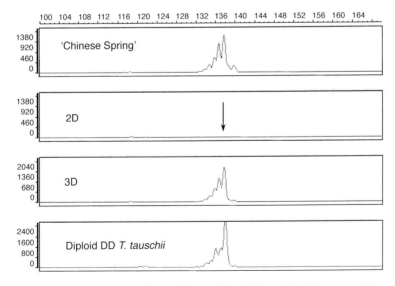

Fig. 10.4. Localization of microsatellite marker WMC144 PCR-amplified fragments (in base pairs) using capillary electrophoresis. The arrows represent the missing peaks. Stutter peaks as illustrated are typically seen in dinucleotide microsatellite repeats.

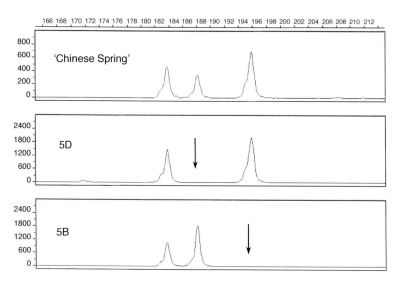

Fig. 10.5. Localization of microsatellite multilocus marker WMC136 PCR-amplified fragments (in base pairs) using capillary electrophoresis. The arrows represent the missing peaks.

one of the PCR products was missing, provided that all of the other aneuploid lines gave the relevant PCR product. Marker WMC144 amplified a single product and the fragment could be localized to chromosome 2D, shown in Fig. 10.4. The chromosomal location of two of the three loci produced by PCR amplification with primer WMC136 was possible using aneuploid lines seen in Fig. 10.5. The chromosome location is indicated by arrows; the smallest allele (183 bp) could not be localized but may be from the missing stock for 4A, the corresponding location to 5 in the A genome.

Discussion

The technique used for enrichment of microsatellites required a high degree of efficiency in order to produce a useful DNA library in hexaploid wheat and to compensate for the relatively low abundance of microsatellites in the plant genome (Edwards *et al.*, 1996). Despite this, 40% of the sequenced clones contained no substantial microsatellite, with two of the 48 clones sequenced revealing no insert at all. The final yield of useful markers was quite low in this study, with only 5% showing polymorphism in hexaploid wheat. Similar results are shown in another recent study on wheat, with only 3% of positive clones giving useful genetic markers (Bryan *et al.*, 1997). Additionally, low levels of polymorphism have been detected using microsatellite probes in bread wheat, reported to be due to their large genome size, polyploid nature and high proportion of repetitive DNA (Varshney *et al.*, 1998).

Marker WMC144 is located on chromosome 2D, as shown in Fig. 10.4. Interestingly, alleles present in the hexaploid cultivars were also present in each of the progenitor species; in particular, accessions of *T. tauschii* (represented in Fig. 10.4) contained a common allele held in the majority of hexaploid cultivars, indicating conservation of the microsatellite repeat regions between the genomes. Another study using 64 microsatellite loci showed that the D genome contains the least microsatellite loci; in particular, the B genome had 45%, followed by the A at 30%, with the D genome at only 25% (Plaschke *et al.*, 1996). This may be because the D genome is the smallest of the genomes in size (Gupta, 1991), which may indicate that there is less repetitive DNA in this genome due to selection pressure. In this study, insufficient markers were used to make these comparisons.

Primer pair WMC136 is a hexameric repeat $(AGGCGG)_4$ and produced three fragments that were monomorphic across the hexaploid varieties in this study. Large microsatellite motifs are rare, thus the occurrence of hexameric repeats and subsequent use in genetic analysis has not been reported previously. Most enrichment techniques do not appear to enrich for larger than tetrameric repeats. Two of the fragments amplified by this unique marker could readily be assigned to the same chromosome, one on each 5B and 5D (Fig. 10.5), and the third is likely to be located on chromosome 4A originally 5A prior to translocation between chromosomes 4A and 5A (Plaschke *et al*, 1996). Detection of homologous loci is common when using RFLP probes; however, little is known of similar detection by microsatellite-specific primer pairs. Amplification of homologous loci offers the advantage of allowing genome comparison studies (Plaschke *et al.*, 1996).

Primer pair WMC145 amplified a compound microsatellite repeat, $GTGCCG(CGG)TGG(CGG)_2$. Although this marker amplified a PCR product of the expected size, the alleles were monomorphic across hexaploid varieties in this study and failed to amplify any easily detectable alleles in any of the wild diploid relatives. It is questionable whether the repeat region flanked by these primers constitutes a microsatellite, as the repeat is very short as well as being imperfect.

Primer pairs WMC141 and WMC146 amplified multiple products, which suggests that the microsatellite flanking regions are conserved and duplicated within and between genera. This has been observed for a number of species such as brassica (Westman and Kresovich, 1998) and potato (Provan *et al.*, 1996). Chromosomal assignment for these markers is made difficult by the possible presence of more than one locus that could be on the same or different chromosomes. Failure to localize microsatellite markers was reported in another study involving around half of the microsatellite loci (Plaschke *et al.*, 1996).

Genetic variability among cultivated hexaploid wheat varieties was low in this study whereas the wild diploid wheat accessions showed a greater degree of polymorphism with a high degree of heterozygosity. Such diversity could potentially be of interest to breeders for genes that can be introgressed, particularly when the diploids possess genomes common to the cultivated wheat (Le Corre and Bernard, 1995). This finding is not surprising in view of the fact that bread wheat is supposed to have been cultivated close to the time of its hybridization and therefore isolated from other species for a long period. This means, therefore, that the potential for genetic variation is lower for bread wheat than its wild progenitors and other relatives (CIMMYT, 1996).

Hexaploid wheat contains only one of several alleles from the B and D genome represented in wild progenitors. This result also suggests that these three loci have not evolved to have multiple alleles in hexaploid wheat, possibly due to the large sizes of the repeat (hexanucleotide). In contrast, the dinucleotide repeat polymorphism of three CA repeat microsatellite markers in hexaploid wheat is consistent with the evolution of diversity occurring since hexaploid formation.

In this study, microsatellite markers designed to amplify in *T. aestivum* have been shown to be useful in amplifying microsatellite loci in related species. These molecular markers have been useful tools in estimating the extent of genetic diversity and relatedness within and between wild diploid progenitors and cultivated hexaploid wheat. The comparative study of diploid wild wheat and cultivated hexaploid wheat species can lead to a greater understanding of evolutionary relationships as well as suggesting a potential source of greater genetic diversity for applications such as wheat breeding (Giorgi *et al.*, 1997).

Acknowledgements

We thank Dr Giovanni Cordeiro and Dr Mervyn Shepherd of the Centre for Plant Conservation Genetics, Lismore, Australia, for helpful suggestions on the manuscript. This work has been supported by the Grains Research Development Corporation.

References

Bryan, G.J., Collins, A.J., Stephenson, P., Orry, A., Smith, J.B. and Gale M.D. (1997) Isolation and characterisation of microsatellites from hexaploid bread wheat. *Theoretical and Applied Genetics* 94, 557–563.

CIMMYT (1996) *World Wheat Facts and Trends: Understanding Global Trends in the Use of Wheat Diversity and International Flows and Wheat Genetic Resources*. CIMMYT, Mexico, DF.

Dvorak, J. (1988) Cytogenetical and molecular inferences about the evolution of wheat. *Proceedings of the 7th International Wheat Symposium*, pp. 187–192.

Dvorak, J., Di Terlizzi, P., Zhang, H.-B. and Resta, P. (1993) The evolution of polyploid wheats: identification of the A genome donor species. *Genome* 36, 21–31.

Dvorak, J., Luo, M.C., Yang, Z.L. and Zhang, H.B. (1998) The structure of the *Aegilops tauschii* genepool and the evolution of hexaploid wheat. *Theoretical and Applied Genetics* 97, 657–670.

Edwards, K.J., Barker, J.H.A., Daly, A., Jones, C. and Karp, A. (1996) Microsatellite libraries enriched for several microsatellite sequences in plants. *Biotechniques* 20, 758–760.

Giorgi, D., D'Ovidio, R., Tanzarella, O.A., Ceoloni, C. and Porceddu, E. (1997) Molecular characterisation of *Aegilops* species belonging to Sitopsis section and their relationships with the B genome of cultivated wheats (abstract). *International Triticeae Mapping Initiative*, June 1997, Public Workshop. P6.01.

Gupta, P.K. (ed.) (1991) Cytogenetics of wheat and its close wild relatives – *Triticum* and *Aegilops*. In: *Chromosome Engineering in Plants: Genetics, Breeding, Evolution*. Elsevier Sciences, New York.

Kihara, H. (1944) Discovery of the DD-analyser, one of the ancestors of *Triticum vulgare*. [In Japanese] *Agric. Hortic.* 19, 13–14.

Kimber, G. and Sears, E.R. (1987) Evolution in the genus *Triticum* and the origin of cultivated wheat. In: Heyne, E.G. (ed.) *Wheat and Wheat Improvement*, 2nd edn. Agronomy. ASA, CSSA, SSSA. No. 13, pp. 154–163.

Le Corre, V. and Bernard, M. (1995) Assessment of the type and degree of restriction fragment length polymorphism (RFLP) in diploid species of the genus *Triticum*. *Theoretical and Applied Genetics* 90, 1063–1067.

Maestra, M. and Naranjo, T. (1998) Homoeologous relationships of *Aegilops speltoides* chromosomes to bread wheat. *Theoretical and Applied Genetics* 97, 181–186.

McFadden, E.S. and Sears, E.R. (1946) The origin of *Triticum spelta* and its free-threshing hexaploid relatives. *Journal of Heredity* 37, 81–89.

Plaschke, J., Borner, A., Wendehake, K., Ganal, M.W. and Röder, M.S. (1996) The use of wheat aneuploids for the chromosomal assignment of microsatellite loci. *Euphytica* 89, 33–40.

Raven, P.H., Evert, R.F. and Eichhorn, S.E. (1986) In: *The Biology of Plants*, 4th edn. Worth Publishing, pp. 580–581.

Röder, M.S., Plaschke, J., Konig, S.U., Borner, A., Sorrells, Tanksley, S.D. and Ganal, M.W. (1995) Abundance, variability and chromosomal location of microsatellites in wheat. *Molecular and General Genetics* 246, 327–333.

Sambrook, J., Fritsch, E.F. and Maniatis, T. (1989) *Molecular Cloning – a Laboratory Manual*, 2nd edn. Cold Spring Harbor Laboratory Press, Cold Spring Harbor, New York.

Somers, D.J., Zhou, Z., Bebeli, P.J. and Gustafson, J.P. (1996) Repetitive genome-specific probes in wheat (*Triticum aestivum*) L.em. Thell amplified with minisatellite core sequences. *Theoretical and Applied Genetics* 93, 982–989.

Varshney, R.K., Sharma, P.C., Gupta, P.K., Balyan, H.S., Ramesh, B., Roy, J.K., Kumar, A. and Sen, A. (1998) Low level of polymorphism detected by SSR probes in bread wheat. *Plant Breeding* 117, 182–184.

Weining, S. and Henry, R.J. (1995) Molecular analysis of the DNA polymorphism of wild barley (*Hordeum spontaneum*) germplasm using the polymerase chain reaction. *Genetic Research and Crop Evolution* 42, 273–281.

Weising, K., Fung, R.W.M., Keeling, J.D., Atkinson, R.G. and Gardner, R.C. (1996) Characterisation of microsatellites from *Actinidia chinesis*. *Molecular Breeding* 2, 117–131.

Westman, A.L. and Kresovich, S. (1998) The potential for cross-taxa simple-sequence repeat (SSR) amplification between *Arabidopsis thaliana* L. and crop brassicas. *Theoretical and Applied Genetics* 96, 272–281.

Xiao, J., Li, J., Yuan, L., McCouch, S.R. and Tanksley, S.D. (1996) Genetic diversity and its relationship to hybrid performance and heterosis in rice as revealed by PCR-based markers. *Theoretical and Applied Genetics* 92, 637–643.

Yang, G.P., Saghai-Maroof, M.A., Xu, C.G., Zhang, Q. and Biyashev, R.M. (1994) Comparative analysis of microsatellite DNA polymorphisms in landraces and cultivars in rice. *Molecular and General Genetics* 245, 187–194.

Chapter 11

Comparison of RFLP and AFLP Marker Systems for Assessing Genetic Diversity in Australian Barley Varieties and Breeding Lines

K.J. CHALMERS, S.P. JEFFERIES AND P. LANGRIDGE

CRC for Molecular Plant Breeding, Department of Plant Science, University of Adelaide, Australia

Introduction

Estimation of genetic diversity in a crop species can assist in the evaluation of germplasm collections as potential gene pools to improve the performance of cultivars. Molecular markers offer the opportunity to assess variation at the DNA sequence level and, as such, have become an important tool in plant genetics (Gebhardt *et al.*,1991). Restriction fragment length polymorphisms (RFLPs) have led to the integration of DNA markers into molecular genetic studies and plant breeding programmes, and have been used extensively in investigations of genetic diveristy in both wild and cultivated barley (*Hordeum* spp.) (Graner *et al.*, 1994; Melchinger *et al.*, 1994). RFLPs are co-dominant and locus-specific markers that are unequalled for many applications. In particular, they are well suited for the construction of linkage maps (Graner *et al.*, 1991; Heun *et al.*, 1991) and, because of their locus specificity, allow for synteny studies (Tanksley *et al.*, 1992; Ahn and Tanksley, 1993). However, the procedures involving RFLPs are labour intensive, expensive and few loci are detected per assay.

Recently a number of locus-specific PCR-based markers such as sequence-tagged sites (Vanichanon *et al.*, 2000) and microsatellites (simple sequence repeats) (Litt and Luty, 1989) have been developed. These marker systems are rapid, technically simple and require only small amounts of DNA. Microsatellites have been used with some success in cereals (Saghai Maroof *et al.*, 1994; Röder *et al.*, 1995) and offer the potential as co-dominant PCR-based markers with uniform genome coverage. Microsatellites show a high level of polymorphism and are particularly suited to identifying a large number of alleles at a specific locus. However, the development of locus-specific markers is time-consuming, expensive and relatively few loci are detected per assay.

Consequently there is a requirement for a PCR-based marker system which is technically simple, highly polymorphic, cost efficient and assays a significant portion of the genome.

Randomly amplified polymorphic DNAs (RAPDs) (Welsh and McClelland, 1990; Williams *et al.*, 1990) have been widely used in studies assessing genetic diversity (Dawson *et al.*, 1993; Heun *et al.*, 1994). RAPDs use a single short primer coupled with a low annealing temperature that allows amplification of multiple loci dispersed throughout the genome and provide a rapid assay for nucleotide sequence polymorphism. Unfortunately, the reproducibility of RAPD analysis is often poor due to low annealing temperatures and other reaction conditions (Yang *et al.*, 1996). RAPDs were also found to be ill-suited for use in cereals as a given primer can amplify a number of non-homologous sequences that may vary between varieties (Devos and Gale, 1992).

Recently, a reliable and efficient method has been developed to identify a large number of molecular markers. The method called amplified fragment length polymorphism (AFLP) is a novel PCR-based technique for DNA fingerprinting. AFLP has found application in germplasm characterization (Hill *et al.*, 1996), genome mapping and marker-assisted breeding programmes (Becker *et al.*, 1995; Van Eck *et al.*, 1995). AFLP relies on PCR amplification of a subset of small restriction fragments. The procedure involves the digestion of genomic DNA with two restriction enzymes, one a frequent cutter and one a rare cutter. Oligonucleotide adaptors are then ligated to the resulting restriction fragments to generate template DNA for PCR. The DNA fragments are then selectively amplified by primers complementary to the oligonucleotide adaptors and up to four selective nucleotides at the 3' end. Polymorphisms are detected as the presence or absence of an amplified restriction fragment and are therefore dominant. AFLP has the capacity to survey a much larger number of loci for polymorphism than other currently available PCR-based methods and has been demonstrated to generate a large number of polymorphisms in barley (*Hordeum vulgare* L.) (Becker *et al.*, 1995).

This study demonstrates the application of barley RFLPs and AFLPs for the differentiation and estimation of genetic relationships between 80 barley accessions that have been widely grown in or have played a major role as progenitors of modern varieties in Australia. We also report the comparison of RFLP and AFLP in their ability to detect polymorphism and in the frequency of polymorphism detected. Phylogenetic trees derived from both sets of data were compared to determine the similarities between the genotypes found using the different marker systems.

Materials and methods

Plant material

Eighty barley cultivars/accessions were screened (Table 11.1). Cultivars were selected on the basis of: (i) historical significance to barley breeding and pro-

Table 11.1. Details of accessions tested, including pedigree, 2/6 row and spring/winter growth habit, where known.

Variety number	Variety name	Country	Pedigree	2/6 Row	Spring/winter
1	Alexis	Germany	St1622 × Triumph		Spring
2	AmajiNijo	Japan		2	
3	Andre	USA	Klages/Zephyr	2	
4	Arapiles	Australia	Noyep/Proctor/Cl2576/Union/4/Kenia/3/Research/2/Noyep/Proctor/5/Domen		Spring
5	Bandulla	Australia	(Prior × Lenta) × (Noyep × Lenta)		
6	Bearpaw	USA	Klaves/Zephyr/Centennial/3/Clark		Spring
7	Blenheim	UK	Triumph/Egmont		
8	Brindabella	Australia	Weeah/Cl7115/HCB27/3/Jadar II/4/Cantala		Spring
9	Carmargue	Germany	49428/68 × (Diamant × 14029/64/6) × KM1192		Spring
10	Caminant	Denmark			Spring
11	Cantala	Australia	Kenia/Erectoides 16		Spring
12	Carina	Germany	(Union × Ackermans WV16) × Volla		Spring
13	Chariot	UK	Dera × CSB626/12		Spring
14	Chebec	Australia	Orge Martin/2/Clipper(86)//Schooner		Spring
15	Cheri	Germany	Triumph × (Medusa × Diamant)		Spring
16	Cl3576	Ethiopia	Landrace		
17	CIMMYT42002	CIMMYT			
18	Clipper	Australia	Proctor/Prior A		Spring
19	Corvette	Australia	Bonus × Cl3576		Spring
20	Dera	UK	Complex Cross		
21	Ellice	Canada	Cl579/Parkland//Betzes/3/Betzes/Piroline/4/Akka/5/Centennial/6/Klages/7/Cambrinus/T		
22	Fergie	UK	Athos/Koru//(Mari/Goldmarker)		Spring
23	Forrest	Australia	Atlas57 × (A16) Prior × Ymer		Spring
24	Franklin	Australia	Shannon/Triumph		Spring
25	Galleon	Australia	(Clipper × Hiproly 3) × (Proctor × Cl3576)		Spring

Continued

Table 11.1. Continued

Variety number	Variety name	Country	Pedigree	2/6 Row	Spring/winter
26	Gilbert	Australia	Reselection from Koru		Spring
27	Gimpel	Germany	[(Proctor × Calsberg II) × (Heine4808 × Stammyl-P)]		Spring
28	GoldenPromise	UK	Gamma-ray mutant of Maythorpe		Spring
29	Grimmett	Australia	Bussell/Zephyr		Spring
30	Grit	Germany	(Ha.st55474/67 × Ha.st.46459/68)		Spring
31	H.Spont.7128448mountain	Israel	H. spontaneum		
32	H.Spont.7128527Desert	Israel	H. spontaneum		
33	H.Spont.771292IsraelCoast	Israel	H. spontaneum		
34	H.Spont.771371ITemperate	Israel	H. spontaneum		
35	Halycon	UK	Warboys/Maris Otter		Winter
36	Harrington	Canada	Klages × ((Gazelle × Betzes) × Centenial)		Spring
37	HarunaNijo	Japan	Satsuko Nijo × (K-3 × G-65)	2	Spring
38	Igri	Germany	820 × 1427 × Ingrid	2	Winter
39	Kaputar	ICARDIA	5604/1025/3/Emir/Shabet/CM67/4	2	Spring
40	KMBR52	Czech Republic			
41	Lara	Australia	Rersearch/Lenta	2	Spring
42	Malebo	Australia	Selection from CPI 11083	6	Spring
43	Maltine	France	Aramir × (Zephyr × Sundance)		
44	Moondyne	Australia	Dampier//(A14)Prior/Yemer/3/Kristina(7052 0.20)/4(73S13)Clipper/Tenn 65-117	2	Spring
45	Morrell	Australia	WUM221/P23822(81S806)/5/(81S719)Forre st/4/(80S64)Psaknom/Dampier/M19(76T1 11)/3/Zephyr	2	Spring
46	Namoi	Australia	Sultan/Nackta//RM1508/Godiva	2	Spring
47	Natasha	France			
48	Norbet	Canada	(((((C15791 × Parkland)/2 × Betzes) × Piroline) × Akka) × Centennial) × Klages	2	Spring

#	Name	Country	Pedigree		Season
49	Noyep	Australia	Single Plant Selection from Prior	2	Spring
50	O'Connor	Australia	Proctor/CI3576(Wl2231)/3/(XBV1212)AILAS5 7//(A14)Prior/Ymer	2	Spring
51	Onslow	Australia	Forrest/Aapo	2	Spring
52	Osiris	France	((Herta × Pallidum191) × Kentia) × ((Monte Cristo × Minerva) × (CI1179 × Deba)) Plumage	6	
53	Parwan	Australia	Archer/Prior//Lenta/3/Research/Lenta	2	Spring
54	Prisma	Netherlands	[Triumph × Cambrinus] × Piccolo	2	Spring
55	Proctor	UK	Kenia/Plumage Archer	2	Spring
56	Puffin	UK	(Athos × Maris Otter) × Igri	2	Winter
57	Research	Australia	Plumage Archer × Prior	2	Spring
58	Richard	Canada		2	Spring
59	Rubin	Czech Republic		2	Spring
60	Sahara3771	North Africa		6	
61	Schooner	Australia	(Proctor × Prior) × (Proctor × CI3576)	2	Spring
62	SH302	Japan	Proctor*4 × CI3208·1	2	Spring
63	Shannon	Australia	(Frankeld × Mona) × Triumph	2	Spring
64	Sissy	USA	(Abed Deba × ((Proctor × CI3576)60 × (CPI-18197 × Beka))/12)22 × ((Clipper × Diamant)/28 × (Proctor × CI3576)/28)	2	Spring
65	Skiff	Australia	(St101 × Aramir) × St210	2	Spring
66	Steffi	Germany	Dampier/((A14)Prior/Yemer/3/Piroline	2	Spring
67	Stirling	Australia	Balder × (Agio × Kenia × 3/Arabische)	2	Spring
68	Sultan	Netherlands	Triumph/Grimmett	6	
69	Tallon	Australia	Diamant × St 1402964/6	2	Spring
70	Triumph	Germany	Warboys/Alpha	2	Spring
71	Ulandra	Australia	Noyep × Prior//CI3576/4/Union//Kenia//Research/3/	2	Winter

Continued

Table 11.1. *Continued*

Variety number	Variety name	Country	Pedigree	2/6 Row	Spring/winter
72	Vic86045B	Australia	Nopyep/Prior/5/Elgina	2	Spring
73	Vic9104	Australia		2	Spring
74	WA73S276	Australia		2	Spring
75	WA83S514	Australia	O'Conner/Yagan	2	Spring
76	Weeah	Australia	Prior/Research	2	Spring
77	WI2868	Australia			Spring
78	WI287522	Australia			Spring
79	Yagan	Australia		2	Spring
80	Yerong	Australia	M22/Malebo	6	Spring

duction in Australia; (ii) accounting for over 2% of deliveries to silos in any Australian state since 1982; (iii) recently released cultivars carrying traits of particular interest; and (iv) significant parents in current breeding programmes.

RFLPs

Seventy-six DNA clones selected to give representative genome coverage were obtained through the Australian Triticeae Mapping Initiative (Langridge *et al.*, 1995). DNA extraction, restriction digestion and Southern blotting and hybridization were carried out as described by Pallotta *et al.* (2000). Total genomic DNA was digested with *Bam*HI, *Dra*I, *Eco*RI, *Eco*RV and *Hin*dIII. DNA membranes of the 80 barley lines were screened with RFLP.

AFLPs

The AFLP method developed by Vos *et al.* (1995) was followed with some modifications. Genomic DNA (1 mg) was digested with the restriction endonucleases *Pst*I and *Mse*I. Double-stranded *Pst*I and *Mse*I adaptors were then ligated to the ends of the restriction fragments followed by ethanol precipitation and resuspension in 60 µl of 0.1 M TE. Pre-amplification was performed using primers specific for the *Pst*I and *Mse*I adaptors including one selective nucleotide, followed by selective amplification using similar primers with three selective bases. The pre-amplification mix was diluted 1:5 in water before being used in the selective amplification step. Pre-amplification PCR consisted of 20 cycles of 94°C for 30 s, 56°C for 1 min and 72°C for 1 min. PCR conditions for selective amplification consisted of one cycle at 94°C for 30 s, 65°C for 30 s and 72°C for 1 min followed by nine cycles over which the annealing temperature was decreased by 1°C per cycle with a final step of 25 cycles of 94°C for 30 s, 56°C for 30 s and 72°C for 1 min. The *Pst*I primer used in selective amplification was end labelled with [γ-^{32}P]ATP (Freinberg and Vogelstein, 1983). Amplified fragments were separated on 6% denaturing polyacrylamide gels. The gels were transferred to 3MM paper for drying and autoradiography was carried out with Fuji RX medical X-ray film at room temperature for 24–48 h.

Data analysis

Genetic distances were calculated between pairs of accessions based on the method of Nei and Li (1979). Genetic distance is defined as the extent of genetic difference between cultivars, as measured by allele frequencies at a sample of loci (Nei, 1987). Genetic similarity is defined as the converse of genetic distance, that is, the extent of similarity among cultivars.

The presence or absence of each specific band was scored for each genotype and for both methods used in this study. AFLP fragments detected by all primer combinations and RFLP fragments detected by all probes for each accession were used to make comparisons with all other accessions. RFLP fragments of the same size for a particular probe/enzyme combination were assumed to identify the same allele, as were AFLP fragments of the same size for a give primer combination. The measure of distance or similarity among cultivars was the covariance of allele frequencies summed for all fragments scored. The distances used in this study were based on RFLP and AFLP fragments scored in each of the cultivars. Parsimony analysis of both RFLP and AFLP data was carried out using PAUP (Version 4.0b3a), a Macintosh computer software package developed by Swaford (1998). The heuristic option of the PAUP program using stepwise addition was used to generate phylogenetic trees.

Results

RFLP

The 80 barley accessions were evaluated using 76 probes with three enzyme combinations. These probes were selected to give a uniform genome coverage based on their map location determined in three barley doubled haploid populations (Langridge et al., 1995). As a consequence of the inclusion of the parents of the three doubled haploid populations, all of the probes were found to be polymorphic with at least one enzyme. A total of 1324 polymorphic fragments were scored, ranging from 4 to 18 polymorphisms per probe/enzyme combination. In order to avoid possible duplication of data, only the most polymorphic probe/enzyme combination at each locus was considered in the analysis. The resulting data set consisted of 595 polymorphic fragments. Each accession was uniquely identified by the marker data, although no single probe discriminated among all accessions.

Genetic distance (GD) based on all possible pairs of lines ranged from 0.026 to 0.360 with a standard error of 3.8×10^{-3}. The mean GD was 0.204, that is, two randomly chosen lines differed on average in 121 of the 595 polymorphic bands scored. The least pairwise GD was between 'Arapiles' and 'Vic86045B' (cv 'Picola') (GD 2.6×10^{-2}) which had 15 polymorphic bands. These were both Victorian accessions derived from the cultivar 'Noyep'. The two most closely related cultivars were 'Research' and its backcross derivative 'Weeah' which differed at only 17 loci (GD 2.8×10^{-2}). The greatest GD was between the *Hordeum spontaneum* accession 771292 and the Australian six-row cultivar, 'Yerong', with 214 polymorphic bands identified between these lines.

AFLP

The 80 barley accessions were evaluated using five AFLP primer combinations. On average, 80 amplification products were identified per primer combination. A total of 243 polymorphic fragments were scored, ranging from 33 to 60 polymorphisms per primer combination. Each accession was uniquely identified by the marker data, although no single primer combination discriminated among all accessions. GD based on all possible pairs of lines ranged from 0.027 to 0.451 with a standard error of 1.1×10^{-3}. The mean GD was 0.234, that is, two randomly chosen lines differed on average in 57 of the 243 polymorphic bands scored. The least pairwise GD was between the two cultivars 'Research' and its backcross derivative 'Weeah' (GD 2.7×10^{-2}) which differed at seven loci. The greatest GD was between the *H. spontaneum* accession 712841 and the accession 'Bandulla'; 214 polymorphic bands were identified between these lines.

Cluster analysis of accessions based on genetic distance

Cluster analysis of accessions based on genetic distance values provides a good estimate of genetic relationships between related and unrelated accessions. The hierarchy clusters shown in Figs 11.1 and 11.2 were consistent with the origin of these lines and their pedigree information as far as was known. Cluster analysis, based on RFLP data and AFLP data, is summarized in Figs 11.3 and 11.4 respectively. In general, cluster analysis of both data sets separated all accessions at two major levels, species and row number. Several other similarities in clustering pattern were also observed.

Four of the 80 accessions included in the study were accessions of the wild ancestor of cultivated barley, *H. spontaneum* (Harlan, 1979). The remaining 76 accessions were commercial cultivars, or breeding lines, of the cultivated barley, *H. vulgare* L. Cluster analysis of both RFLP and AFLP data sets clearly separated these two species groups. The four *H. spontaneum* accessions produced the four highest mean RFLP genetic distance estimates (mean of all pairwise estimates) (Table 11.2). While mean genetic distance estimates between the two species groups were very high (0.31–0.33), genetic distance estimates between the four individual *H. spontaneum* accessions were also high (0.24–0.3).

Cluster analysis of both RFLP and AFLP data sets clearly separated the North African six-row landrace, 'Sahara 3771', from other *H. vulgare* accessions. While the AFLP data set separated 'Sahara 3771' uniquely, the RFLP data set grouped 'Sahara 3771' with a two-row CIMMYT introduction, 'CIMMYT 42002', and also uniquely separated the CIMMYT introduction, 'Namoi', from other *H. vulgare* accessions. The clear separation of 'Sahara 3771' from other *H. vulgare* accessions is reflected in its mean RFLP genetic distance estimate over all pairwise comparisons (0.30) (Table 11.2). In addition, 'Sahara 3771' was shown to be genetically divergent from the four *H. spontaneum* accessions (RFLP genetic distance estimates of 0.35–0.36).

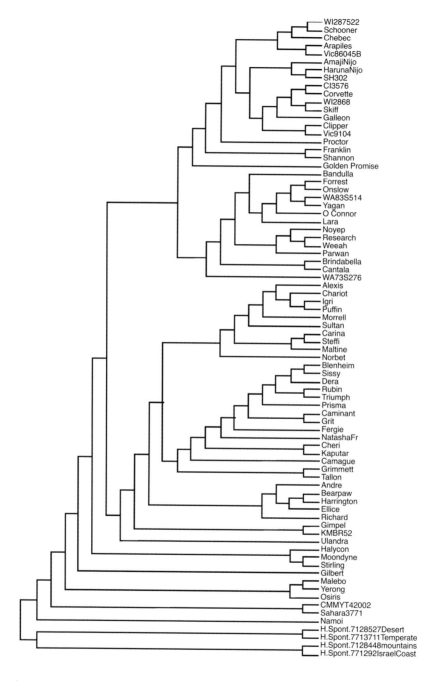

Fig. 11.1. Dendogram showing a cluster analysis of the 80 barley accessions based on restriction fragment length polymorphism analysis. A total of 1324 polymorphic fragments were used in the analysis.

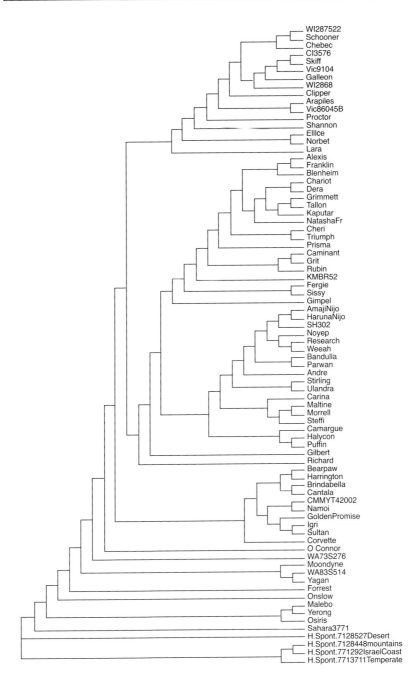

Fig. 11.2. Dendogram showing a cluster analysis of the 80 barley accessions based on amplified fragment length polymorphism analysis. A total of 595 polymorphic fragments were used in the analysis.

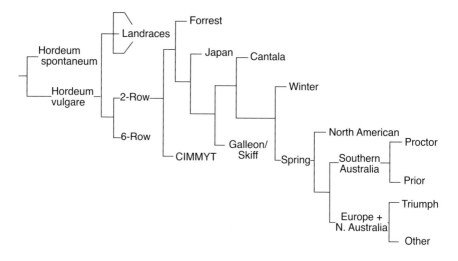

Fig. 11.3. Summary of major restriction fragment length polymorphism phylogram groupings.

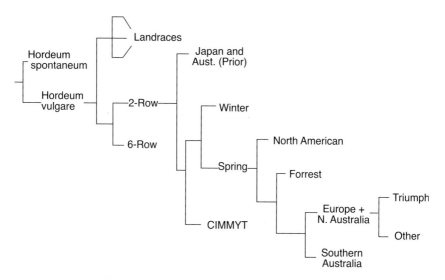

Fig. 11.4. Summary of major amplified fragment length polymorphism phylogram groupings.

The remaining *H. vulgare* accessions were separated into six-row and two-row groups in cluster analysis performed on both RFLP and AFLP data sets (Figs 11.3 and 11.4). The Australian six-row varieties 'Malebo' and 'Yerong' cluster together ('Yerong' is derived from a cross involving 'Malebo') and, in turn, this group clusters with the European six-row variety 'Osiris'.

The clear separation of the two species groups, the landrace 'Sahara 3771', and six-row accessions from two-row accessions is consistent between cluster

Table 11.2. Mean pairwise (mean genetic distance between a selected accession and all other accessions) restriction fragment length polymorphism genetic distance estimates (only accessions with mean GD estimates greater than 0.21 included).

0.31–0.33	0.28–0.30	0.25–0.27	0.22–0.24
H. spontaneum – mountains	Sahara	CI3576	Amaji Nijo
H. spontaneum – desert		Osiris	Corvett
H. spontaneum – coast		Malebo	CIMMYT 42002
H. spontaneum – temperate		Yerong	Forrest
			Namoi
			Skiff
			Yagan

analysis performed on both RFLP and AFLP data sets. It is beyond this level of differentiation that major RFLP and AFLP clustering patterns become less consistent. The RFLP data set clearly separates two-row *H. vulgare* accessions into two groups:

1. Accessions selected in the southern hemisphere, but also including Japanese accessions;
2. Accessions selected in the Northern hemisphere, but also including northern Australian accessions.

The only exception to this RFLP clustering pattern is the UK-bred malting quality variety 'Proctor', which clusters with Australian and Japanese accessions.

The RFLP data set separates the southern hemisphere (not including northern Australian cultivars) cluster into two groups:

1. Accessions with strong pedigree link to 'Proctor';
2. Accessions with strong pedigree link to 'Prior'.

The RFLP data set separates the northern hemisphere cluster into two clear selection environment and/or gene pool groups:

1. European accessions but also including northern Australian-bred cultivars;
2. North American-bred accessions.

The North American group all have a clear pedigree link to the North American malting quality variety 'Betzes' (not included in this study). The European and northern Australian RFLP cluster group separates into two cluster groups which are generally based (some exceptions) on pedigree links to either of the European malting quality varieties, 'Triumph' or 'Aramir' (not included in this study).

In contrast to the RFLP data set, the AFLP data do not separate the two-row *H. vulgare* accessions into distinct northern and southern hemisphere-derived cluster groups. The AFLP data set identifies a 'Proctor' group very similar to the RFLP 'Proctor' group but excludes the Japanese subgroup and the southern Australian cultivar 'Corvette' and includes two North American cultivars

'Ellice' and 'Norbet'. Similarly, cluster analysis of the AFLP data set identifies a 'Triumph'-based subgroup (very similar to the RFLP 'Triumph' group) and an 'Aramir'-based European (including northern Australian selected cultivars) subgroup. In addition, cluster analysis of the AFLP data set forms a 'Prior'-based subgroup which also includes the three Japanese accessions but does not include the southern Australian cultivars 'Lara', 'Cantala' and 'Brindabella' nor the Western Australian feed varieties 'Forrest', 'Onslow', 'Mundah', 'Yagan' and 'O'Connor'. Analysis of the RFLP data set clusters the Western Australian feed quality cultivars within the 'Prior' group due to their strong pedigree link to 'Forrest' which is derived from 'Prior'. 'Prior' and its landrace progenitor, 'Chevallier', are both common parents in the pedigrees of the three Japanese accessions included in the study. In addition, the Japanese accessions have pedigree links to 'Plumage Archer', a parent of 'Proctor', and 'Golden Melon', which is likely to also have pedigree links to 'Proctor'.

Cluster analysis of the RFLP data set clearly separates five North American cultivars 'Andre', 'Bearpaw', 'Harrington', 'Ellice' and 'Richard' into a unique group within the northern hemisphere major group. The only exception was 'Norbet' which clustered with European two-row cultivars. In contrast, cluster analysis of AFLP data paired closely related North Amercian cultivars but failed to cluster them all into a single group.

Discussion

Russell *et al.* (1997) derived relationships among 18 cultivated barley accessions using RFLP, AFLP, RAPD and SSR data and utilized principal coordinate analysis on the combined data. They found accessions clearly separated into spring and winter types. Among the winter types, the two-rowed and six-rowed varieties formed two distinct groups, with the two-rowed types forming an intermediate group between the spring and six-rowed winter types. Similar arrangements were observed for both AFLP and RFLP data. Melchinger *et al.* (1994), using RFLP data on 48 European barley cultivars, also observed a clear separation between spring and winter types, subgrouping of two-row and six-row types, and further subgrouping of cultivars with similar pedigrees.

In this study, cluster analysis of RFLP and AFLP data also produced similar clustering patterns. However, in contrast to both Russell *et al.* (1997) and Melchinger *et al.* (1994), barley accessions were not separated into winter and spring types. Pairs of closely related winter types were widely dispersed throughout clusters of spring types. Only five winter types of relatively diverse background were included in this study and could account for this difference. This study did, however, separate barley accessions into two-rowed and six-rowed types, with the AFLP data providing a more distinct separation at this level.

Cluster analysis of both RFLP and AFLP data sets clearly separated the two species groups represented in the study. In addition, the four *H. spontaneum* accessions produced the four highest mean RFLP genetic distance estimates (mean of all pairwise estimates) (Table 11.2). Nevo (1986) compared allozymic diversity of 19 wild barleys from Iran, Turkey and Israel with landraces from Iran, Europe, Egypt and India and two composite crosses. Allozymic diversity was lowest in the composite crosses, intermediate in the landraces and highest in the wild germplasm, with significant differences between centres of origin of wild germplasm. Similarly, chloroplast DNA diversity in 11 lines of *H. spontaneum* was significantly higher than in nine lines of *H. vulgare* (Clegg et al., 1984).

The four *H. spontaneum* accessions were selected to represent four distinct ecogeographical regions of the Near East Fertile Crescent, particularly Israel. This area has been recognized as the centre of origin and diversity of the wild progenitors of cultivated barley (Nevo, 1991). The accessions were selected to represent mountainous, desert, coastal and temperate regions of Israel. A number of studies have described the basis of diversity in *H. spontaneum* in Israel (Nevo et al., 1979; Nevo, 1991; Chalmers et al., 1992; Dawson et al., 1993). The patterns of variation appear to be related to ecological, climatic and geographic factors (Chalmers et al., 1992). In this study, relatively high RFLP genetic distance values (0.24–0.33) were observed between the four *H. spontaneum* accessions (Table 11.3), with greatest genetic distance between accessions selected in the desert and coastal regions. These results add further support to observations on the effect of ecogeographic factors on genetic diversity in *H. spontaneum*.

In this study, mean RFLP genetic distance estimates were in the following order: *H. spontaneum* accessions > landraces ('Sahara 3771' and 'CI3576') > six-row cultivars ('Malebo', 'Yerong' and 'Osiris') > CIMMYT introductions ('Namoi', 'Yagan', 'CIMMYT 42002') > other two-row cultivated barley accessions (Table 11.2). Modern plant breeding practices have reduced the range of genetic variability available in the cultivated *H. vulgare* gene pool. The pattern of mean RFLP genetic distance estimates observed in this study confirms the observations of the previous studies and demonstrates the potential value of *H. spontaneum*, landraces and even the CIMMYT gene pool as genetic resources for cultivated barley improvement in Australia.

Cluster analysis of RFLP and AFLP data produced many similar groupings. However, cluster analysis of RFLP data clearly differentiated southern Australian germplasm with pedigree links to 'Prior', 'Proctor' and 'CI3576' from European, North American and northern Australian germplasm. A number of individual accessions in the 'Proctor' group also have pedigree links to

Table 11.3. Restriction fragment length polymorphism genetic diversity estimates between *H. spontaneum* accessions.

	Mountains	Desert	Coast
Desert	0.30		
Coast	0.24	0.33	
Temperate	0.31	0.31	0.31

'Prior' and many in the 'Prior' group also have pedigree links to 'Proctor'. The foundation for this major clustering appears to be based on the relative proportion of 'Prior' or 'Proctor' in the pedigree, and/or strong pedigree links to the North African landrace CI 3576 ('Proctor' group) or the Western Australian feed variety 'Forrest' ('Prior' group). This provides evidence that the southern Australian cluster is a relatively unique germplasm pool. This clustering distinction is evident but not as clear in the analysis of AFLP data.

Melchinger *et al.* (1994) reported one major pedigree-based separation and clustering within and among European barley cultivars. They found cultivars related to 'Aramir' clustered separately from varieties closely related to 'Triumph', supporting their conclusion of these two unique germplasm sources. Cluster analysis of both RFLP and AFLP data sets, in this study, also successfully separated European germplasm into groups with strong pedigree relatedness to either 'Triumph' or 'Aramir'. Cluster analysis of AFLP data, however, clustered the 'Aramir' group with a 'Prior'-based southern Australian subgroup and then, at a secondary clustering level, with the 'Triumph' group.

In conclusion, the results of this study, in agreement with other studies, indicate that both RFLP and AFLP markers provide similar and generally consistent data that are suitable for barley germplasm identification. This is not altogether surprising as both techniques are based on identifying restriction site changes. However, cluster analysis of RFLP data appeared to differentiate two-rowed *H. vulgare* accessions into more logical pedigree and ecogeographic groupings. Considering the two techniques, based on the time and labour requirements to generate roughly equivalent amounts of data, the main conclusion is that AFLP profiling techniques offer a significant time and cost advantage over RFLPs. Furthermore, the potential to automate the AFLP technique indicates that this is a suitable candidate to replace the latter technique for genetic relatedness studies where large numbers of samples are to be processed on a routine basis.

References

Ahn, S. and Tanksley, S.D. (1993) Comparative linkage maps of rice and maize genomes. *Proceedings of the National Academy of Sciences USA* 90, 7980–7984.

Becker, J., Vos, P., Kuiper, M., Salamini, F. and Heun, M. (1995) Combined mapping of AFLP and RFLP markers in barley. *Molecular and General Genetics* 249, 65–73.

Chalmers, K.J., Waugh, R., Waters, J., Forster, B.P., Nevo, E., Abbott, R.J. and Powell, W. (1992) Grain isozyme and ribosomal DNA variability in *Hordeum spontaneum* populations from Israel. *Theoretical and Applied Genetics* 84, 313–322.

Clegg, M.T., Brown, A.H.D. and Whitefeld, P.R. (1984) Chloroplast DNA diversity in wild and cultivated barley: implications for genetic conservation. *Genetical Research* 43, 339–343.

Dawson, I.K., Chalmers, K.J., Waugh, R. and Powell, W. (1993) Detection of genetic variation in *Hordeum spontaneum* populations from Israel using RAPD markers. *Molecular Ecology* 2, 151–159.

Devos, K.M. and Gale, M.D. (1992) The use of random amplified DNA markers in wheat. *Theoretical and Applied Genetics* 84, 567–572.

Freinberg, A.P. and Vogelstein, B. (1983) A technique for radiolabelling DNA restriction fragments to a high specific activity. *Analytical Biochemistry* 132, 6–13.

Gebhardt, C., Ritter, E., Barone, A., Debener, T., Walkameier, B., Schachtschabel, U., Kaufmann, H., Thompson, R., Bonierbale, M., Ganal, M., Tanksley, S. and Salamini, F. (1991) RFLP maps of potato and their alignment with the homeologous tomato genome. *Theoretical and Applied Genetics* 83, 49–57.

Graner, A., Jahoor, A., Schondelmaier, J., Siedler, H., Pillen, K., Fishbeck, G. and Wenzel, G. (1991) Construction of a RFLP map of barley. *Theoretical and Applied Genetics* 83, 250–256.

Graner, A., Ludwig, W.F. and Melchinger, A.E. (1994) Relationships among European barley germplasm. 2. Comparison of RFLP and pedigree data. *Crop Science* 34, 1199–1205.

Harlan, J.R. (1979) Barley. In: Simmonds, N.W. (ed.) *Evolution of Crop Plants*. Longman, Harlow, UK, pp. 93–98.

Heun, M., Kennedy, A.E., Anderson, J.A., Lapitan, N.L.V., Sorrells, M.E. and Tanksley, S.D. (1991) Construction of a restriction fragment length polymorphism map for barley (*Hordeum vulgare*). *Genome* 34, 437–447.

Heun, M., Murphy, J.P. and Phillips, T.D. (1994) A comparison of RAPD and isozyme analyses for determining the genetic relationships among *Avena sterilis* L accessions. *Theoretical and Applied Genetics* 87, 689–696.

Hill, M., Witsenboer, H., Zabeau, M., Vos, P., Kesseli, R. and Michelmore, R. (1996) PCR-based fingerprinting using AFLPs as a tool for studying genetic relationships in *Lactuca* spp. *Theoretical and Applied Genetics* 93, 1202–1210.

Langridge, P., Karakousis, A., Collins, N., Kretschmer, J. and Manning, S. (1995) A consensus linkage map of barley. *Molecular Breeding* 1, 389–395.

Litt, M. and Luty, J.A. (1989) A hypervariable microsatellite revealed by *in vitro* amplification of a dinucleotide repeat within the cardiac muscle actin gene. *American Journal of Human Genetics* 44, 397–401.

Melchinger, A.E., Graner, A., Singh, M. and Messmer, M. (1994) Relationships among European barley germplasm: I. Genetic diversity among winter and spring cultivars revealed by RFLPs. *Crop Science* 34, 1191–1199.

Nei, M. and Li, W.H. (1979) Mathematical model for studying genetic variation in terms of restriction endonucleases. *Proceedings of the National Academy of Sciences USA* 76, 5269–5273.

Nei, M. (1987) *Molecular Evolutionary Genetics*. Columbia University Press, New York.

Nevo, E. (1986) Genetic resources of wild cereals and crop improvement: Israel, a natural laboratory. *Israel Journal of Botany* 35, 255–278.

Nevo, E. (1991) Origin, evolution, population genetics and resources for breeding of wild barley, *Hordeum spontaneum*, in the fertile crescent. In: Shewry, P. (ed.) *Genetics, Biochemistry, Molecular Biology and Biotechnology*. CAB International, Wallingford, UK, pp. 19–43.

Nevo, E., Zohary, D., Brown, A.H.D. and Haber, M. (1979) Genetic diversity and environmental associations of wild barley, *Hordeum spontaneum*, in Israel. *Evolution* 33, 815–833.

Pallotta, M.A., Graham, R.D., Langridge, P., Sparrow, D.H.B. and Barker, S.J. (2000) RFLP mapping of manganese efficiency in barley. *Theoretical and Applied Genetics* (in press).

Röder, M.S., Plaschke, J., König, S.U., Börner, A., Sorrels, M.E., Tanksley, S.D. and Ganal, M.W. (1995) Abundance, variability and chromosomal location of microsatellites in wheat. *Molecular and General Genetics* 246, 327–333.

Russell, J.R., Fuller, J.D., Maccaulay, M., Hatz, B.G., Jahoor, A., Powell, W. and Waugh, R. (1997) Direct comparison of levels of genetic variation among barley accessions detected by RFLPs, AFLPs, SSRs and RAPDs. *Theoretical and Applied Genetics* 95, 714–722.

Saghai Maroof, M.A., Biyashev, R.M., Yang, G.P., Zhang, Q. and Allard, R.W. (1994) Extraordinarily polymorphic microsatellite DNA in barley: species diversity, chromosomal locations and population dynamics. *Proceedings of the National Academy of Sciences USA* 91, 5466–5470.

Swofford, D.L. (1998) PAUP*. Phylogenetic Analysis Using Parsimony (*and Other Methods). Version 4. Sinauer Associates, Sunderland, Massachusetts.

Tanksley, S.D., Ganal, M.W., Prince, J.P., de Vincente, M.C., Bonierbale, M.W., Broun, P., Fulton, T.M., Giovannoni, J.J., Grandillo, S., Martin, G.B., Messeguer, R., Miller, L., Patterson, A.H., Pineda, O., Röder, M.S., Wing, R.A., Wu, W. and Young, N.D. (1992) High-density molecular linkage maps of the tomato and potato genomes. *Genetics* 132, 1141–1160.

Van Eck, H.J., van der Voort, J.R., Draaistra, J., Vanenckervort, E., Segers, B., Peleman, J., Jacobsen, E., Helder, J. and Bakker, J. (1995) The inheritance and chromosomal localisation of AFLP markers in a non-inbred potato offspring. *Molecular Breeding* 1, 397–410.

Vanichanon, A., Blake, N.K., Martin, J.M. and Talbert, L.E. (2000) Properties of sequence-tagged site primer sets influencing repeatability. *Genome* 43, 47–52.

Vos, P., Hogers, R., Bleeker, M., Reijans, M., van de Lee, T., Hornes, M., Frijters, A., Pot, J., Peleman, J., Kuiper, M. and Zabeau, M. (1995) AFLP: a new technique for DNA fingerprinting. *Nucleic Acids Research* 23, 4407–4414.

Welsh, J. and McClelland, M. (1990) Fingerprinting genomes using PCR with arbitrary primers. *Nucleic Acids Research* 18, 7213–7218.

Williams, J.G.K., Kubelic, A.R., Livak, K.J., Rafalski, J.A. and Tingey, S.V. (1990) DNA polymorphisms amplified by arbitrary primers are useful as genetic markers. *Nucleic Acids Research* 18, 7213–7218.

Yang, W., de Oliveira, A.C., Goodwin, I., Schertz, K. and Bennetzen, J.L. (1996) Comparison of DNA marker technologies in characterizing plant genome diversity: variability in Chinese sorghums. *Crop Science* 36, 1669–1676.

Chapter 12

Discovery and Application of Single Nucleotide Polymorphism Markers in Plants

D. BHATTRAMAKKI[1] AND A. RAFALSKI[2]

[1]*Pioneer Hi-Bred International Inc., Johnston, Iowa, USA;*
[2]*DuPont Agricultural Products – Genomics, Delaware Technology Park, Newark, Delaware, USA*

Introduction

In crop plants, genetic markers are useful for the creation of genetic maps, for map-based cloning, and for many breeding applications, including marker-assisted selection, backcross conversion and genotyping. They can also be used for population studies and genetic resource management. Morphological or phenotypic markers are traditional markers that are recognized by visual observation of the phenotype in the field or laboratory. Only a few morphological markers are available and often they are greatly influenced by the environment. Allozymes are another type of markers that take advantage of differential mobility of the enzymes originating from different alleles of a particular gene.

DNA molecular markers have gained popularity in the last decade. Among them the restriction fragment length polymorphism (RFLP) markers (see Chapter 11, this volume) are the markers of choice for comparative genomic studies. Although RFLPs are one of the early markers still used in many laboratories, examining numerous marker–genotype combinations using RFLPs is laborious and time-consuming. Random amplified polymorphic DNA markers (see Chapter 3, this volume) do not require *a priori* sequence information and are advantageous in terms of simplicity and cost and are, therefore, frequently used for species in which no other markers have been developed. Amplified fragment length polymorphism markers (see Chapters 3 and 11, this volume) can be used to quickly create well-saturated genetic maps. Microsatellites or the simple sequence repeat (SSR) markers (see Chapter 2, this volume) are the markers that are currently most widely used. Single nucleotide polymorphisms (SNPs, pronounced as 'snips') are rapidly becoming the markers of choice in human genetic studies because of their high frequency in the genome and low

© CAB *International* 2001. *Plant Genotyping: the DNA Fingerprinting of Plants* (ed. R.J. Henry)

mutation rates compared with microsatellites. They are also amenable to automated analysis in a high throughput manner. Single nucleotide polymorphisms are DNA sequence variations between individuals. Not many systematic efforts to discover SNPs have yet been undertaken in plants although there has been a big effort in the pharmaceuticals sector. In this chapter we will discuss the approaches we have taken to discover SNPs in a large number of maize (*Zea mays*) loci. Before describing SNPs, we will briefly discuss SSRs.

DNA molecular markers

SSRs: the widely used markers

Simple sequence repeats are co-dominant markers that are routinely used in many industrial and academic labs. They occur at high frequency and appear to be distributed throughout the genomes of higher plants. SSRs display a high level of polymorphism even among closely related accessions, and are amenable to simple and inexpensive PCR-based assays (for reviews, see Brown *et al.*, 1996; Holton, Chapter 2, this volume). SSR loci have been genetically mapped in many important crop plants including wheat (Röder *et al.*, 1998), rice (Cho *et al.*, 1998), soybean (Cregan *et al.*, 1994), maize (Taramino and Tingey, 1996) and barley (Ramsey *et al.*, 2000).

The use of SSR markers involves the isolation of SSR-containing DNA clones from enriched genomic DNA libraries, synthesizing primer sets to amplify the SSR contained region, and mappping SSR loci that are polymorphic. Although many improved procedures are now available to construct SSR-enriched libraries and subsequently to sequence positive clones, the isolation of SSRs is still a time-consuming and expensive process. The cost of developing a substantial number of robust SSR markers for use in genotyping applications involving thousands of individuals is often prohibitive. Moreover, even in the 'dense maps' containing many SSRs, there are many regions of the map that are completely devoid of any SSR marker. Although they are abundant and may occur with a frequency of one SSR for every 30-kb region of plant genome, the realization of that density on a genetic map has not been achieved yet in any crop species. Some SSRs can also be identified by searching expressed sequence tag (EST) databases. As these SSRs are likely to be within or adjacent to coding sequences, they may be less polymorphic than SSRs derived from non-coding regions.

SNPs: the widely prevalent polymorphisms

Single nucleotide polymorphisms are the most common form by far of DNA polymorphisms in a genome. A large amount of SNP data is available in humans, but very limited data are available on SNPs in plants. Just as with other genetic marker systems, SNPs can be used for germplasm fingerprinting, marker-assisted backcross conversion and marker-assisted breeding. SNPs are highly amenable to automation and can potentially be used to create a very high-density genetic map required for association mapping (Wang *et al.*, 1998). Interestingly, some of the SNPs in the coding region (cSNPs) may have functional significance if the resulting amino acid change causes the altered phenotype. Detection of SNPs does not require DNA fragment length measurement, thus allowing one to design high throughput, automated assays, without separating DNA by size. While SSRs can often represent many alleles, SNPs are biallelic in nature. However, multiple haplotypes can be distinguished by combining the analysis of closely linked SNP alleles.

Variations in maize: SNPs and indels

In an early study in maize, a high level of sequence variation was noticed in *Shrunken-1* locus (*Sh1*) (Werr *et al.*, 1985). This variation amounts to one difference in 33 bases (3%) in the coding region and one in 27 (3.7%) bases in the untranslated 3′ region of the gene. In another study, Shattuck-Eidens *et al.* (1990) studied four loci for variations due to SNPs and insertions/deletions (indels) in seven yellow dent genotypes. A very high frequency of variants was noticed, one variant occurring for every 13 bases. A measure of polymorphism (θ) varied from 0.006 to 0.040 for base substitutions and from 0.002 to 0.023 for indels. Recently, Selinger and Chandler (1999) sequenced a 594-bp region of a regulatory gene of maize (*b* gene) in 18 genotypes consisting of 11 cultivated and seven ancestral genotypes. They found 116 SNPs plus 30 small insertions of 1–15 bp.

Doebley and colleagues (Wang *et al.*, 1999) used SNP data from the so-called 'evolutionary locus' *teosinte branched1* (*tb1*) gene of maize in an attempt to answer some questions pertaining to maize domestication. They sampled around 2.9 kb region of the transcribed and non-transcribed upstream regions of *tb1* from 17 genotypes of maize (*Zea mays mays*), 12 genotypes of teosinte (*Z. mays parviglumis*, ancestor of maize) and several genotypes of *Z. mays mexicana* (another species of teosinte). They concluded that there was not much reduction in variation in the part of the gene that codes for the *tb1* protein. In contrast, in the promoter region, the changes were dramatic. Maize represented only about 3% of the variations observed in teosinte, and all cultivated maize variants were grouped in a single haplotype. This study suggested that during maize domestication, selection affected the promoter region more strongly than the coding region, thus leading to a change in phenotype.

The common theme that arises from the above studies is that all of them detected a great deal of DNA sequence polymorphism in maize even though only a few loci were examined. It will be interesting to examine the level of SNPs in other plant species also.

Other plant SNPs: how variable are they?

As opposed to maize, melon exhibited a much lower variation, the range being 0–0.002 among three loci between eight cultivars, even though the germplasm was much broader than those represented in maize (Shattuck-Eidens *et al.*, 1990). Grimm *et al.* (1999) assayed 18 North American soybean germplasms at 12 different loci, for which information is available in GenBank. This work suggested that the SNPs in the soybean genome occur at a frequency of 3.4 per kbp. Interestingly, the frequency of SNPs was equal in SSR-flanking regions and in non-SSR-containing regions.

In the three genomes of wheat (A, B and D), the frequency of SNPs among three genes involved in starch biosynthesis was one in every 15 bases (W. Powell, personal communication). All observed indels were in the non-coding region with a frequency of one in every 83 bases. This abundant SNP variation can be utilized to design genome-specific primers, which could in turn be used to genotype wheat of different ploidy levels.

The utility of SNPs for genotyping different tree species was recently demonstrated (Germano and Klein, 1999). Twelve nuclear and 13 chloroplast interspecific SNPs were identified in spruce tree. Two species-specific SNPs distinguished black spruce (*Picea mariana*) from red (*Picea rubens*) and white spruce (*Picea glauca*). These SNPs were consistent among 96–100% of the trees surveyed. In addition, five SNPs that distinguished white spruce from red and black spruce were consistent among 100% of surveyed trees.

Savolainen *et al.* (2000) studied nucleotide variation at the alcohol dehydrogenase locus in the outcrossing *Arabidopsis lyrata*, a close relative of the selfing *Arabidopsis thaliana*. The estimated diversity (θ) was found to be 0.0038 and was lower than that previously reported in *A. thaliana* (0.0069) at the same locus (Innan *et al.*, 1996). The authors concluded that outcrossing alone does not increase the polymorphism between the populations as opposed to selfing. However, the completely different pattern of variation within the *Adh* locus between these two species was partly attributed to the mating system.

Large-scale discovery of maize SNPs: SNPs everywhere

A pilot study was conducted (Rafalski *et al.*, 1999) in our laboratory to determine the frequency of SNPs in the maize genome and to demonstrate the utility of SNPs as tools for mapping and as genetic markers. The study was later expanded. A large number of SNPs were identified using several maize genotypes, which capture more than 90% of the allelic diversity present in the cultivated maize germplasm that was used for the pilot study. The strategy we used was essentially a 'resequencing' approach similar to that described for humans (Wang *et al.*, 1999). Sequences in the DuPont/Pioneer EST database served as initial templates for primer design in our study. Primer pairs were designed mainly from 3'-untranslated regions to maximize the frequency of SNPs identified. The 3' regions of genes were amplified from genomic DNA extracted from pre-selected inbreds. PCR products, usually around 350 bases in length, were sequenced using dideoxy terminator chemistry. The sequences were aligned using Phred/Phrap suite (Ewing *et al.*, 1998; Ewing and Green, 1998) and viewed through Consed (Gordon *et al.*, 1998). SNPs and small indels were identified and a file of polymorphic sites generated in a semi-automated fashion. We catalogued all sequence polymorphisms including single nucleotides, indels and miniature inverted repeat transposable elements. Some of these polymorphisms, especially the short indels we encountered frequently, can probably be attributed to transposable element footprints.

Our large-scale study has confirmed the previously observed high frequency of polymorphism occurrence in the maize genome. Results from the analysis of several hundred loci in the eight maize inbreds used in the study indicate the average SNP frequency to be one SNP polymorphism for every 83 bases and an indel for every 250 bases. The high frequency of SNPs observed here is not surprising since maize is an open pollinated species. It has been previously demonstrated that the type of mating system is a major determinant of the distribution of genetic variation within and between populations (Allard *et al.*, 1968). Abundant RFLP variation was also previously observed in outcrossing species like maize as compared with self-pollinated species like melon and tomato (Helentjaris *et al.*, 1985). Our SNP data suggest that the present day maize is much more genetically diverse than many other crop plants studied to date.

In our study, we observed that only a few haplotypes are found, as depicted in Table 12.1 for maize embryo storage protein gene *globulin1* (*glb1*). Selinger and Chandler (1999) also noted three clades (haplotypes) consisting of three, five and eight genotypes with distinct indel polymorphisms. Alleles within each clade were much more closely related than alleles in different clades. In this study, two genotypes did not match to any of the three common haplotypes.

Table 12.1. The major haplotypes found in the region spanning intron IV of the *globulinI-S* locus among the eight genotypes analysed for SNP.

		Polymorphic base position						
Haplotype	Genotype name/no.	57	130	165	232	236	274	374
		Exon IV			Intron IV			Exon V
H1	H_4	T	T	A	T	T	ND	C
	H_3	T	T	A	T	T	ND	C
	H_1	T	T	A	T	T	ND	C
	E_2	T	T	A	T	T	ND	C
	B73	T	T	A	T	T	ND	C
H2	MO17	C	T	A	C	C	D	C
	B_1	C	T	A	C	C	D	C
H3	H60	C	C	T	T	T	D	T

Note: (i) Sequence length covered: 403 bp. (ii) Only the polymorphic positions are shown. (iii) ND, no deletion; D, deletion.

This is consistent with our observation of occasional rare haplotypes. Three major haplotypes were found in both cultivated and ancestral genotypes by Selinger and Chandler (1999). In a similar study from our lab with many more samples, we also observed some conservation of haplotypes in the present day maize and their ancestors (A.-J. Rafalski, unpublished results).

Discovery of SNPs via data mining: an alternative approach

In the last 5 years we have seen an explosion of DNA sequence information from many organisms. The human and *Arabidopsis* genome sequencing is complete and there is considerable progress towards the completion of the rice genome. In other plant species, sequencing efforts have been sporadic and often limited to the partial sequencing of a few ESTs. Many biotechnology companies created proprietary databases of ESTs for crop plants, including maize and soybean. Frequently, several genotypes of diverse origin are represented in the EST databases. Therefore, the databases are a rich source for polymorphisms that exist between the populations. The current challenge is to fully utilize this resource to mine SNPs. Specialized computer software can be used to identify and align the sequences from the same locus of different genotypes, and catalogue the resultant SNPs in an automated manner. Identifying SNPs via data mining may be an alternative, economical approach wherever a large amount of diverse data already exists. SNPs identified through data mining must be verified empirically. Already data mining has been successfully used to identify human SNPs (Picoult-Newberg *et al.*, 1999) and specialized software called POLYBAYES was developed and used to discover SNPs (Marth *et al.*, 1999).

SNP genotyping: myriad of assay choices

There are several competing technologies available to assay SNPs in an automated manner, as listed in Table 12.2. Although much has been promised by the commercial vendors, ultimately cost and the ease of operation will drive the choice of a particular platform. Eventually SNPs should allow for higher throughput, low cost multiplexed genotyping for molecular breeding, genetic diagnostics and research applications. The review of assay technologies presented below is by no means exhaustive and is weighted by our own experiences and interests.

Allele-specific PCR

Appropriately designed PCR primers can be used to discriminate SNP alleles. In the assay developed by See *et al.* (2000) in barley, two primers are labelled with different fluorophores at their 5′ nucleotides with their 3′ termini matching

Table 12.2. SNP assays compared with respect to equipment needed to perform the assays.

Assay	Equipment needed
PCR dependent	
Single-stranded conformational polymorphism detection	DNA sequencer
Fluorescent capillary sequencing	DNA sequencer
Pyrosequencing	Pyrosequencer
Temperature-modulated heteroduplex assay	DHPLC machine
TaqMan	Fluorescence monitoring thermocycler
Molecular beacons	Fluorescence monitoring thermocycler
Affymetrix, custom chip	Chip scanner
Allele-specific hybridization on nylon arrays	Low resolution spotter and phosphorimager
Allele-specific hybridization on glass arrays	High resolution spotter and scanner
Dynamic allele-specific hybridization	Fluorescence monitoring thermocycler
Allele-specific ligation, on plates/arrays	Fluorescence plate reader or scanner
Allele-specific ligation, sequencer detection	DNA sequencer
Single base extension, on plates or arrays	Fluorescence plate reader or scanner
Single base extension, on Affymetrix flexarrays	Chip scanner
Single base extension, fluorescent polarization detection	Fluorescence polarization plate reader
Single base extension, four colour gel based	DNA sequencer
Base extension, single colour, multiplexed	DNA sequencer
Base extension, gel detection	DNA sequencer
Base extension, mass spectrophotometer detection	MALDI mass spectrophotometer
PCR independent	
Invader assay	Fluorescence plate reader
Padlock probes	Fluorescence plate reader

DHPLC, denaturing high-performance liquid chromatography; MALDI, matrix-assisted laser desorption/ionization.

each of the SNP alleles. The PCR is performed using two labelled forward primers and an unlabelled, common reverse primer. A separate pre-amplification step reduces the complexity manyfold and may be a necessary step in large genomes. Each primer perfectly matches one of the two available alleles and the alleles can be scored based on fluorescence spectrum or size of the PCR product. Although the technique is simple, the throughput is not very high.

Allele-specific hybridization

In the allele-specific oligonucleotide hybridization technique the target PCR product is immobilized and denatured to a membrane and hybridized with allele-specific oligonucleotides. An oligonucleotide that is complementary to one of the alleles will hybridize to that allele and the other allelic variant will hydridize with its specific complementary probe. The detection of the hybridized probe is by radiolabel, fluorophore or biotin assay. In a variation of this assay, oligonucleotides can be immobilized (instead of amplified targets) and probed with labelled PCR products of the samples.

Fluorescence resonance energy transfer-based methods

The TaqMan (PE Biosystems) and molecular beacons (MBs) are the homogeneous SNP genotyping assays that depend on fluorescence energy transfer. The TaqMan assay uses the exonuclease activity of *Taq* DNA polymerase to discriminate between perfectly matched and mismatched oligonucleotides (Heid *et al.*, 1996). In the TaqMan assay, fluorogenic oligonucleotide probes are synthesized with a fluorescent reporter dye at the 5′ terminus, and the 3′ terminus contains a blocking group to prevent probe extension and a quencher that inhibits the fluorescence of the reporter. The *Taq* DNA polymerase, during its polymerization step in PCR, encounters the annealed probe and begins to displace it. This leads to clipping of the probe by the nuclease activity of the enzyme and results in increased fluorescence. The presence of an allele is deciphered by monitoring increase of the fluorescence resulting from the separation of fluorophore from the quencher. Hundreds of samples can be analysed simultaneously, and there is no need for downstream electrophoresis.

MBs (Tyagi and Kramer, 1996) are short oligonucleotide probes of a stem-and-loop configuration with a fluorophore attached to the 5′ end and a non-fluorescent quencher attached to the 3′ end. The fluorophore and the quencher are held in close proximity through a short segment of complementary 5′ and 3′ ends. When the loop segment hybridizes to the target sequence, the quencher and fluorophore become dissociated and fluorescence is detected. The genotyping assay is performed with separate PCRs, each containing a pair of allele-specific MBs, and each containing a different fluorophore. With the commercial fluorescence imaging system, several thousand samples can be read simultaneously. Multiplexing can be achieved by using MBs with different fluorescence emission maxima. Published data (Täpp *et al.*, 2000) conclude that this system provides a rapid and inexpensive genotyping technique and has a good dynamic range.

Both Taqman and MB methods could be employed when a few SNP loci have to be analysed for a large number of samples. The signals are detected after the probe degradation in the TaqMan assay, whereas in the MB assay the probe is still intact when the fluorescent detection is performed. The MB assays could be used in an immobilized format, so potentially could be used in a high throughput manner. These two assays are comparable with respect to cost of labour and consumables (Täpp *et al.*, 2000).

Pyrosequencing

Pyrosequencing allows short segments of sequence, typically of 20 nucleotides, and possibly up to 100 nucleotides, to be obtained in an automated manner. In the present configuration, up to 96 different templates can be sequenced simultaneously in 15 min after template preparation. Pyrosequencing relies on the step-wise addition of individual deoxynucleotide triphosphates (dNTPs) and sequencing-by-synthesis (Nyren *et al.*, 1993). The template-guided incorporation of dNTPs into the growing DNA chain is monitored via luminescent detection of released pyrophosphate from the incorporation reaction. Genotyping of previously identified SNPs requires only a small stretch of sequence beyond the primer binding site, and pyrosequencing handles this very efficiently. The procedure involves designing sequencing primers close to the identified SNP sites, PCR amplifying the SNP loci, obtaining single-stranded template, and sequencing several bases including the target SNP site using Luc96 pyrosequencer.

Like most other genotyping methods, pyrosequencing requires a PCR step before the SNP assay. The need to generate a single-stranded sequencing template adds to the cost and time required for this assay. With the available pyrosequencing machine and the protocol, the pyrosequencing can be used for medium throughput applications in genotyping and EST mapping, without incurring additional assay development costs. In maize, the major advantage of the sequencing approach, including pyrosequencing, is the ability to identify haplotypes through the identification of several adjacent SNP alleles. This increases the information content of the marker system. The results from our preliminary experiments (Ching and Rafalski, 2000) using this method for genotyping and mapping maize SNPs are encouraging.

Third Wave Technology

Third Wave Technologies, Inc. developed an enzyme-based system of genetic identification that utilizes the property of cleavase enzyme (Lyamichev *et al.*, 1999). The assay is known as cleavase fragment length polymorphism and it makes use of the specific sequence-dependent secondary structures containing duplexed and single-stranded regions. The cleavase recognizes these sequences and produces fragments after cleaving the junction of the duplexed region. This technology does not involve a PCR amplification step and thus reduces assay costs and the artefacts that can be introduced during PCR (Mein *et al.*, 2000).

Array-based hybridization

The SNP genotyping can be performed using very high-density gene chips. The user-defined chips are available for human SNP analysis from a variety of manufacturers. The DNA chip or SNP chip of Affymetrix, for example, contain precisely ordered arrays of oligonucleotides synthesized *in situ* on a (glass or) silica-wafer. This can accommodate as many as 60,000 oligonucleotide probes and can be used to screen as many as 1500 human SNPs simultaneously (Lipshutz *et al.*, 1999). The procedure involves PCR amplification of the target, hybridization to the oligonucleotides on the chip, scanning the chip to see which probe produces the signal, and analysing the data. The genotype is determined based on the probe sequences that show the strongest hybridization signal, according to a proprietary algorithm. Highly multiplex PCR is necessary to take full advantage of the capacity of chips to assay multiple loci.

The model plant *Arabidopsis* whose genome has now been published is a potential target for manufacturing universal chips that can be used for experimental purposes. Another potential target is maize, for which a set of unigenes will be available to design chip-based assays. However, the need for highly multiplexed template PCR, hybridization, non-flexible assay format, and the high cost may limit the use of chips.

Single base extension – fluorescent polarization assay

The fluorescent polarization assay method for detecting SNPs is a variation of the template-directed dye terminator incorporation assay, which is detected using fluorescence polarization and was developed recently (Chen *et al.*, 1999). This method involves an oligonucleotide probe that hybridizes immediately upstream of the SNP site. All the four dideoxynucleotide triphosphates (ddNTPs), each labelled with a different fluor, are added followed by DNA polymerase and the probe is allowed to extend by a single base. Fluorescence depolarization is then used to determine which ddNTP was incorporated. The advantages of this method are the speed and accuracy of SNP detection, the low cost and the ability to genotype many targets rapidly.

Denaturing high-performance liquid chromatography (DHPLC)

DHPLC is a mismatch detection technology that relies on differences in physical properties between DNA homoduplexes and mismatched heteroduplexes formed during the annealing of wild-type and mutant DNA (Oefner and Underhill, 1998). The procedure is also called temperature-modulated heteroduplex assay (TMHA) since the method involves heat denaturation of the DNA and the subsequent slow cooling at an empirically determined optimal temperature. It is during the cooling that heteroduplexes are formed. Homo- and heteroduplexes are resolved using a proprietary separation matrix. This method does not need any *a priori* information about the SNP, but it only detects the presence or absence of a mutation, not the nature and location of the mutation. The major advantage of TMHA is that it does not need modified PCR primers, detection labels or any sample pre-treatment and still allows some multiplexing and a degree of automation. However, the optimum assay temperature has to be empirically determined for each of the targets and may take considerable time. This method will be useful for the identification of SNP-containing DNA segments prior to sequencing. In highly polymorphic species, like maize, direct sequencing of most DNA segments is likely to detect SNPs, and DHPLC may not be necessary for routine SNP discovery.

Linkage disequilibrium (LD) study: what is the right number?

Estimating the extent of LD – the non-random association of alleles at linked loci – will offer a method of calculating the minimum number of SNP markers that can be used to tag all the loci in the genome. Markers closest to the candidate gene will manifest higher levels of disequilibrium than those that are farther away. This phenomenon could potentially be used to identify associations between phenotypes and marker loci without reliance on specially developed mapping populations. LD depends on the recombination frequency. It is well established that the recombination rates are not uniformly distributed along the chromosomes, and generally are low around the centromere and increase towards the telomere. In addition, there may be many recombination hot (and cold) spots distributed throughout the genome. We are also aware of the limitations of the study arising from the history of the material studied and the specific region of the chromosome in question. More studies are necessary to determine the extent of linkage disequilibrium in the maize germplasm, in different regions of the genome, and consequently to determine the number of molecular markers necessary for the association studies in maize.

Issues: what is the winning strategy?

The history of SNPs as a research tool is very short, but the interest in the scientific community is mounting. Although plant scientists are excited about the opportunity of making use of SNPs, at present there is not much literature available. Several issues must be resolved before SNPs can be routinely used as molecular markers. First, the number of SNPs required for genome-wide association studies is not known. An estimate of this could be obtained by understanding the level of linkage disequilibrium for a few select loci. Efforts to saturate the genome with SNPs have to be undertaken. Secondly, SNPs must be mapped, preferably in a high throughput, cost-effective, accurate and sensitive manner. Although many technologies are available, none has emerged as dominant. Lastly, SNP research is dependent on large-scale data handling and specialized software, some of which are not currently widely available. Resolving these issues is essential if SNPs are to be used as genetic markers in agriculture.

Conclusion

Essential to genetics research is the association of sequence variations with heritable phenotypes. The most common type of sequence variation is SNPs, occurring approximately once every 700–1000 bp in humans and with much higher frequency in maize, as seen from our work. Because SNPs are expected to facilitate large-scale association genetics studies, there has recently been great interest in their discovery and detection.

Most molecular marker analysis methods used today are time, labour and cost intensive, involve electrophoresis, thereby preventing truly high throughput analyses of breeding populations and germplasm. An ideal marker analysis method would be inexpensive, highly automated, and preferably not based on DNA separation. SNP markers have the potential to fit these requirements and undoubtedly represent one of the most powerful tools for analysis of genomes in plants.

Acknowledgements

The authors are grateful to Dr Karin Lohman for critical reading of the manuscript and many useful suggestions. We also thank Dr Scott Tingey for his support for this work.

References

Allard, R.W., Jain, S.K. and Workman, P. (1968) The genetics of inbreeding species. *Advanced Genetics* 14, 55–131.

Brown, S.M., Hopkins, M.S., Mitchell, S.E., Senior, M.L., Wang, T.Y., Duncan, R.R., Gonzalez-Candelas, F. and Kresovich, S. (1996) Multiple methods for the identification of polymorphic simple sequence repeats (SSRs) in sorghum [*Sorghum bicolor* (L.) Moench]. *Theoretical and Applied Genetics* 93, 190–198.

Chen, X., Levine, L. and Kwok, P.-Y. (1999) Fluorescence polarization in homogenous nucleic acid analysis. *Genome Research* 9, 492–498.

Ching, A. and Rafalski, A.J. (2000) Rapid EST mapping and genotyping of SNP loci using pyrosequencing. *Proceedings of Plant and Animal Genome Conference VIII*, San Diego, California, p112. http://www.intl-pag.org/pag/8/abstracts/pag8419.html

Cho, Y.G., McCouch, S.R., Kupier, M., Kang, M.-R., Pot, J., Groenen, J.T.M. and Eun, M.Y. (1998) Integrated map of AFLP, SSLP, and RFLP markers using a recombinant inbred population of rice (*Oryza sativa* L.). *Theoretical and Applied Genetics* 97, 370–380.

Cregan, Y.G., Bhagwat, A.A., Akkaya, M.S. and Rongwen, J. (1994) Microsatellite fingerprinting and mapping of soybean. *Methods in Molecular and Cellular Biology* 5, 49–61.

Ewing, B. and Green, P. (1998) Base calling sequencer traces using *Phred*. II. Error probabilities. *Genome Research* 8, 186–194.

Ewing, B., Hillier, L., Wendl, M.C. and Green, P. (1998) Base calling sequencer traces using *Phred*. I. Accuracy assessment. *Genome Research* 8, 175–185.

Germano, J. and Klein, A.S. (1999) Species-specific nuclear and chloroplast single nucleotide polymorphisms to distinguish *Picea glauca*, *P. mariana* and *P. rubens*. *Theoretical and Applied Genetics* 99, 37–99.

Gordon, D., Abajian, C. and Green, P. (1998) *Consed*: a graphical tool for sequence finishing. *Genome Research* 8, 175–185.

Grimm, D.A., Denesh, D., Mudge, J., Young, N.D. and Cregan, P.B. (1999) Assessment of single nucleotide polymorphisms (SNPs) in soybean. *Proceedings of Plant and Animal Genome Conference VII*, San Diego, California, p. 140. http://www.intl-pag.org/pag/7/abstracts/pag7784.html

Heid, C.A., Stevens, J., Livak, K.J. and Williams, P.M. (1996) Real time quantitative PCR. *Genome Research* 6, 986–994.

Helentjaris, T., King, G., Slocum, M., Siedenstrang, C. and Wegman, S. (1985) Restriction fragment polymorphisms as probes for plant diversity and their development as tools for applied plant breeding. *Plant Molecular Biology* 5, 109–118.

Innan, H., Tajima, F., Terauchi, R. and Miyashita, N.T. (1996) Intragenic recombination in the *Adh* locus of the wild plant *Arabidopsis thaliana*. *Genetics* 143, 1761–1770.

Lipshutz, R.J., Fodor, S.P.A., Gingeras, T.R. and Lockhart, D.J. (1999) High density synthetic oligonucleotide arrays. *Nature Genetics* Supplement 21, 20–24.

Lyamichev, V., Mast, A.L., Hall, J.G., Prudent, J.R., Kaiser, M.W., Takova, T., Kwiatkowski, R.W., Sander, T.J., de Arruda, M., Arco, D.A., Neri, B.P. and Brow, M.A. (1999) Polymorphism identification and quantitative detection from genomic DNA by invasive cleavage of oligonucleotide probes. *Nature Biotechnology* 17, 292–296.

Marth, G.T., Korf, I., Yandell, M.D., Yeh, R.T., Gu, Z., Zakeri, H., Stitziel, N.O., Hillier, L., Kwok, P.-K. and Gish, W.R. (1999) A general approach to single-nucleotide polymorphism discovery. *Nature Genetics* 23, 452–456.

Mein, C.A., Barratt, B.J., Dunn, M.G., Siegmund, T., Smith, A.N., Esposito, L., Nutland, S., Stevens, H.E., Wilson, A.J., Phillips, M.S., Jarvis, N., Law, S., Arruda, M. and Todd, J.A. (2000) Evaluation of single nucleotide polymorphism typing with invader on PCR amplicons and its automation. *Genome Research* 10, 330–343.

Nyren, P., Pettersson, B. and Uhlen, M. (1993) Solid phase DNA minisequencing by an enzymatic luminometric inorganic pyrophosphate detection assay. *Analytical Biochemistry* 208, 171–175.

Oefner, P.J. and Underhill, P.A. (1998) DNA mutation detection using denaturing high performance liquid chromatography (DHPLC). In: Dracopoli, N.C., Hines, J., Korf, B.R., Morton, C., Seidman, C.E., Seidman, J.G., Mair, D.T. and Smith, D.R. (eds) *Current Protocols in Human Genetics* Suppl. 19, Wiley, New York, 7.10.1–7.10.12.

Picoult-Newberg, L., Ideker, T.E., Pohl, M.G., Taylor, S.L., Donaldson, M.A., Nickerson, D.A. and Boyce-Jacino, M. (1999) Mining SNPs from EST databases. *Genome Research* 9, 167–174.

Rafalski, A., Ching, A., Bhattramakki, D., Henderson, K., Jung, M., Morgante, M., Dolan, M., Register, J., Smith, O. and Tingey, S. (1999) Single nucleotide polymorphisms (SNPs) in the 3'-untranslated flanks of maize genes reveal conserved ancestral haplotypes. *Cold Spring Harbor Meeting on Genome Sequencing and Biology*, Cold Spring Harbor, New York.

Ramsey, L., Macaulay, M., degli Ivahissevich, S., MacLean, K., Cardle, L., Fuller, J., Edwards, K.J., Tovesson, S., Morgante, M., Massari, A., Maestri, E., Marmiroli, M., Sjakste, T., Ganal, M., Powell, W. and Waugh, R. (2000) A simple sequence repeat based linkage map of barley. *Genetics* 150, 1997–2005.

Röder, M.S., Korzun, V., Wendehake, K., Plaschke, J., Tixier, M.-H., Leroy, P. and Ganal, M.W. (1998) A microsatellite map of wheat. *Genetics* 149, 2007–2023.

Savolainen, O., Langley, C.H., Lazzaro, B.P. and Helene, F. (2000) Contrasting patterns of nucleotide polymorphism at the alcohol dehydrogenase locus in the outcrossing *Arabidopsis lyrata* and the selfing *Arabidopsis thaliana*. *Molecular Biology and Evolution* 17, 645–655.

See, D., Kanazin, V., Talbert, H. and Blake, T. (2000) Electrophoretic detection of single-nucleotide polymorphisms. *BioTechniques* 28, 710–716.

Selinger, D.A. and Chandler, V.C. (1999) Major recent and independent changes in levels and patterns of expression have occurred at the *b* gene, a regulatory locus in maize. *Proceedings of the National Academy of Sciences USA* 96(26), 15007–15012.

Shattuck-Eidens, D.M., Bell, R.N., Neuhausen, S.L. and Helentjaris, T. (1990) DNA sequence variation within maize and melon: observations from polymerase chain reaction amplification and direct sequencing. *Genetics* 126, 207–217.

Täpp, I., Malmberg, L., Rennel, E., Wik, M. and Syvänen, A.-C. (2000) Homogenous scoring of single-nucleotide polymorphisms: comparison of the 5'-nuclease TaqMan assay and molecular beacon probes. *BioTechniques* 28, 732–738.

Taramino, G. and Tingey, S. (1996) Simple sequence repeats for germplasm analysis and mapping in maize. *Genome* 39, 277–287.

Tyagi, S. and Kramer, F.R. (1996) Molecular beacons: probes that fluoresce upon hybridization. *Nature Biotechnology* 14, 533–536.

Wang, D.G., Fan, J.B., Siao, C., Berno, A., Young, P., Sapolsky, R., Ghandour, G., Perkins, N., Winchester, E., Spencer, J., Kruglyak, L., Stein, L., Hsie, L., Topaloglou, T., Hubbell, E., Robinson, E., Mittmann, M., Morris, M.S., Shen, N., Kilburn, D., Rioux, J., Nusbaum, C., Rozen, S., Hudson, T.J., Lipshutz, R., Chee, M. and Lander, E.S. (1998) Large-scale identification, mapping, and genotyping of single-nucleotide polymorphisms in the human genome. *Science* 280, 1077–1082.

Wang, R.-L., Syec, A., Hey, J., Lukens, L. and Doebley, J. (1999) The limits of selection during maize domestication. *Nature* 398, 236–239.

Werr, W., Frommer, W.-B., Maas, C. and Starlinger, P. (1985) Structure of the sucrose synthase gene in on chromosome 9 of *Zea mays* L. *EMBO Journal* 4, 1373–1380.

Chapter 13
Producing and Exploiting Enriched Microsatellite Libraries

T.L. MAGUIRE

Department of Botany, The University of Queensland, Brisbane, Queensland, Australia

Introduction

Microsatellites, or simple sequence repeats, are stretches of DNA containing tandem repeats of di-, tri- or tetranucleotide units, which are evenly distributed throughout eukaryotic genomes (Tautz and Renz, 1984). Microsatellites are an abundant source of variation in many organisms (Weber and May, 1989; Weber, 1990). The high information content, definitive allele assignment and rapid analysis by PCR make microsatellites well poised for exploitation in plant breeding applications such as linkage analysis, agronomic trait selection, germplasm assessment, varietal identification, as well as population genetic applications in both natural and breeding populations. The widespread use of microsatellites, however, is hindered by the difficulty and expense of isolating microsatellite loci from the species of interest.

Most microsatellite loci are identified from small insert (< 1000 bp) genomic libraries using oligonucleotide probes. Standard methods for the isolation of plant microsatellite loci involve digestion of genomic DNA, ligation into a plasmid vector, which is then transformed into *Escherichia coli*, followed by hybridization with a labelled microsatellite oligonucleotide probe, DNA sequencing of positive clones, primer design, locus-specific PCR amplification and identification of polymorphisms (Fig. 13.1). The number of positive clones containing microsatellites is a function of the abundance of the target microsatellite repeats in the genome under study. This is especially problematic in plants, where there is a tenfold reduction in the frequency of dinucleotide repeats compared with humans (Powell *et al.*, 1996). The relatively low frequency of microsatellites in plant genomes presents some technical problems for the large-scale isolation of microsatellites. To overcome this problem,

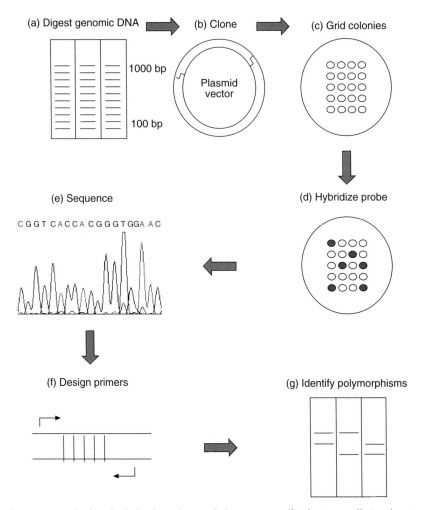

Fig. 13.1. Standard methods for the isolation of plant microsatellite loci generally involve (a) digestion of genomic DNA, (b) ligation into a plasmid vector, (c) transformation into *Escherichia coli* and grid colonies, (d) hybridization with a labelled microsatellite oligonucleotide probe, (e) DNA sequencing of positive clones, (f) primer design and locus-specific PCR amplification, and (g) identification of polymorphisms.

various enrichment methods have been developed, which will be reviewed in this chapter.

Traditional methods of obtaining microsatellites

Approaches for the isolation of new microsatellite loci are generally based on searching known sequences in public databases, or the identification of clones containing microsatellite loci by screening genomic libraries. Brown *et al.* (1996) compared three commonly used approaches for isolating plant microsatellite loci: (i) searching public databases; (ii) cross-species amplification; and (iii) screening genomic libraries.

Database searching was found to be the least costly in terms of time and resources for obtaining new microsatellite loci (Brown *et al.*, 1996). It has been the starting point for many plant species, providing there are sufficient entries in the database. However, database searching alone is unlikely to provide sufficient markers in any plant species for mapping or breeding applications, which require in excess of 500 markers. This will change in the future as large sequencing and expressed sequence tag (EST) projects generate large databases, allowing more microsatellite loci to be found by database searches. Cho *et al.* (2000) compared the variability characteristics of microsatellites obtained from EST databases with those obtained from genomic libraries. It was found that database-derived microsatellites had lower values than genomic library microsatellites. Similar observations have also been reported in other plant species (Becker and Huen, 1995; Chin *et al.*, 1996). Microsatellites isolated from EST sequences are generally trinucleotide repeats, specifically GC-rich repeats. These EST-derived repeats are generally less polymorphic than dinucleotide repeats. However, it was found that some trinucleotide repeats isolated from genomic clones were highly variable, indicating that particular trinucleotide repeats may be as variable as dinucleotide repeats (Cho *et al.*, 2000).

The application of cross-species transfer of microsatellites was difficult to predict (Brown *et al.*, 1996). The taxonomic distance of the species of interest and conservation of the flanking sequence determines whether the correct region is amplified, and how variable the microsatellite loci are. Often, the reaction conditions need to be optimized and the products sequenced to verify the presence of a microsatellite region. Microsatellites have been transferred between closely related plant species, but the wider application between genera has not been explored fully. Many heterologous primer pairs may need to be screened to find useful primers for the species of interest. This may be cost-effective if access to many primer pairs is granted through collaboration; however, synthesizing primers from published sequences may not be an efficient approach. Library screening, despite the effort, may be the best option for isolating new microsatellite loci.

Library screening is an inherently inefficient method. In a typical experiment using an un-enriched library, such as Brown *et al.* (1996), approximately 0.2% of the library clones hybridized to a mixed microsatellite probe. A second round of hybridization was then required to screen for false positives. This greatly reduces the amount of time wasted on sequencing clones that do not contain microsatellite regions. Following sequencing, it was found that up to 70% of the sequenced clones cannot be used for primer design, due to insufficient flanking sequence, clones that contain too few repeats, severely imperfect repeats, or absence of microsatellite repeats. Following primer synthesis, up to 65% of the primer pairs failed to produce any polymorphic products from an array of plant genotypes.

Methods for producing genomic DNA libraries

There are different types of genomic libraries, categorized by the type of cloning vector used. Different cloning vectors carry different amounts of DNA, so the choice of vector for library construction depends on the insert size. Plasmid and phage vectors carry short fragments, so they are suitable for small insert libraries. Bacterial artificial chromosomes (BACs) and yeast artificial chromosomes (YACs) carry longer fragments, so they are suitable for large insert libraries. Ease of manipulation is also another important consideration in choosing a cloning vector.

Small or large insert libraries?

There are two main approaches for the isolation of microsatellite loci from genomic libraries. One approach is to screen a large insert genomic library with an end-labelled microsatellite oligonucleotide probe. Clones that hybridize to the probe are then purified and divided into subclones, which are then screened by hybridization for fragments containing microsatellites. Selected clones are then sequenced, and the flanking regions of the microsatellite repeat are used to design PCR primers. The drawbacks of this approach are the requirement for many blot hybridizations, and the difficulty of sequencing the relatively large subclones. The alternative is to produce a small insert (< 1000 bp) genomic library constructed in a plasmid or phage vector. Small insert genomic libraries are directly suitable for sequencing the entire insert. They can also be highly enriched for the desired microsatellite repeats using enrichment strategies (Edwards *et al.*, 1996; Maguire *et al.*, 2000).

The limitation of small insert libraries is, however, that the inserts are too small for some applications, such as physical mapping by *in situ* hybridization. In this case, large insert libraries may be advantageous. Baron *et al.* (1992) used large insert libraries when isolating microsatellites from chromosomes purified by flow cytometry. To avoid subcloning into small insert libraries, an alternative method was described, which was based on the combination of dinucleotide repeat-specific probing, and on the production of nested-deleted clones using exonuclease III, to detect microsatellite-containing clones that allow direct sequencing.

Large insert libraries may also be of advantage in other applications, such as the targeted isolation of microsatellite loci. Due to the random distribution of microsatellite loci, large genomic regions can remain untagged with microsatellites. In order to target microsatellite marker development to specific regions of the genome, large insert libraries such as BAC libraries can be used. Using this approach, microsatellite loci can be isolated which are associated with previously characterized restriction fragment length polymorphism (RFLP), amplified fragment length polymorphism (AFLP) or other markers isolated using BAC clones. BAC libraries have already been developed for many plant species (see for example the available listing from the Clemson University Genome Institute http://www.clemson.genome.edu or the Texas A&M University BAC Centre), and the insert size tends to be quite large, up to 150 kb (Cregan *et al.*, 1999). These libraries can be easily screened for microsatellites. BAC clones have several advantages over other large insert libraries such as YAC clones: (i) they can be isolated and manipulated simply using basic plasmid technology; (ii) they form fewer hybrid inserts than YAC clones; and (iii) they have a reduced possibility of contamination with other eukaryotic DNA, which may also contain microsatellite loci.

Using this approach, Cregan *et al.* (1999) screened BAC libraries with several RFLP markers known to map to a particular linkage group of soybean. Positive clones were then verified and linkage mapping demonstrated that the BAC clones were located at the expected chromosomal positions. Selected BAC clones were then used as a source of DNA for the construction of small insert libraries. This method involves preparing BAC DNA, followed by digestion with restriction enzymes. The resulting fragments were ligated into a plasmid vector and transformed into *E. coli*. Colonies were screened with oligonucleotide probes complementary to microsatellite repeats. This was then repeated to avoid false positives. Positive clones were then screened by PCR for insert size and selected clones were sequenced. Those clones containing microsatellite regions and suitable flanking sequence were then used to design PCR primers. Primers were tested for polymorphism and placed on medium density DNA marker maps of soybean, thus confirming that the microsatellite loci mapped to the correct regions. Using this approach, nearly 800 microsatellite loci were placed on the map, allowing detection by PCR instead of the more costly RFLP methods. Despite the abundance of markers, several genome regions remain unmarked. For example, a region of at least 30 cM next to the supernodulation gene (*nts-1*) (Kolchinsky *et al.*, 1997) did not contain any microsatellite markers mapped by Cregan *et al.* (1999).

Cloning with plasmid vectors or λ bacteriophage vectors?

In principle, cloning with plasmid vectors is straightforward. The plasmid DNA is cut with a restriction enzyme and ligated to a foreign DNA fragment. The resulting recombinant plasmid is then used to transform bacteria. Selectable markers on the plasmid allow only those bacteria that have been transformed to survive. Further selection of bacteria containing plasmids with inserts is achieved by blue/white screening for insertional inactivation of the β-galactosidase (*lacZ*) gene. Because of their reliability and ease of handling, plasmids have become the workhorses of molecular cloning. However, the generation of large, representative libraries in plasmid vectors can be technically demanding and tends to select for smaller inserts.

Cloning in λ bacteriophage vectors, however, enables efficient production of very large libraries that are not size-selective. For example, cloning methods using the λ Zap vector combine the high efficiency of λ library construction, with the convenience of a plasmid vector. Clones in λ vectors can be screened with microsatellite oligonucleotide probes at greater efficiencies than in *E. coli* and, following selection, the pBluescript phagemid can be rapidly excised *in vivo*. This allows further characterization of the clones in a plasmid system. The pBluescript phagemid contains universal priming sites for sequencing as well as the bacteriophage f1 origin of replication, allowing rescue of ssDNA, which can be used for sequencing or other applications. The pBluescript phagemid also contains the *lacZ* gene for α-complementation in the host bacterial strain for blue/white colony screening.

Enrichment methods for obtaining microsatellites

Various enrichment methods have been developed to increase the efficiency of microsatellite loci isolation from genomic DNA libraries. They can be broadly categorized by the mode of enrichment such as: (i) enrichment by colony/plaque hybridization; (ii) enrichment by primer extension; (iii) enrichment by hybridization; and (iv) enrichment by screening random amplified polymorphic DNA (RAPD) profiles.

Enrichment by colony/plaque hybridization

Standard methods for the isolation of plant microsatellite loci involve screening colonies/plaques with oligonucleotide probes complementary to microsatellite repeats. Similarly, enrichment methods involve colony/plaque screening, prior to sequencing, to select further for clones containing microsatellite regions. This increases the overall efficiency of microsatellite loci isolation, following enrichment, by reducing the cost of sequencing clones that do not contain microsatellite regions. For example, Scott *et al.* (1999) screened libraries that had low levels of enrichment, when sequencing random clones. It was found that, following screening with microsatellite oligonucleotides and sequencing of positive clones, nearly all of the clones contained a microsatellite region, therefore increasing the efficiency of microsatellite loci isolation.

Enrichment by primer extension

Enrichment methods by primer extension can be further categorized by the type of primer used, such as: (i) primers complementary to microsatellite oligonucleotides; and (ii) degenerate oligonucleotide primers.

Primers complementary to microsatellite oligonucleotides

One of the earliest methods for producing enriched microsatellite libraries was described by Ostrander *et al.* (1992). This method involved digestion of genomic DNA, followed by ligation into a plasmid vector, which was then transformed into *E. coli*. Portions of the ligation mixture were then transformed into λ phagemid vectors and propagated in a bacterial strain deficient in dUTPase (*dut* gene) and uracil N glycosylase (*ung* gene). Following infection with a *dut ung* strain with M13 helper phage, circular ssDNA molecules were isolated. The circular ssDNA molecules were then converted to circular dsDNA molecules by *in vitro* primer extension, using microsatellite oligonucleotides and *Taq* DNA polymerase. The resulting products were then transformed into an *E. coli* strain with the wild-type alleles of *dut* and *ung*. This strongly favoured the replication of clones containing primer-extended products, leading to a highly enriched microsatellite library. Colonies were then screened for the presence of microsatellites using labelled oligonucleotide probes. When comparing the original un-enriched library with the enriched library, it was found that the enriched library had approximately 50-fold more positive clones (Ostrander *et al.*, 1992), demonstrating the effectiveness of the enrichment method.

An enrichment method described by Paetkau (1999) involved a traditional (non-enriching) method followed by enrichment of the clones, rather than selection of the genomic fragments, before ligation into a vector. The selection process involves two steps whereby microsatellite oligonucleotides were annealed to and extended on clones containing the appropriate microsatellite sequence. After the first strand extension reaction with biotintylated microsatellite oligonucleotides, clones were selected using streptavidin-coated magnetic beads. Single-stranded DNA was then eluted from the bound molecules and the extension reaction repeated, resulting in dsDNA fragments. After the second extension reaction, aliquots were transformed into phagemids. The second strand reaction increased the efficiency with which microsatellite clones were transformed. Phage stocks were then tested for the presence of microsatellite regions using PCR with microsatellite oligonucleotides and a universal sequencing primer. Clones that produced PCR products were then sequenced for the presence of microsatellite regions. Of 24 positive clones, all contained microsatellite regions. However, positive clones from a library of a different species revealed that of 12 clones, only one contained a microsatellite region.

Degenerate oligonucleotide primers

An enrichment method described by Fisher *et al.* (1996) involved a 5′ anchoring procedure that consistently anchors PCR primers at the 5′ ends of microsatellites, therefore amplifying two close and inverted microsatellites and the region between them. A degenerate primer was designed to anneal to a specific dinucleotide repeat, which also had an additional seven nucleotides, forming the 'anchor'. The two most 5′ nucleotides were designed to anneal to any nucleotide, with G used to pair with C or T, and T used to pair with A or G. The next five nucleotides were designed so that they would not pair with the dinucleotide repeat, but were as redundant as possible. These redundancies ensured that the primer was complementary to one in six of all possible random sequences adjacent to the specific dinucleotide repeat. PCR products were amplified using this approach and ligated into a plasmid vector and transformed into *E. coli*. Selected colonies were then isolated and sequenced to reveal the presence of microsatellite regions at each end of the insert. Screening the library with oligonucleotide probes complementary to the original dinucleotide repeat identified further positive clones. Fifteen positive clones were isolated, and of these 13 were unique (Fisher *et al.*, 1996). Primers were then designed and tested for polymorphism. It was found that the primer pairs were polymorphic, and were reliably anchored at the 5′ end of the microsatellite. The advantage of this method is that only one additional primer is needed to amplify the terminal repeats, thereby reducing the cost of obtaining microsatellite loci. This approach is also applicable to other di-, tri- and tetranucleotide repeats, and to any genome containing microsatellite loci.

Koblizkova *et al.* (1998) described an enrichment method that involved a degenerate oligonucleotide-primed PCR prior to enrichment. Following PCR amplification of DNA, the fragments were bound to biotintylated microsatellite oligonucleotides captured on streptavidin-coated paramagnetic beads. After washing, the fragments were eluted and re-amplified using the degenerate primers, and the enrichment process was repeated. The resulting fragments were then ligated into a plasmid vector, and transformed into *E. coli*. Colonies were screened with labelled oligonucleotide probes for the presence of microsatellites. After sequencing, it was found that the flanking regions of the microsatellites were very similar between clones. This was because two or more fragments always shared the whole sequence region either before or after the microsatellite region. It was thought that the fragments originated as amplification artefacts by recombination of different PCR products within the microsatellite region. It was hypothesized by Koblizkova *et al.* (1998) that traces of the microsatellite oligonucleotide left in the PCR after the subtraction step could prime synthesis of incomplete products, which were subsequently converted to dsDNA fragments containing terminal microsatellite sequence. These fragments can then anneal to any complementary microsatellite sequence, resulting in artificial products composed of two unrelated sequences surrounding the microsatellite. To overcome this problem, a modified 3′ oligonucleotide was used (3-hydroxypropyl phosphate), which prevents extension of the oligonucleotide by *Taq* DNA polymerase. Of 19 positive clones sequenced, all contained microsatellite regions.

It should be noted that the occurrence of PCR artefacts in microsatellite-enriched libraries has seldom been reported to date. This may be because the partial homology of flanking regions can be easily overlooked. The result of this is that PCR primers designed from these flanking regions will most probably not function, because the corresponding targets are not close enough in the genomic DNA. In practice, this results in a failed PCR or a smear on a gel.

Enrichment by hybridization

Enrichment methods by hybridization can be further categorized by the type of support used such as: (i) streptavidin-coated magnetic beads; and (ii) Nylon membranes.

Streptavidin-coated magnetic beads

Streptavidin-coated magnetic beads are uniform, paramagnetic beads, covalently coupled with purified streptavidin. Streptavidin is a protein that has a high affinity for biotin. The high binding affinity of streptavidin for biotin allows the rapid and efficient isolation of biotin-labelled target molecules. The stability and strength of the interaction enable DNA manipulations such as strand melting, hybridization and elution to be performed without affecting the immobilization of the DNA on the coated magnetic beads. Due to these unique properties, streptavidin-coated magnetic beads have been used in various enrichment methods for isolating microsatellite loci.

Fischer and Bachmann (1998) described an enrichment method that involved the affinity capture of single-stranded restriction fragments annealed to biotintylated microsatellite oligonucleotides, followed by PCR. Genomic DNA was digested with a restriction enzyme, followed by ligation to a double-stranded adaptor. The constructs were then heated and allowed to hybridize to biotintylated microsatellite oligonucleotides. These hybrids were then bound to streptavidin-coated magnetic beads. After washing, the fragments were eluted from the beads. The resulting fragments were then PCR amplified using the adaptors as PCR primers. The amplification products were then digested and ligated into a plasmid vector, which was then transformed into *E. coli*. Selected clones were sequenced for the presence of microsatellite regions. Of 48 positive clones, 29 contained microsatellite regions. A similar method was also described by Prochazka (1996), which involved an additional PCR step after ligation of the adaptors, followed by hybridization to a pool of biotintylated microsatellite oligonucleotides. In addition, colonies were further screened for the presence of microsatellite regions using labelled oligonucleotide probes. Sequence analysis of nine positive clones revealed that five contained microsatellite regions.

Similar methods were described by Brondani *et al.* (1998), Milbourne *et al.* (1998) and White and Powell (1997). These methods involved an additional size selection step following restriction digestion of the genomic DNA. In contrast, fragments were cloned using λ phagemids. Plaques were further screened for the presence of microsatellite regions using labelled oligonucleotide probes, and plasmids were excised from positive clones. In the method of Brondani *et al.* (1998), the presence of the microsatellite region and its position within the insert was also determined by an anchored PCR strategy, based on the method of Rafalski *et al.* (1996). Selected clones were then isolated and sequenced, confirming the presence of microsatellite regions. Of the 207 positive clones, 180 contained suitable microsatellite regions.

An earlier report by Kijas *et al.* (1994) describes another, similar method. Genomic DNA was digested, size-selected and ligated into a plasmid vector. Predominantly single-stranded copies of the inserts were obtained using asymmetric PCR, using universal sequencing primers. Biotintylated microsatellite probes were attached to streptavidin-coated magnetic beads and the single-stranded products were allowed to hybridize to the probe–bead complex. After washing, fragments were eluted and PCR amplified. PCR products were then separated on agarose gels and transferred on to Nylon membranes, which were then screened for microsatellites. Those PCR products that were enriched for microsatellite sequence (revealed by a strong smear of hybridization) were then digested and ligated into a plasmid vector, which was transformed into *E. coli*. Colonies were further screened for microsatellites using labelled oligonucleotide probes. There were at least 20% more positive clones in the enriched library compared with the original library, which had no detectable positive clones. This method was reported to be more suited to the enrichment of lower copy number repeats such as tri- and tetranucleotide repeats.

A high throughput approach for isolating microsatellite loci from enriched libraries was described by Connell *et al.* (1998). This method involved nebulization of genomic DNA, followed by adaptor ligation and pre-selective PCR amplification, with a low number of thermocycling steps, to obtain usable 5'- and 3'-flanking sequence of microsatellite loci. This was followed by hybridization to a biotintylated microsatellite oligonucleotide, which was then captured with streptavidin-coated paramagnetic beads. Eluted fragments were then ligated into a plasmid vector and transformed into *E. coli*. Colonies were screened for microsatellites using a high-density colony screening approach. Selected colonies were sequenced for the presence of microsatellite regions. Of 12 positive clones, eight contained microsatellite regions. There are two main advantages of this method: (i) nebulization of the genomic DNA generates fragments of the expected size range, without bias according to sequence context; and (ii) minimal thermocycling reduces the bias in the library towards very small or highly repetitive inserts.

Nylon membranes

Most of the enrichment methods described so far in this chapter have focused on the isolation of a single type of microsatellite repeat. Since little is known about the level of polymorphism of different types of microsatellite repeats, it is possible that the sequences chosen do not detect the desired level of polymorphism. Furthermore, it may require the production of additional libraries in the future.

To overcome this potential limitation, an enrichment method was described by Edwards *et al.* (1996), which results in libraries containing a variety of microsatellite repeats. This method involved digestion of genomic DNA by restriction enzymes, followed by ligation of an adaptor sequence. Simultaneous enrichment for pools of microsatellite repeats was carried out by hybridizing fragments to a Nylon filter with many bound microsatellite oligonucleotides.

After washing, the bound fragments were eluted and amplified by PCR, using the adaptor sequence as a primer. The enrichment procedure was then repeated. The resulting fragments were digested, size-selected and ligated into a plasmid vector, and transformed into *E. coli*. Selected colonies were isolated and sequenced. Sequencing revealed that the libraries were highly enriched for microsatellites, where 50–70% of the clones (depending on the species tested) contained microsatellite sequences (Edwards *et al.*, 1996). There are several advantages to this method: (i) it is applicable to many plant species; (ii) it is quick and relatively inexpensive in that it does not require biotintylation of the microsatellite oligonucleotides; and (iii) it results in the production of large numbers of clones containing many different microsatellite repeats, thus eliminating the need for further library construction with different microsatellite oligonucleotides.

When using subtractive hybridization (such as streptavidin-coated magnetic beads or Nylon membranes) as the method of enrichment, it is also important to consider the early stages of adaptor ligation to the restriction-digested genomic DNA. Following ligation of the adaptor, the genomic DNA is usually enriched for microsatellites and then made double stranded by PCR amplification, using the adaptor sequence as a priming site.

Hamilton *et al.* (1999) reported that the adaptor sequence which is ligated to the restriction-digested DNA in the first steps of most enrichment processes can greatly influence the quality and enrichment rate of microsatellite libraries. It was thought that unless a high efficiency of adaptor ligation was obtained, a large proportion of genomic DNA fragments may not be recovered during the enrichment process. The design of the adaptor sequence is, therefore, essential. An ideal adaptor sequence must: (i) ligate to the genomic DNA fragments; (ii) serve as a priming site; and (iii) contain restriction sites for subsequent ligation into a vector. If the adaptor sequence has an overhang for ligation, then this limits the adaptor to certain restriction enzymes. This can be disadvantageous as single restriction enzyme size selection can overlook many potential microsatellite loci, or cut too close to a microsatellite sequence resulting in little or no flanking sequence. Using more than one restriction enzyme results in more fragments in the desired size range (< 1000 bp for small insert libraries). A blunt-ended adaptor, however, can ligate to genomic DNA (after treatment with single strand-specific nuclease) generated by any restriction enzyme or combination of enzymes. This results in a larger proportion of the genome sampled for microsatellites.

Hamilton *et al.* (1999) described an adaptor sequence containing two restriction sites for ligation into a vector, where an *Xmn*I site is created if two adaptors ligate to form a dimer. The presence of *Xmn*I during the ligation reaction drives the ligation of adaptor sequence to the genomic DNA fragments. This overcomes the inefficiency of blunt-end ligation and the tendency of adaptors to self-ligate, due to their higher molar end concentration. A similar strategy was used to prevent dimerization of inserts, during ligation of the insert into the vector. Of 120 positive clones sequenced, all contained microsatellite regions. A similar adaptor design strategy was reported by Edwards *et al.* (1996).

Enrichment by screening RAPD profiles

RAPD techniques have previously been identified as a rich source of microsatellites and other repetitive elements, probably because they must necessarily include inverted repeats which are themselves associated with repeat duplication processes (Ender *et al.*, 1996). Therefore, microsatellite loci can theoretically be isolated by screening RAPD profiles, to avoid the time-consuming screening of genomic DNA libraries.

Ueno *et al.* (1999) described an enrichment method that used this approach. This method involved first amplifying RAPD fragments from genomic DNA samples, followed by separation on agarose gels. Fragments were then transferred on to Nylon membranes by Southern transfer. Labelled microsatellite oligonucleotides were hybridized to membranes to screen for positive bands. Positive bands were then purified and cloned into plasmid vectors, which were transformed into *E. coli*. Colonies were subsequently screened by PCR for insert size. Selected inserts were then PCR amplified and sequenced to test for the presence of microsatellite regions. Sequencing revealed that among 30 clones, 21 contained microsatellite regions.

Lunt *et al.* (1999) reported a similar approach that builds on previously described RAPD enrichment procedures (Ender *et al.*, 1996), but develops the use of a repeat-specific PCR to detect microsatellites. This method involved RAPD amplification of genomic DNA, followed by ligation into a plasmid vector, and transformation into *E. coli*. Colony PCR was performed in duplicate, using two sets of PCR primers: M13 forward and reverse primers, and both M13 primers and a repeat-specific primer. Samples containing an extra amplification product were sequenced. Sequencing of PCR products revealed that twice as many microsatellites were isolated (of 14 positive clones, 12 contained microsatellite regions) when compared with traditional screening approaches using hybridization (of 14 positive clones, six contained microsatellite regions). The advantages of this method are: (i) it is cheap and efficient; (ii) it requires minimum specialized equipment; (iii) it requires no radioactive hybridization techniques; and (iv) it produces templates suitable for direct PCR sequencing.

Trouble-shooting enriched microsatellite libraries

Sometimes, the recovery of microsatellite loci from an enriched library is poor. This may be due to the comparatively low density of the particular sequence in the genome, or the association of these sequences with other repetitive DNA. The sequence to marker success rate can be considerably improved through the application of screening methods to identify non-informative clones prior to sequencing. For example, hybridization with labelled genomic DNA can identify clones containing repetitive DNA (Smith and Devey, 1994). PCR screening of the inserts using various combinations of universal or microsatellite-specific primers can also identify clones which contain microsatellite regions too close to the end of a fragment. In addition, prior to microsatellite library construction, the choice of microsatellite repeat can be screened by Southern hybridization, as well as various restriction enzyme combinations, in order to optimize the choice of enzymes and oligonucleotide probe combinations. Other factors may also be important contributors to the low recovery of microsatellite loci from enriched libraries. For example, PCR amplification may bias the enriched fragment population to contain shorter fragments. Also, optimization of conditions for each microsatellite oligonucleotide probe is important to maximize the final recovery of fragments. For example, Cordeiro *et al.* (1999) found that optimizing hybridization conditions for the enrichment step significantly improved (40% increase) the recovery of microsatellite loci from an enriched library. Optimization is especially important in plants because some microsatellite repeats such as AT/TA (which are among the most abundant in plants) are difficult to isolate from genomic libraries. This is thought to be due to the palindromic nature of the sequences, and the low T_m of the probe, making hybridization conditions difficult.

Other methods which exploit the variability associated with microsatellites

Alternative methods have been described that do not require the construction of microsatellite libraries, for those laboratories which do not have the resources to carry out such labour-intensive cloning and sequencing efforts. These methods utilize microsatellite oligonucleotides without the need to determine the unique sequence flanking the microsatellite repeats. For example, microsatellite oligonucleotides have been applied to RFLP fingerprinting of plants, in a procedure termed hybridization-based microsatellite fingerprinting (Weising and Kahl, 1997). This method involved digestion of genomic DNA by restriction enzymes, followed by separation in agarose gels. Fragments were then transferred to Nylon membranes, which were allowed to hybridize with a labelled microsatellite oligonucleotide probe. Fragments were detected using radiography or non-radioactive detection, producing complex fingerprint-like patterns. Using this approach, high levels of polymorphism were detected between related genotypes.

Another method that exploits the variability associated with microsatellites was described by Zietkiewicz *et al.* (1994). This method involved 5'- or 3'-anchored dinucleotide repeats, which serve as single PCR primers. The resulting amplification products were then separated on agarose or acrylamide gels. This revealed complex, fingerprint-like patterns that were polymorphic among genotypes. Although microsatellite repeats were used as priming sites in this method, the amplified fragments do not contain true microsatellite sequences. Therefore, they are more like other random amplified methods such as RAPD, AFLP, arbitrarily primed PCR and DNA amplification fingerprinting (DAF).

Concluding comments

Given the myriad of different microsatellite enrichment methods (Table 13.1), an obvious problem that has to be faced is how to choose the most appropriate method for a specific investigation, especially whenever new attractive enrichment methods are described. Two major deciding factors for the choice of enrichment method are: (i) the question being addressed; and (ii) the number of microsatellite loci required. For example, if microsatellites are needed for a large mapping programme, a high throughput approach for the isolation of many microsatellite loci is appropriate. In contrast, if microsatellites are needed for population genetic applications, then fewer microsatellite loci are needed. Technological and resource issues may also determine the choice of enrichment

Table 13.1. Summary of microsatellite enrichment methods.

Enrichment method	Level of enrichment	Reference
Enrichment by primer extension		
Microsatellite oligonucleotides	50-fold compared with un-enriched	Ostander *et al.* (1992)
	24 positive clones sequenced, all contained microsatellites	Paetkau (1999)
Degenerate oligonucleotides	15 positive clones sequenced, 13 contained microsatellites	Fisher *et al.* (1996)
	19 positive clones sequenced, all contained microsatellites	Koblizkova *et al.* (1998)
Enrichment by hybridization		
Streptavidin-coated magnetic beads	48 positive clones sequenced, 29 contained microsatellites	Fisher and Bachmann (1998)
	9 positive clones sequenced, 5 contained microsatellites	Prochazka (1996)
	207 positive clones sequenced, 180 contained microsatellites	
	20% positive clones compared with un-enriched with no detectable positive clones	Kijas *et al.* (1994)
	12 positive clones sequenced, 8 contained microsatellites	Connell *et al.* (1998)
	120 positive clones sequenced, all contained microsatellites	Hamilton *et al.* (1999)
Nylon membranes	50–70% of clones randomly sequenced contained microsatellites	Edwards *et al.* (1996)
Enrichment by screening RAPD profiles	30 positive clones sequenced, 21 contained microsatellites	Ueno *et al.* (1999)
	14 positive clones sequenced, 12 contained microsatellites	Lunt *et al.* (1999)

RAPD, random amplified polymorphic DNA.

method. Some laboratories may not have access to specialized equipment or facilities. In addition, cost may be an important factor, excluding some enrichment methods.

References

Baron, B., Poirier, C., Simon-Chazottes, D., Barnier, C. and Guenet, J.L. (1992) A new strategy useful for rapid identification of microsatellites from DNA libraries with large size inserts. *Nucleic Acids Research* 20, 3365–3369.

Becker, J. and Huen, M. (1995) Barley microsatellites: allele variation and mapping. *Plant Molecular Biology* 27, 835–845.

Brondani, R.P.V., Brondani, C., Tarchini, R. and Grattapaglia, D. (1998) Development, characterisation and mapping of microsatellite markers in *Eucalyptus grandis* and *E. urophylla*. *Theoretical and Applied Genetics* 97, 816–827.

Brown, S.M., Hopkins, M.S., Mitchell, S.E., Senior, M.L., Wang, T.Y., Duncan, R.R., Gonzalez-Candelas, F. and Kresovich, S. (1996) Multiple methods for the identification of polymorphic simple sequence repeats (SSRs) in sorghum [*Sorghum bicolour* (L.) Moench]. *Theoretical and Applied Genetics* 93, 190–198.

Chin, E.C.L., Senior, M.L., Shu, H. and Smith, J.S.C. (1996) Maize simple repetitive DNA sequences: abundance and allele variation. *Genome* 39, 866–873.

Cho, Y.G., Ishii, T., Temnykh, S., Chen, X., Lipovich, L., McCouch, S.R., Park, W.D., Ayres, N. and Cartinhour, S. (2000) Diversity of microsatellites derived from genomic libraries and GenBank sequences in rice (*Oryza sativa* L.). *Theoretical and Applied Genetics* 100, 713–722.

Connell, J.P., Pammi, S., Iqbal, M.J., Huizinga, T. and Reddy, A.S. (1998) A high throughput procedure for capturing microsatellites from complex plant genomes. *Plant Molecular Biology Reporter* 16, 341–349.

Cordeiro, G.M., Maguire, T.L., Edwards, K.J. and Henry, R.J. (1999) Optimisation of a microsatellite enrichment technique in *Saccharum* spp. *Plant Molecular Biology Reporter* 17, 225–229.

Cregan, P.B., Mudge, J., Fickus, E.W., Marek, L.F., Danesh, D., Denny, R., Shoemaker, R.C., Matthews, B.F., Jarvick, T. and Young, N.D. (1999) Targeted isolation of simple sequence repeat markers through the use of bacterial artificial chromosomes. *Theoretical and Applied Genetics* 98, 919–928.

Edwards, K.J., Barker, J.H.A., Daly, A., Jones, C. and Karp, A. (1996) Microsatellite libraries enriched for several microsatellite sequences in plants. *BioTechniques* 20, 759–760.

Ender, A., Schwenk, K., Stadler, T., Streit, B. and Schierwater, B. (1996) RAPD identification of microsatellites in *Daphnia*. *Molecular Ecology* 5, 437–441.

Fischer, D. and Bachmann, K. (1998) Microsatellite enrichment in organisms with large genomes (*Allium cepa* L.). *BioTechniques* 24, 796–802.

Fisher, P.J., Gardner, R.C. and Richardson, T.E. (1996) Single locus microsatellites isolated using 5′ anchored PCR. *Nucleic Acids Research* 24, 4369–4371.

Hamilton, M.B., Pincus, E.L., DiFiore, A. and Fleischer, R.C. (1999) Universal linker and ligation procedures for construction of genomic DNA libraries enriched for microsatellites. *BioTechniques* 27, 500–507.

Kijas, J.M.H., Fowler, J.C.S., Garbett, C.A. and Thomas, M.R. (1994) Enrichment of microsatellites from the citrus genome using biotintylated oligonucleotide sequences bound to streptavidin-coated magnetic particles. *BioTechniques* 16, 656–662.

Koblizkova, A., Dolezel, J. and Macas, J. (1998) Subtraction with 3' modified oligonucleotides eliminates amplification artefacts in DNA libraries enriched for microsatellites. *BioTechniques* 25, 32–38.

Kolchinsky, A., Landau-Ellis, D. and Gresshoff, P.M. (1997) Map order and linkage distances of molecular markers close to the supernodulation (*nts*-1) locus of soybean. *Molecular and General Genetics* 254, 29–36.

Lunt, D.H., Hutchinson, W.F. and Carvalho, G.R. (1999) An efficient method for PCR based isolation of microsatellite arrays (PIMA). *Molecular Ecology* 8, 891–894.

Maguire, T.L., Edwards, K.J., Saenger, P. and Henry, R. (2000) Characterisation and analysis of microsatellite loci in a mangrove species *Avicennia marina* (Forsk.) Vierh. (Avicenniaceae). *Theoretical and Applied Genetics* 101, 279–285.

Milbourne, D., Meyer, R.C., Collins, A.J., Ramsay, L.D., Gebhardt, C. and Waugh, R. (1998) Isolation, characterisation and mapping of simple sequence repeat loci in potato. *Molecular and General Genetics* 259, 233–245.

Ostrander, E.A., Jong, P.M., Rine, J. and Duyk, G. (1992) Construction of small-insert genomic DNA libraries highly enriched for microsatellite repeat sequences. *Proceedings of the National Academy of Sciences USA* 89, 3419–3423.

Paetkau, D. (1999) Microsatellites obtained using strand extension: an enrichment protocol. *BioTechniques* 26, 690–697.

Powell, W., Machray, G.C. and Provan, J. (1996) Polymorphism revealed by simple sequence repeats. *Trends in Plant Science* 1, 215–222.

Prochazka, M. (1996) Microsatellite hybrid capture technique for simultaneous isolation of various STR markers. *Genome Research* 6, 646–649.

Rafalski, J.A., Vogel, J.M., Morgante, M., Powell, W., Andre, C. and Tingey, S.V. (1996) Generating and using DNA markers in plants. In: Birren, B. and Lai, E. (eds) *Analysis of Non-mammalian Genomes – a Practical Guide*. Academic Press, New York, pp. 75–134.

Scott, L.J., Cross, M., Shepherd, M., Maguire, T. and Henry, R.J. (1999) Increasing the efficiency of microsatellite discovery from poorly enriched libraries in coniferous forest species. *Plant Molecular Biology Reporter* 17, 351–354.

Smith, D.N. and Devey, M.E. (1994) Occurrence and inheritance of microsatellites in *Pinus radiata*. *Genome* 37, 977–983.

Tautz, D. and Renz, M. (1984) Simple sequences are ubiquitous repetitive components of eukaryotic genomes. *Nucleic Acids Research* 12, 4127–4138.

Ueno, S., Yoshimaru, H., Tomaru, N. and Yamamoto, S. (1999) Development and characterisation of microsatellite markers in *Camellia japonica* L. *Molecular Ecology* 8, 335–346.

Weber, J.L. (1990) Informativeness of human (dC-dA)n (dG-dT)n polymorphisms. *Genomics* 7, 524–530.

Weber, J.L. and May, P. (1989) Abundant class of human DNA polymorphisms which can be typed using the polymerase chain reaction. *American Journal of Human Genetics* 44, 388–396.

Weising, K. and Kahl, G. (1997) Hybridisation-based microsatellite fingerprinting of plants and fungi. In: Caétano-Anolles, G. and Gresshoff, P.M. (eds) *DNA Markers: Protocols, Applications, and Overviews*. Wiley-Liss, New York, pp. 27–54.

White, G. and Powell, W. (1997) Isolation and characterisation of microsatellite loci in

Swietenia humilis (Meliaceae): an endangered tropical hardwood species. *Molecular Ecology* 6, 851–860.

Zietkiewicz, E., Rafalski, A. and Labuda, D. (1994) Genome fingerprinting by simple sequence repeat (SSR)-anchored polymerase chain reaction amplification. *Genomics* 20, 176–183.

Chapter 14
Sourcing of SSR Markers from Related Plant Species

M. ROSSETTO

Centre for Plant Conservation Genetics, Southern Cross University, Lismore, Australia

Microsatellite technology

As microsatellites, or simple sequence repeats (SSRs), have been discussed at length in previous chapters, an in-depth description of this technique is not necessary. Their abundant and uniform distribution throughout the genome, codominant inheritance, simple screening requirements and reproducibility ensure that microsatellites are valuable for a multitude of applications. One of the greatest advantages of these markers is a rate of polymorphism which is higher than in other common techniques such as restriction fragment length polymorphism and random amplified polymorphic DNA (Powell et al., 1996b). Slipped-strand mispairing during DNA replication is thought to be the predominant mode of mutation for microsatellites (Levinson and Gutman, 1987; Schlötterer and Tautz, 1992; Wierdl et al., 1997). Such mutational events accumulate more rapidly than point mutations and insertions/deletions (indels); rates are estimated at 10^{-2} to 10^{-3} per locus, per gamete, per generation (Weber and Wong, 1993). Some questions remain about the predominant mutation pattern of these rapid modifications, the main alternatives being single-step or random mutation events. A number of theoretical studies supporting one model or the other (as well as combinations of both) have been suggested (see Shriver et al., 1993; Di Rienzo et al., 1994; Estoup et al., 1995; Goldstein and Pollock, 1997 for example).

The potential of microsatellites as useful markers for plant studies was promptly recognized, resulting in their successful characterization and application in a number of species. Most early studies were based on crops such as rice (Wu and Tanksley, 1993), barley (Saghai Maroof et al., 1994) and wheat (Röder et al., 1995). More recently, the ability of these hypervariable regions to reveal high allelic diversity and delimit fine genetic structure has resulted in an

© CAB *International* 2001. *Plant Genotyping: the DNA Fingerprinting of Plants* (ed. R.J. Henry)

increase in elaborate population studies (Gupta *et al.*, 1996; Jarne and Lagoda, 1996; Powell *et al.*, 1996a; Goldstein and Pollock, 1997; Rossetto *et al.*, 1999). Unfortunately, this is not a universal technique, and the availability of species-specific SSR loci for only a small number of plant taxa is a major constraint to their ubiquitous adoption. Many laboratories have sufficient resources and expertise for conducting SSR-based research but not for characterizing new loci. Even with the development of highly efficient enrichment procedures (Edwards *et al.*, 1996; Powell *et al.*, 1996a; Cordeiro *et al.*, 1999; Maguire, Chapter 13 this volume), the characterization of SSR markers for new species is still a laborious and costly exercise, in particular if research is focused on more than one species. An alternative to the development of specific SSR libraries can be to search through published sequences and expressed sequence tag databases, an option discussed further by Scott (Chapter 15, this volume). Such luxury, however, is only possible for a small minority of taxa.

If a fraction of existing microsatellite libraries could be at least transferred across related species, one of the major limitations of this technology would be circumvented. This approach would make SSRs available to a variety of species with minimal resource expenditure. Empirical studies have shown that an inverse relationship exists between primer site conservation and evolutionary relationship between test taxa, with the threshold distance after which no amplification can be expected being lower in plants than in animals. However, there is still an insufficient number of studies investigating the use of heterologous SSR primers in plants. The current understanding on interspecific amplification of microsatellites in plants is the main focus of this chapter.

The challenges of cross-species amplification

Microsatellite primers developed for one species can be used to detect polymorphism at homologous sites in related species. For this to be possible, the repeat sequence and the flanking regions containing the selected priming sites must be conserved across taxa. The success of heterologous PCR amplification will depend upon the evolutionary distance between the source and the target species. Higher genomic homology is likely to translate into greater conservation of SSR-flanking regions and, as a result, in transferability of primer pairs. Therefore, similar sized amplification products should be obtained with DNA from related species. However, a cautious approach is required when comparing similar PCR products obtained across different species, as various factors can cause size homoplasy. Over long evolutionary times, the interspecific allelic differences at one locus are often more complex than simple changes in repeat number. Two equally sized products amplified in different species might include mutations, rearrangements and duplications in the flanking region and/or changes in the repeat (Peakall *et al.*, 1998; Sefc *et al.*, 1999; Rossetto *et al.*, 2000). These variations hinder the use of SSRs in phylogenetic studies and sometimes, if undetected, can lead to misinterpretation of the data.

Possible problems include false positives (the microsatellite repeat is absent from the fragment amplified) and false negatives (or null alleles, the microsatellite repeat is present but not amplified). Null alleles are caused by sequence variations at the primer target site that will prevent primer annealing and will result in no amplification. An individual with a null allele can be incorrectly scored as homozygous at that specific locus, with the obvious implications for population and parentage studies. Some examples of null alleles detected in target but not in source species are those of Fisher *et al.* (1998) (two loci in pines) and Sefc *et al.* (1999) (two loci in *Vitis* hybrids). Nevertheless, null alleles can also be detected in the species for which the SSRs were developed and their frequencies can be statistically calculated (Brookfield, 1996). Other potential problems include the duplication of the flanking regions, which can result in multiple banding patterns (Fisher *et al.*, 1998), and their duplication and inversion which can result in the amplification of single-primed fragments (of particular concern for taxa known to include numerous duplications and rearrangements such as the brassicas; Westman and Kresovich, 1998). Duplications can cause serious interpretation problems but, with the use of sequencing and pedigree analysis to clarify their origin, they can become advantageous (Fisher *et al.*, 1998).

Little attention has been given to the patterns of microsatellite evolution across species and, unfortunately, most cross-species studies have been conducted using a limited number of individuals. There are, however, some recognized trends and expectations. For instance, changes in repeat sequences (especially interruptions of perfect repeats) are likely to reduce polymorphism (Gupta *et al.*, 1996). The reason for this is that long continuous perfect repeats are more unstable (and therefore more variable) than shorter interrupted repeats (Wierdl *et al.*, 1997). As a result, it is usually expected that loci incorporating long repeats in the source species will be less informative in target species because of the potential repeat interruptions. There are, however, recorded cases of loci more polymorphic across target species than within the source species (Weising *et al.*, 1996; Smulders *et al.*, 1997).

Sourcing of SSR markers from related species: current status

A brief summary of cross transferability in animals

Microsatellites were initially described in humans (Litt and Luty, 1989; Weber and May, 1989). Similar findings quickly followed in other mammalian species such as mice (Love *et al.*, 1990), pigs (Johansson *et al.*, 1992) and cattle (Kemp *et al.*, 1993). In the zoological world the potential of interspecific amplification was realized very quickly. As early as 1991, Schlötterer *et al.* successfully tested SSR primers developed for the long-finned pilot whale (*Globicephala melas*) on a range of whales representative of the major cetacean radiations. Four of these loci were further tested across a number of individuals from two *Balenoptera* species, showing polymorphism in all but one locus/species combination. This study was quickly followed by numerous other reports of interspecific amplification between animal species. Deka *et al.* (1994) reported the successful transfer of human SSRs to chimpanzee, with very similar allelic diversity between the two species as well as successful amplification in gorillas and orang-utans. Various degrees of interspecific amplification and polymorphism were also reported on such diverse groups as artiodactyls (Engel *et al.*, 1996) and pinnipeds (Gemmell *et al.*, 1997). Despite the fact that many studies reported a negative relationship between amplification success and evolutionary distance, Primmer and Ellegren (1998) described a locus conserved between birds species that had diverged an estimated 60 million yeas ago, and Fitzsimmons *et al.* (1995) reported locus conservation between turtles that diverged 300 million years ago. In view of the success of cross-species amplification in animals, most of these investigations propose that this is an adequate approach for population genetics studies. As a result, research based on the use of SSRs sourced from closely related animal species is becoming increasingly frequent. For example, Roy *et al.* (1994) used domestic dog microsatellites to study genetic differentiation and hybridization patterns in grey wolf (*Canis lupus*), coyote (*Canis latrans*) and red wolf (*Canis rufus*). Similarly, cattle and horse microsatellites were used, respectively, to study genetic variability and population structure in African buffalo (*Syncerus caffer*, Simonsen *et al.*, 1998) and Catalonian donkey (*Equus asinus*, Jordana *et al.*, 1999). Thus, despite the prevailing tendency for developing species-specific microsatellites (for example, a representative journal such as *Molecular Ecology* reported the characterization of microsatellites for 47 and 56 animal species and six and 13 plant species in 1998 and 1999, respectively), sourcing SSRs from related animal species still represents an attractive option.

Cross transferability in plants

It has been estimated that microsatellite regions are less frequent in plants than in animals. Various figures have been published, ranging from five times (Lagercrantz *et al.*, 1993) to ten times (Powell *et al.*, 1996a) less abundant in plants than in animals. Other reports have detected differences between major groupings, with SSRs being three times more frequent in dicots than in monocots (Wang *et al.*, 1994). Such lower frequency, if real and not related to the bias in sequence information available, should be a further reason for assessing cross-species transferability of SSR primer pairs.

The characterization of microsatellites in plants has been particularly popular in agriculture and forestry. The greater availability of SSRs in certain groups of species has increased interest in primer transferability at least across closely related taxa. As a result, microsatellite amplification has been successful across pine species that have diverged over 100 million years ago (Karhu *et al.*, 2000). Table 14.1 illustrates many of the plant microsatellite cross-species studies available from the scientific literature at the time of publication (data from a number of studies on cultivars, landraces and variants are not included as they represent the same species and transfer success is generally 100%). The source species include a range of monocots, dicots and gymnosperms, mostly with commercial relevance. The target species include a variety of related taxa usually selected for their commercial and/or conservation significance. The number of target species tested is high; 239; 57.7% belong to the same genera as the source species and 42.3% belong to different genera. Polymorphism has not been evaluated in all studies and, in most cases, only a few individuals for each species have been analysed. As a result, in many instances the data presented represent an estimation of potential, rather than a final verdict.

Similar data are presented in Table 14.2 for unpublished research currently being conducted at the Centre for Plant Conservation Genetics. Table 14.3 condenses the data from Table 14.1 by grouping the findings according to evolutionary relationship between the source and target species. This summary shows that closely related species are more likely to share SSR priming sites than more distantly related ones. Close to 90% of species/primer combinations tested within subgenera were successful, a much higher success rate than between genera attempts (just over 35%, Table 14.3). Table 14.3 also shows that when priming sites did transfer across species, the rate of polymorphism was still relatively high, even between evolutionarily distant species. These results suggest that it is possible to obtain functional SSR primers even from more distantly related species.

Plant microsatellites sourced from related species have been applied to various types of research. For example, in agricultural research, interspecific microsatellite-based genetic maps have been developed in rice. A map based on a *Oryza sativa* × *Oryza glaberrima* cross has been produced using 159 *O. sativa* SSRs and is in agreement with previous maps (Lorieux *et al.*, 2000). Such studies can provide useful data on breeding and selection, marker-assisted localization of genes and quantitative trait loci and the development of interspecifc introgression lines. As in the zoological world, plant population studies have also relied on the application of SSR primer pairs characterized in related species. Strieff *et al.* (1998) successfully studied fine-scale genetic structure in populations of *Quercus petrea* and *Quercus robur* using a total of six SSR loci characterized in *Q. petrea* and *Quercus macrocarpa*. The authors obtained an average of 21.7 alleles per locus and verified Mendelian inheritance using known crosses. A similar study by Degen *et al.* (1999) investigated genetic structure in *Q. robur* stands using four SSRs developed for *Q. petrea*.

Despite their lower frequency in organelle DNA (Wang *et al.*, 1994), microsatellites have been characterized from the chloroplast genome and have proved to be a potentially useful source of polymorphic markers across 13 *Glycine* taxa (Powell *et al.*, 1995). It was initially suggested that such markers would have greater potential for phylogeographic and phylogenetic studies; however, further experimentation demonstrated that two of these SSR loci did not successfully represent relationships among genomes (Doyle *et al.*, 1998). Microsatellites have also been characterized from the mitochondrial genome and have been successfully transferred across 15 species (Soranzo *et al.*, 1999).

Table 14.1. Summary of the major published studies on cross transferability of plant microsatellites. The source species are indicated as well as their taxonomic divergence with the target species. Transfer success and polymorphism of the loci successfully transferred are also indicated.

Source species	No. of loci tested	No. of target species tested	Taxonomic divergence	Transfer success (%)	Polymorphism of transferred loci (%)	Reference
Actinidia chinensis	16	8	Same genus	78.9	—	Weising et al., 1996
Arabidopsis thaliana	30	6	Diff. genera	66.7	55.4	Westman and Kresovich, 1998
Arabidopsis thaliana	30	3	Diff. genera	30	60	van Treuren et al., 1997
Arachis hypogaea	6	3	Same genus	83.3	—	Hopkins et al., 1999
Brassica napus	17	2	Same genus	76.5	100	Szewc-McFadden et al., 1996
Caryocar brasiliense	10	5	Same genus	100	100	Garcia Collevatti et al., 1999
Citrus × *Poncirus*	2	10	Same genera	100	100	Kijas et al., 1995
Cucumis melo/sativus	7	8	Diff. genera	71.4	100	Katzir et al., 1996
Eucalyptus nitens	4	6	Same genus	66.7	100	Byrne et al., 1996
Eucalyptus grandis/urophylla	100	4	3 same genus 1 diff. genus	65 20	— —	Brondani et al., 1998
Glycine max	31	9	3 same genus 6 diff. genera	57 10.3	— —	Peakall et al., 1998
Grevillea macleayana	7	8	Same genus	46.2	—	England et al., 1999
Helianthus annuus	13	25	3 same genus 22 diff. genera	79.5 21.5	62.5 —	Whitton et al., 1997
Lycopersicum esculentum	44	6	4 same genus 2 diff. genera	88.6 28.4	82.1 —	Smulders et al., 1997
Magnolia obovata	11	10	5 same genus 5 diff. genera	90.9 33.3	— —	Isagi et al., 1999
Medicago sativa	3	16	Same genus	100	100	Diwan et al., 1997
Melaleuca alternifolia	35	7	2 same genus 5 diff. genera	72.9 38.3	66.7 65.7	Rossetto et al., 2000
Mimulus guttatus	6	3	Same genus	100	100	Awadalla and Ritland, 1997
Oryza sativa	8	4	Same genus	90.6	—	Wu and Tanksley, 1993

Species				Reference
Paspalum vaginatum	10	1	Diff. genus	0	—	Brown et al., 1996
Phitecellobium elegans	6	13	Diff. genera	42.3	43.2	Dayanandan et al., 1997
Picea sitchensis	5	4	2 same genus	100	—	van de Ven and McNicol, 1996
			2 diff. genus	20	—	
Pinus radiata	7	16	10 same genus	62.9	72.7	Fisher et al., 1998
			6 diff. genera	10.2	0	
Pinus strobus	15	9	Same genus	51.5	71	Echt et al., 1999
Prunus persica	2	1	Same genus	50	100	Downey and Iezzoni, 2000
P. cerasus	1	1	Same genus	100	100	
P. avium	5	1	Same genus	60	67	
Quercus myrsinifolia	9	11	8 same genus	94	—	Isagi and Suhandono, 1997
			3 diff. genera	22	—	
Quercus petraea	17	7	5 same genus	71.8	82	Steinkellner et al., 1997
			2 diff. genera	35.3	—	
Swietenia humilis	11	10	Diff. genus	50.9	—	White and Powell, 1997
Triticum aestivum	15	3	Diff. genera	63.3	68.4	Roder et al., 1995
Vitis riparia	18	6	Same genus	93.5	100	Sefc et al., 1999
Vitis vinifera	5	7	Same genus	100	100	Thomas and Scott, 1993
Vitis vinifera	16	3	1 same genus	62.5	—	Scott et al., 2000
			2 diff. genera	15.6	—	
Zea mays	67	1	Diff. genus	64.2	27.9	Brown et al., 1996
Zostera marina	12	2	Same genus	4.7	0	Reusch, 2000

Table 14.2. Summary of unpublished data from research conducted at the Centre for Plant Conservation Genetics.

Source species	No. of loci tested	No. of target species tested	Taxonomic divergence	Transfer success (%)	Polymorphism of transferred loci (%)	Reference
Eucalyptus cloeziana/grandis/ variegata	5	5	4 diff. subgenera	90	94.4	R. Mellick, personal communication
Pinus caribea	2	3	1 diff. genus	20	100	M. Cross, personal communication
P. elliottii	1	3	Same genus	100	100	
P. radiata	8	3	Same genus	33.3	100	
P. strobus	5	4	Same genus	70	100	
P. taeda	26	2	Same genus	60	0	
Sorghum bicolor	20	25	Same genus	71.2	67.6	S. Dillon, personal communication
Vitis vinifera	9	27	Diff. genera	71	66.7	M. Rossetto (unpublished)
				39.5	47.9	

Table 14.3. Cross transferability and polymorphism of SSR primers within plant subgenera, genera, alliances and families. These data summarize the information contained in Table 14.1.

Taxonomic divergence	Transferability % (N)		Polymorphism % (N)	
Same subgenus	89.8	(521)	78.3	(299)
Same genus	76.4	(1800)	86.0	(773)
Same alliance	45.4	(403)	50.4	(141)
Same family	35.2	(1683)	58.4	(363)

(N) represents the number of species/primer tests attempted, and the measure of polymorphism is calculated only from the loci that were successfully transferred.

Some practical suggestions

Usually modifications of PCR protocols are necessary for the successful transfer of microsatellite primer pairs between species. Not all the studies cited in Table 14.1 describe PCR protocols in detail and, of those that do, approximately 35% used unchanged conditions. In general, as the evolutionary distance between the source and the target species increases, the PCR primer annealing temperature is lowered by 2–5°C or a touchdown protocol is used. A lowering of annealing temperature should compensate for potential differences between the primer and the priming site in the target species. Dayanandan *et al.* (1997) found that high GC content in primers ensured greater transfer success without modification of the protocol, with lower GC contents requiring modifications of the PCR protocols.

It is important to emphasize, however, that amplification of a PCR product does not necessarily imply success. Therefore, since the generation of SSRs can be expensive and complex, it is important to ensure that the fragments amplified are those anticipated. This is particularly true as the evolutionary gap between source and target species increases. There are various ways to assess the presence of the expected SSR within the PCR fragments obtained. Stutter bands associated with the amplification products are usually considered as a sign that the product obtained is indeed an SSR. However, better verification techniques such as hybridization with a selective probe and sequencing are more reliable in detecting the presence of the expected repeats within amplification products (Westman and Kresovich, 1998). Sequencing is arguably the best approach but it is also potentially complex and expensive, as in most cases it will require cloning.

Sequence verification is important in a study that is fully reliant on microsatellites sourced from other species, as product size and variation alone are not always good predictors of SSR presence. Once their presence has been verified, cross-sourced SSRs can be applied to a variety of research fields. It is possible that the development of a specific SSR database listing all available loci for selected groups of species may facilitate cross-species experimentation. The main restriction will then be the number of loci available in the source species.

Applications such as mapping might be only possible for taxa closely related to major crops or forestry species for which very large SSR libraries exist. However, SSR-based population genetics studies, which generally rely on a smaller number of loci, can become accessible to a much larger number of taxa by using heterologous primers.

References

Awadalla, P. and Ritland, K. (1997) Microsatellite variation and evolution in the *Mimmulus guttatus* species complex with contrasting mating systems. *Molecular Biology and Evolution* 14, 1023–1034.

Brondani, R.P.V., Brondani, C., Tarchini, R. and Grattapaglia, D. (1998) Development, characterisation and mapping of microsatellite markers in *Eucalyptus grandis* and *E. urophylla*. *Theoretical and Applied Genetics* 97, 816–827.

Brookfield, J.F.Y. (1996) A simple new method for estimating null allele frequency from heterozygote deficiency. *Molecular Ecology* 5, 453–455.

Brown, S.M., Hopkins, M.S., Mitchell, S.E., Senior, M.L., Wang, T.Y., Duncan, R.R., Gonzalez-Candelas, F. and Kresovich, S. (1996) Multiple methods for the identification of polymorphic simple sequence repeats (SSRs) in sorghum [*Sorghum bicolor* (L.) Moench]. *Theoretical and Applied Genetics* 93, 190–198.

Byrne, M., Marquez-Garcia, M.I., Uren, T., Smith, D.S. and Moran, G.F. (1996) Conservation and genetic diversity of microsatellite loci in the genus *Eucalyptus*. *Australian Journal of Botany* 44, 331–341.

Cordeiro, G.M., Maguire, T.L., Edwards, K.J. and Henry, R.J. (1999) Optimisation of microsatellite enrichment technique in *Saccharum* spp. *Plant Molecular Biology Reporter* 17, 225–229.

Dayanandan, S., Bawa, K.S. and Kesseli, R. (1997) Conservaton of microsatellites among tropical trees (Leguminosae). *American Journal of Botany* 84, 1658–1663.

Degen, B., Streiff, R. and Ziegenhagen, B. (1999) Comparative study of genetic variation and differentiation of two pedunculate oak (*Quercus robur*) stands using microsatellite and allozyme loci. *Heredity* 83, 597–603.

Deka, R., Shriver, M.D., Yu, L.M., Jin, L., Aston, C.E., Chakraborty, R. and Ferrell, R.E. (1994) Conservation of human chromosome 13 polymorphic microsatellite (CA)n repeats in chimpanzees. *Genomics* 22, 226–230.

Di Rienzo, A., Peterson, A.C., Garza, J.C., Valdes, A.M., Slatkin, M. and Freimer, N.B. (1994) Mutational processes of simple-sequence repeat loci in human populations. *Proceedings of the National Academy of Sciences USA* 91, 3166–3170.

Diwan, N., Bhagwat, A.A., Bauchan, G.B. and Cregan, P.B. (1997) Simple sequence repeat DNA markers in lucerne and perennial and annual *Medicago* species. *Genome* 40, 887–895.

Downey, S.L. and Iezzoni, A.F. (2000) Polymorphic DNA markers in black cherry (*Prunus serotina*) are identified using sequences from sweet cherry, peach, and sour cherry. *Journal of the American Society for Horticultural Sciience* 125, 76–80.

Doyle, J.J., Morgante, M., Tingey, S.V. and Powell, W. (1998) Size homoplasy in chloroplast microsatellites of wild perennial relatives of soybean (*Glycine* Subgenus *Glycine*). *Molecular Biology and Evolution* 15, 215–218.

Echt, C.S., Vendramin, G.G., Nelson, C.D. and Marquardt, P. (1999) Microsatellite DNA as shared genetic markers among conifer species. *Canadian Journal of Forest Research* 29, 365–371.

Edwards, K.J., Barker, J.H.A., Daly, A., Jones, C. and Karp, A. (1996) Microsatellite libraries enriched for several microsatellite sequences in plants. *BioTechniques* 20, 759–760.

Engel, S.R., Linn, R.A., Taylor, J.F. and Davis, S.K. (1996) Conservation of microsatellite loci across species of Artiodactyls: implications for population studies. *Journal of Mammalogy* 77, 504–518.

England, P.R., Ayre, D.J. and Whelan, R.J. (1999) Microsatellites in the Australian shrub *Grevillea macleayana* (Proteaceae). *Molecular Ecology* 8, 689–690.

Estoup, A., Garnery, L., Solignac, M. and Cornuet, J.M. (1995) Microsatellite variation in Honey Bee (*Apis mellifera* L.) populations: hierarchical genetic structure and test of the infinite allele and stepwise mutation models. *Genetics* 140, 679–695.

Fisher, P.J., Richardson, T.E. and Gardner, R.C. (1998) Characteristics of single- and multi-copy microsatellites in *Pinus radiata*. *Theoretical and Applied Genetics* 96, 969–979.

Fitzsimmons, N.N., Moritz, C. and Moore, S.S. (1995) Conservation and dynamics of microsatellite loci over 300 million years of marine turtle evolution. *Molecular Biology and Evolution* 12, 432–440.

Garcia Collevatti, R., Vianello Brondani, R. and Grattapaglia, D. (1999) Development and characterisation of microsatellite markers for genetic analysis of a Brazilian endangered tree species *Caryocar brasiliense*. *Heredity* 83, 748–756.

Gemmell, N.J., Allen, P.J., Goodman, S.J. and Reed, J.Z. (1997) Interspecific microsatellite markers for the study of pinniped populations. *Molecular Ecology* 6, 661–666.

Goldstein, D.B. and Pollock, D.D. (1997) Launching microsatellites: a review of mutation process and methods of phylogenetic inference. *Journal of Heredity* 88, 335–342.

Gupta, P.K., Balyan, P.C., Sharma, P.C. and Ramesh, B. (1996) Microsatellites in plants: a new class of molecular markers. *Current Science* 70, 45–54.

Hopkins, M.S., Casa, A.M., Wang, T., Mitchell, S.E., Dean, R.E., Kochert, G.D. and Kresovich, S. (1999) Discovery and characterisation of polymorphic simple sequence repeats (SSRs) in peanut. *Crop Science* 39, 1243–1247.

Isagi, Y. and Suhandono, S. (1997) PCR primers amplifying microsatellite loci of *Quercus myrsinifolia* Blume and their conservation between oak species. *Molecular Ecology* 6, 897–899.

Isagi, Y., Kanazashi, T., Suzuki, W., Tanaka, H. and Abe, T. (1999) Polymorphic microsatellite DNA markers for *Magnolia obovata* Thunb. and their utility in related species. *Molecular Ecology* 8, 698–700.

Jarne, P. and Lagoda, P.J.L. (1996) Microsatellites from molecules to populations and back. *Trends in Ecology and Evolution* 11, 424–429.

Johansson, M., Ellegren, H. and Andersson, L. (1992) Cloning and characterisation of highly polymorphic porcine microsatellites. *Journal of Heredity* 83, 196–198.

Jordana, J., Folch, P. and Sanchez, A. (1999) Genetic variation (protein markers and microsatellites) in endangered Catalonian donkeys. *Biochemical Systematics and Ecology* 27, 791–798.

Karhu, A., Dieterich, J.H. and Savolainen, O. (2000) Rapid expansion of microsatellite sequences in pines. *Molecular Biology and Evolution* 17, 259–265.

Katzir, N., Daninpoleg, Y., Tzuri, G., Karchi, Z., Lavi, U. and Cregan, P.B. (1996) Length polymorphism and homologies of microsatellites in several Cucurbitaceae species. *Theoretical and Applied Genetics* 93, 1282–1290.

Kemp, S.J., Brezinsky, L. and Teale, A.J. (1993) A panel of bovine, ovine and caprine polymorphic microsatellites. *Animal Genetics* 24, 363–365.

Kijas, J.M.H., Fowler, J.C.S. and Thomas, M.R. (1995) An evaluation of sequence tagged microsatellite site markers for genetic analysis within *Citrus* and related species. *Genome* 38, 349–355.

Lagercrantz, U., Ellegren, H. and Andersson, L. (1993) The abundance of various polymorphic microsatellite motifs differs between plants and vertebrates. *Nucleic Acids Research* 21, 1111.

Levinson, G. and Gutman, G.A. (1987) High frequencies of short frameshifts in poly-CA/TG tandem repeats borne by bacteriophage M13 in *Escherichia coli* K-12. *Nucleic Acids Research* 15, 5323–5338.

Litt, M. and Luty, J.A. (1989) A hypervariable microsatellite revealed by *in vitro* amplification of a dinucleotide repeat within the cardiac muscle actin gene. *American Journal of Human Genetics* 44, 397–401.

Lorieux, M., Ndjiondjop, M.N. and Ghesquiere, A. (2000) A first interspecific *Oryza sativa* × *Oryza glaberrima* microsatellite-based genetic linkage map. *Theoretical and Applied Genetics* 100, 593–601.

Love, J.M., Knight, A.M., McAleer, M.A. and Todd, J.A. (1990) Towards construction of a high-resolution map of the mouse genome using PCR-analysed microsatellites. *Nucleic Acids Research* 21, 1111–1115.

Peakall, R., Gilmore, S., Keys, W., Morgante, M. and Rafalski, A. (1998) Cross species amplification of soybean (*Glycine max*) simple sequence repeats (SSRs) within the genus and other legume genera: implications for the transferability of SSRs in plants. *Molecular Biology and Evolution* 15, 1275–1287.

Powell, W., Morgante, M., Andre, C., McNicol, J.W., Machray, G.C., Doyle, J.J., Tingey, S.V. and Rafalski, J.A. (1995) Hypervariable microsatellites provide a general source of polymorphic DNA markers for the chloroplast genome. *Current Biology* 5, 1023–1029.

Powell, W., Machray, G.C. and Provan, J. (1996a) Polymorphism revealed by simple sequence repeats. *Trends in Plant Science* 7, 215–222.

Powell, W., Morgante, M., Andre, C., Hanafey, M., Vogel, J., Tingey, S. and Rafalski, A. (1996b) The comparison of RFLP, RAPD, AFLP and SSR (microsatellite) markers for germplasm analysis. *Molecular Breeding* 2, 225–238.

Primmer, C.R. and Ellegren, H. (1998) Patterns of molecular evolution in avian microsatellites. *Molecular Biology and Evolution* 15, 997–1008.

Reusch, T.B.H. (2000) Five microsatellite loci in eelgrass *Zostera marina* and a test of cross-species amplification in *Z. noltii* and *Z. japonica*. *Molecular Ecology* 9, 365–378.

Röder, M.S., Plaschke, J., Konig, S.U., Borner, A., Sorrells, M.E., Tanksley, S.D. and Ganal, M.W. (1995) Abundance, variability and chromosomal location of microsatellites in wheat. *Molecular and General Genetics* 246, 327–333.

Rossetto, M., Slade, R.W., Baverstock, P.R., Henry, R.J. and Lee, L.S. (1999) Microsatellite variation and assessment of genetic structure in tea tree (*Melaleuca alternifolia* – Myrtaceae). *Molecular Ecology* 8, 633–643.

Rossetto, M., Harriss, F.C.L., McLauchlan, A., Henry, R.J., Baverstock, P.R. and Lee, L.S. (2000) Interspecific amplification of tea tree (*Melaleuca alternifolia* – Myrtaceae) microsatellite loci – potential implications for conservation studies. *Australian Journal of Botany* 48, 367–373.

Roy, M.S., Geffen, E., Smith, D., Ostrander, E.A. and Wayne, R.K. (1994) Patterns of differentiation and hybridization in North American wolflike canids, revealed by analysis of microsatellite loci. *Molecular Biology and Evolution* 11, 553–570.

Sagai Maroof, M.A., Biyashev, R.M., Yang, G.P., Zhang, Q. and Allard, R.W. (1994) Extraordinarily polymorphic microsatellite DNA in barley: species diversity, chromosomal locations, and population dynamics. *Proceedings of the National Academy of Sciences USA* 91, 5466–5470.

Schlötterer, C. and Tautz, D. (1992) Slippage synthesis of simple sequence DNA. *Nucleic Acids Research* 20, 211–215.

Schlötterer, C., Amos, B. and Tautz, D. (1991) Conservation of polymorphic simple sequence loci in cetacean species. *Nature* 354, 63–64.

Scott, K.D., Eggler, P., Seaton, G., Rossetto, M., Ablett, E.M., Lee, L.S. and Henry, R.J. (2000) Analysis of SSRs derived from grape ESTs. *Theoretical and Applied Genetics* 100, 723–726.

Sefc, K.M., Regner, F., Turetschek, E., Glossl, J. and Steinkellner, H. (1999) Identification of microsatellite sequences in *Vitis riparia* and their applicability for genotyping of different *Vitis* species. *Genome* 42, 367–373.

Shriver, M.D., Jin, L., Chakraborty, R. and Boerwinkle, E. (1993) VNTR allele frequency distributions under the stepwise mutation model: a computer simulation approach. *Genetics* 134, 983–993.

Simonsen, B.T., Siegismund, H.R. and Arctander, P. (1998) Population structure of African buffalo inferred from mtDNA sequences and microsatellites loci: high variation but low differentiation. *Molecular Ecology* 7, 225–237.

Smulders, M.J.M., Bredemeijer, G., Rus-Kortekaas, W., Arens, P. and Vosman, B. (1997) Use of short microsatellites from database sequences to generate polymorphism among *Lycopersicon esculentum* cultivars and accessions of other *Lycopersicon* species. *Theoretical and Applied Genetics* 97, 264–272.

Soranzo, N., Provan, J. and Powell, W. (1999) An example of microsatellite length variation in mitochondrial genome of conifers. *Genome* 42, 158–161.

Steinkellner, H., Lexer, C., Turetschek, E. and Gossel, J. (1997) Conservation of (GA)n microsatellite loci between *Quercus* species. *Molecular Ecology* 6, 1189–1194.

Streiff, R., Labbe, T., Bacilieri, R., Steinkellner, H., Glossl, J. and Kremer, A. (1998) Within-population genetic structure in *Quercus robur* L. and *Quercus petrea* (Matt.) Liebl. assessed with isozymes and microsatellites. *Molecular Ecology* 7, 317–328.

Szewc-McFadden, A.K., Kresovich, S., Bliek, S.M., Mitchell, S.E. and McFerson, J.R. (1996) Identification of polymorphic, conserved simple sequence repeats (SSRs) in cultivated *Brassica* species. *Theoretical and Applied Genetics* 93, 534–538.

Thomas, M.R. and Scott, N.S. (1993) Microsatellite repeats in grapevine reveal DNA polymorphisms when analysed as sequence-tagged sites (STSs). *Theoretical and Applied Genetics* 86, 985–990.

van Treuren, R., Kuittinen, H., Karkkainen, K., Baena-Gonzales, E. and Savolainen, O. (1997) Evolution of microsatellites in *Arabis petrea* and *Arabis lyrata*, outcrossing relatives of *Arabidopsis thaliana*. *Molecular Biology and Evolution* 14(3), 22–229.

van de Ven, W.T.G. and McNicol, R.J. (1996) Microsatellites as DNA markers in Sitka spruce. *Theoretical and Applied Genetics* 93, 613–617.

Wang, Z., Weber, J.L., Zhong, G. and Tanksley, S.D. (1994) Survey of plant short tandem DNA repeats. *Theoretical and Applied Genetics* 88, 1–6.

Weber, J.L. and May, P.E. (1989) Abundant class of human DNA polymorphisms which can be typed using polymerase chain reaction. *American Journal of Human Genetics* 44, 388–396.

Weber, J.L. and Wong, C. (1993) Mutation of human short tandem repeats. *Human Molecular Genetics* 2, 1123–1128.

Weising, K., Fung, R.W.M., Keeling, R.G., Atkinson, R.G. and Gardner, R.C. (1996) Characterisation of microsatellites from *Actinidia chinensis*. *Molecular Breeding* 2, 117–131.

Westman, A.L. and Kresovich, S. (1998) The potential for cross-taxa simple-sequence repeat (SSR) amplification between *Arabidopsis thaliana* L. and crop brassicas. *Theoretical and Applied Genetics* 96, 272–281.

White, G. and Powell, W. (1997) Cross-species amplification of SSR loci in the Meliaceae family. *Molecular Ecology* 6, 1195–1197.

Whitton, J., Rieseberg, L.H. and Ungerer, M.C. (1997) Microsatellite loci are not conserved across the Asteraceae. *Molecular Biology and Evolution* 14, 204–209.

Wierdl, M., Dominska, M. and Petes, T.D. (1997) Microsatellite instability in yeast: dependence on the length of the microsatellite. *Genetics* 146, 769–779.

Wu, K.S. and Tanksley, S.D. (1993) Abundance, polymorphism and genetic mapping of microsatellites in rice. *Molecular and General Genetics* 241, 225–235.

Chapter 15

Microsatellites Derived from ESTs, and their Comparison with those Derived by Other Methods

K.D. SCOTT

Centre for Plant Conservation Genetics, Southern Cross University, Lismore, Australia

Introduction

Microsatellites or simple sequence repeats (SSRs) are usually single locus markers characterized by their hypervariability, abundance and reproducibility. Microsatellites are five times less abundant in plant genomes than in mammals (Lagercrantz *et al.*, 1993). Estimates of the frequency of microsatellites in plants ranges from one every 3.3 kb in barley (Becker and Heun, 1995) to 1.2 Mb for GA/CT and GT/CA repeats in tomato (Broun and Tanksley, 1996). On average, the estimates for microsatellite occurrence is one every 21.2 kb in dicotyledonous plants and one every 64.6 kb in monocots (Wang *et al.*, 1994). Applications of microsatellites are in the construction of molecular maps, cultivar identification, cultivar parentage assessment, for the purpose of obtaining patents, plant variety rights and for population and evolutionary studies. For evolutionary studies there is a role for both highly polymorphic and more stable microsatellite markers, for recent and more ancient evolutionary studies, respectively (Cho *et al.*, 2000).

In the past, microsatellites have been expensive to develop. Enrichment protocols have reduced this cost of development; however, recently a new source of large numbers of microsatellites has become available. The microsatellites are available from expressed sequence tag (EST) databases, and are effectively a free by-product of the currently expanding EST research. These microsatellites are obviously limited to those species for which this type of database exists. To date, these include some important species, such as *Arabidopsis* (Delseny *et al.*, 1997) rice (Sasaki *et al.*, 1994; Yamamoto and Sasaki, 1997) and maize (Wang and Bowen, 1998). EST studies are not only becoming more common, but are also including species from the smaller agronomic crops. A recent example of the effectiveness of grape EST-derived SSRs is outlined later in this chapter.

Microsatellites can be sourced by many methods, which include derivation from enriched genomic libraries, by the screening of genomic libraries, screening of bacterial artificial chromosome libraries, screening of cDNA libraries, from public databases such as GenBank, from related species and from EST databases. Given that currently there is a rapidly expanding number of EST databases, the exploration of SSRs available from this resource (at little cost) is warranted, as they may provide another important and viable source of many SSRs for essential agronomic applications. In this chapter, ESTs as a source of microsatellites will be compared with the other sources of microsatellites, and the relative advantages and disadvantages of EST-derived microsatellites discussed.

Microsatellites sourced from libraries

Genomic libraries

Genomic libraries can be a source of microsatellites. To develop microsatellites from genomic libraries, the library clones are screened with repetitive probes. Positive clones are then sequenced for verification and primer design. Examples of the successful use of this approach are in wheat (Ma *et al.*, 1996), *Pinus* (Kostia *et al.*, 1995), sorghum (Brown *et al.*, 1996), grapes (Bowers *et al.*, 1996), soybean (Akkaya *et al.*, 1992) and *Beta vulgaris* (Mörchen *et al.*,1996). In comparison with the other methods for obtaining microsatellites, this method can be laborious, particularly if many microsatellites are required. In an example in *Pinus*, 6000 clones were screened to obtain eight useful microsatellites (Kostia *et al.*, 1995), and in sorghum only 0.2% of clones contained SSRs, with even fewer than this being useful (Brown *et al.*, 1996). An advantage of the approach is that the method is technically simple, and thus more available to all laboratories, unlike, for example, microsatellites from EST. Microsatellites from genomic libraries can be representative of the repeat types depending on the selection of probes, and should not result in a high proportion of redundant clones which can occur in enrichment techniques which may rely on the inclusion of a PCR step during microsatellite isolation.

Enriched libraries

Microsatellites can be isolated with better efficiency from libraries if the libraries are enriched for the microsatellites prior to screening or analysis. The enrichment for microsatellites can occur before or after cloning. Enrichment before cloning is achieved by fragmenting the DNA by sonication or restriction, followed by the addition of known sequences at the ends of the DNA. This addition of known sequences is often acheived through the addition of adaptors, and can sometimes be followed by a PCR. Microsatellites are then enriched through hybridization on either membranes (Edwards *et al.*, 1996) or magnetic beads (Kijas *et al.*, 1994; Refseth *et al.*, 1997) before cloning and sequencing.

Enrichment for microsatellites after the cloning step can be accomplished by triplex formation (Ito *et al.*, 1992; Milbourne *et al.*, 1998), or by using single strand DNA and an oligonucleotide repeat to generate the second strand which then retrieves the double-stranded product (Ostrander *et al.*, 1992; Bryan *et al.*, 1997).

Enrichment protocols can be very effective for producing large numbers of microsatellites; however, there can be several associated difficulties with this approach. There can be problems with bias representation of repeat types, in some cases there can be high copy numbers of the same microsatellite clone, and sometimes totally ineffectual or low enrichment levels. Some of these problems can make the time costs of an enrichment strategy prohibitive. The problems of bias representation and significant redundancy do not apply to microsatellites from existing EST databases.

BAC/YAC libraries

Microsatellites derived from either BAC (bacterial artificial chromosome) or YAC (yeast artificial chromosome) libraries are primarily a method of targeting the isolation of microsatellites to regions of genomes that are deficient in SSR markers. Microsatellite targeting is important for good genome coverage and optimal utility of genetic maps. Large insert libraries such as BACs and YACs have not been used very often in plants for microsatellite isolation, as there are only a few plant species for which large insert libraries are available, and for which there are maps containing large numbers of other markers. BACs have been used successfully for this purpose in soybean (Cregan *et al.*, 1999), and there is an example of the use of YACs in fungi (Chen *et al.*, 1995). The relative disadvantage of YACs over BACs is the potential for contamination with eukaryotic DNA that might contain microsatellite sequences (Cregan *et al.*, 1999).

In the study by Cregan *et al.* (1999), BACs that covered areas of the soybean genome deficient in SSR markers, were identified by the use of restriction fragment length polymorphism (RFLPs) within the deficient region. The BAC clones which were positive for these RFLPs were then subcloned and screened as for other small insert libraries (e.g. genomic or enriched SSR libraries). As BAC clones are large (> 100 kb), a number of microsatellites can often be isolated from a target region using this approach.

cDNA libraries

Microsatellites derived from cDNA libraries are equivalent to those derived from EST databases as an EST is a sequenced cDNA. Thus, microsatellite lengths, the mixtures of repeat types and the polymorphic nature of the microsatellites from cDNA and ESTs are identical. Microsatellites from cDNA libraries may be either filtered out from sequences in a database (i.e. an EST), or isolated by the physical screening of library clones by hybridization for inserts containing microsatellites. The details outlined in this section will refer to the process by which microsatellites are obtained by screening cDNA clones before sequencing them. Those cDNAs which are databased and searched electronically for the presence of microsatellites will be discussed under the section specifically on EST database-derived microsatellites. The advantages and drawbacks to the use of cDNA as a source of SSRs will also be discussed in the same section.

Microsatellites obtained by screening cDNA library clones with oligo probes is an approach which has often been utilized in human and animal research (Davis and Maddox, 1997; Ruyter-Spira *et al.*, 1998) and used to a lesser degree in plants. Examples of its use in plants are in *Arabidopsis* (Depeiges *et al.*, 1995), potato (Milbourne *et al.*, 1998) and rice (Panaud *et al.*, 1995).

Microsatellites sourced from databases

GenBank and other public sequence databases

Some of the first studies on microsatellites relied on the availability of microsatellites in sequences from public databases such as EMBL and GenBank. More recent examples of the use of microsatellites sourced from databases have been in potato (Milbourne *et al.*, 1998), sorghum (Brown *et al.*, 1996), barley (Becker and Heun, 1995), tomato (Smulders *et al.*, 1997), soybean (Akkaya *et al.*, 1992) and many other species; reviewed in Gupta *et al.* (1996). Extraction of microsatellites from these databases covers all available sequences and may often include cDNA or EST type data. The characteristics of microsatellites solely derived from cDNA or ESTs are discussed in more detail below, so this section will address the isolation and application of microsatellites from mixed sequences found in public databases.

Microsatellites from databases such as EMBL and GenBank, like EST database-derived SSRs, are easily identified through electronic sorting, and only require primer design for flanking sequence before application. This makes database obtained microsatellites a low labour, low cost approach, although highly reliant on previous research. The criteria which can be set for the electronic sorting can differ, but as an example Becker and Heun (1995) searched for all possible mononucleotide repeats with $n \geq 10$ (poly(A) tails of cDNA were ignored), for all possible dinucleotide repeats with $n < 5$, for all possible trinucleotide repeats with $n \geq 4$ and for two tetranucleotide repeats with $n \geq 3$. The significance of the length of the microsatellite searched for is that longer repeats are correlated with increased polymorphism.

The primary disadvantage of the derivation of microsatellites from public databases is that often only a small number of microsatellites are available for any one species. Given that the number of microsatellites that will be found in the database will be proportional to the length of sequence in the database for the species of interest, this approach is really only a viable source for significant numbers of microsatellites, in species for which there is a large number of database entries, such as for *Arabidopsis*, rice and maize. While it is possible to use database sequence from a related species, it does introduce some of the additional difficulties of microsatellites developed from related species (see below).

A search of the public databases for microsatellites in the *Solanaceae* (Smulders *et al.*, 1997) found that 42% of the SSRs were upstream or downstream from a gene, 26% were in introns, 22% in cDNA and only 10% in coding DNA. Similar results were found for database microsatellites from potato (Milbourne *et al.*, 1998). Smulders *et al.* (1997) also found that the occurrence of repeat types varied according to microsatellite location. Upstream or downstream from the genes and in intronic DNA, 61% of the repeats were dinucleotide. In cDNAs only 37% of the repeats were dinucleotide and in exons only 13%. For trinucleotides, Smulders *et al.* (1997) generally found the opposite trend although it did depend on the particular motif type.

EST databases

Microsatellites from ESTs is an approach that has been used in humans (Haddad *et al.*, 1997) that is becoming available in plants as EST databases become more common. To date, this specific approach has been published for rice (Miyao *et al.*, 1996; Cho *et al.*, 2000) and grapes (Scott *et al.*, 2000). EST microsatellites are functionally identical to cDNA-derived microsatellites, the only defining difference being the electronic searching of sequences from an EST database, compared with hybridization of cDNA clones followed by sequencing for some cDNA-derived SSRs. EST-derived microsatellites have some intrinsic advantages in that they are quick to elucidate (by electronic sorting), abundant, un-biased in repeat type, present in gene-rich areas and highly transferable (Cho *et al.*, 2000; Scott *et al.*, 2000).

The disadvantages of EST-derived microsatellites in comparison with some of the other approaches discussed is that they rely on the prior existence of sequence data, and may be slightly less polymorphic than random microsatellites. EST-derived microsatellites will be slightly less polymorphic than genomic library-derived microsatellites, as there is pressure for sequence conservation in gene regions reducing polymorphism. However, EST microsatellites still have useful levels of polymorphism for mapping and fingerprinting applications (Miyao *et al.*, 1996; Cho *et al.*, 2000; Scott *et al.*, 2000), and may also have some speciality applications in ancient evolutionary studies (Meyer *et al.*, 1995; Cho *et al.*, 2000). Similar to other database-sourced microsatellites, the criteria for the electronic sorting of the SSRs can differ. In the example of the grape ESTs (Scott *et al.*, 2000), all possible mononucleotides of $n \geq 10$, dinucleotide repeats with $n \geq 7$ and all possible trinucleotide repeats with $n \geq 5$ were extracted.

Results from the study of Scott et al. (2000) illustrated that EST-derived microsatellites are highly transferable and polymorphic.

Microsatellites sourced from related species

Testing microsatellites in a species related to the one from which the microsatellites were characterized is a desirable approach, as it does not require the high levels of labour and costs associated with direct development. Examples of microsatellites successfully transferred to related species include: those used in sorghum but developed from *Zea mays* and *Paspalum vaginatum* (Brown et al., 1996); those used in crop brassicas with primers designed from *Arabidopsis* (Westman and Kresovich, 1998); those used in several *Citrus* species although developed from only one *Citrus* species (Kijas et al., 1995); and primers designed in *Swietenia humilis* that have been useful in many *Meliaceae* species (White and Powell, 1997).

The primary limitations in the use of microsatellites from related species is that only a portion of microsatellites from another species will be informative, and the number and species for which these microsatellites are already developed may be limited. Often the use of heterologous primers also requires more optimization than homologous microsatellite primers. Heterologous primers are also more likely to produce products of unexpected sizes, or produce products of expected sizes that are not SSRs. Products resulting from heterologous primers should be verified through hybridization, sequencing or controlled testing (Westman and Kresovich, 1998), before use in significant studies. The transfer efficiency of microsatellites may be determined by the genetic relatedness of the species, and may be affected by the position of the microsatellite. For example, microsatellites from ESTs are more likely to be transferable across to less related species, than those from non-gene regions, as ESTs and more specifically the coding regions will be more conserved.

The use of microsatellites from related species has particular application to the assessment of genetic diversity among wild relatives of agronomic species. Evaluation of wild relatives is important due to the reduction in species diversity when crops are grown in monoculture.

The utility of EST-derived microsatellites: an example from grapes

A study on grape EST-derived microsatellites (Scott et al., 2000) assessed this type of SSR for levels of polymorphism, transferability across related species and genera, and compared EST-derived microsatellites with enriched library-derived microsatellites.

In the example of EST-derived SSRs (Scott et al., 2000), mononucleotides of ten or more repeats, dinucleotides of seven or more repeats, and trinucleotides of five or more repeats were extracted from an EST database from Chardonnay grapes. These microsatellites constituted 2.5% of the total population of ESTs in the database. The 2.5% included 46 non-redundant dinucleotide and 78 non-redundant trinucleotide repeats, from the initial EST database of 5000 sequences. This clearly illustrated the high abundance of microsatellites from the EST database.

The characteristics of the EST-derived microsatellites showed 60.6% of the microsatellites to be in 5'-untranslated regions (5'UTRs), 15.2% in 3'-untranslated regions (3'UTRs) and 24.2% in the coding region. The fact that most microsatellites occurred in the 5'UTRs is consistent with the results in rice (Miyao *et al.*, 1996). This abundance of 5'UTR SSRs may be partly due to the usual 5' sequencing approach to ESTs. The proportions of EST microsatellite repeat types also differed between the regions. No dinucleotides occurred in the coding region, 80% of the dinucleotides occurred in the 5'UTRs and 20% in the 3'UTRs. Of the occurrences of trinucleotides, 44.5% were in 5'UTRs, 11.1% in the 3'UTRs and 44.4% in the coding region.

Sixteen primer pairs were designed and tested from the 124 microsatellites available (Fig. 15.1) (Scott *et al.*, 2000). The 16 primers represented dinucleotide and trinucleotide repeats located in 5'untranslated regions, 3' untranslated regions and trinucleotides from coding sequence (CDS) regions (Table 15.1). The assignment of SSR location to a gene region was based on high homology (> 60% identity) of the ESTs to genes in other species. The genotypes used to screen primers in the study were: two *Vitis vinifera* cultivars (Cabernet Sauvignon and Riesling), three non-*vinifera* cultivars (Ramsey, Riparia and 1616C) and two non-*Vitis* genera from within the *Vitaceae* (*Cissus cardiophylla* and *Cayratia japonica*).

In grapes, the EST SSR primer pairs were shown in most cases to produce products of the expected size, and were polymorphic and transferable. Half of the SSRs examined were useful for differentiating two *V. vinifera* cultivars, while

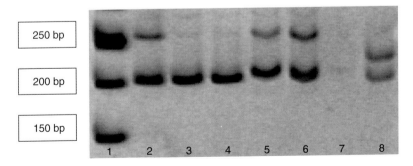

Figure 15.1. Example of expressed sequence tag simple sequence repeat amplification products for primer scu15vv separated on an 8% polyacrylamide gel. Lane 1, 50 bp ladder; 2, Ramsey; 3, 1616C; 4, Riparia; 5, Riesling; 6, Cabernet Sauvignon; 7, *Cissus cardiophylla*; 8, *Cayratia japonica*.

Table 15.1. Simple sequence repeats (SSRs) developed from the *Vitis vinifera* Chardonnay expressed sequence tag (EST) database. For each SSR, the repeat unit, length and location are shown. Range of allele length (base pairs) of the EST SSRs. Polymorphisms (Poly) between cultivars (C), species (Spp) and genera (G) are shown, in addition to the transferability (Trans) of the markers. No amplification products is shown by (—). See Scott *et al.* (2000) for primer sequences and scoring.

Primer	Repeat	Position	Poly	Trans	Range of allele lengths
scu01vv	$(CT)_9 X_{60}(CT)_{10}AT(CT)_5$	5′UTR	—	—	—
scu02vv	$(GA)_9$	5′UTR	—	—	—
scu03vv	$(GA)_8$	5′UTR	—	—	—
scu04vv	$(CT)_8$	5′UTR	C	Spp	171–175
scu05vv	$(AT)_{13}$	3′UTR	C, Spp	Spp	160–168
scu06vv	$(AT)_8$	3′UTR	C, Spp, G	Spp, G	155–191
scu07vv	$(ACC)_5$	CDS	Spp	Spp	203–235
scu08vv	$(GGT)_5$	5′UTR	Spp	Spp	159–180
scu09vv	$(GGT)_5$	3′UTR	—	—	—
scu10vv	$(CAA)_6$	5′UTR	C, Spp, G	Spp, G	205–307
scu11vv	$(CTT)_8$	5′UTR	Spp	Spp	242–248
scu12vv	$(TCT)_{10}$	5′UTR	—	—	—
scu13vv	$(CTT)_6$	CDS	—	—	—
scu14vv	$(GAA)_6$	3′UTR	C, Spp	Spp	167–188
scu15vv	$(GAA)_6$	CDS	C, Spp, G	Spp, G	201–252
scu16vv	$(GAA)_5$	CDS	G	Spp, G	170–179

eight out of the ten SSRs were able to discriminate rootstock species. The level of transferability of the EST SSRs was high, with four of the ten primer pairs amplifying alleles from a related genus.

The polymorphism detected by grape microsatellites from differing EST regions (coding sequence, 5′ untranslated region and 3′ untranslated region) varied at the different taxonomic levels (Scott *et al.*, 2000). In the 5′ untranslated region, microsatellite polymorphism was present at all taxonomic levels; however, the majority of 5′ untranslated region microsatellites varied between species. The 5′ untranslated region SSRs transferred to other species and genera from the *Vitaceae*. All 3′ untranslated region microsatellites were variable between the cultivars and between the species, with some variation also evident across genera. The 3′ untranslated region SSRs were as transferable as the 5′ untranslated region SSRs. The SSRs within the coding sequence were all trinucleotides, and variation was present at all taxonomic levels. However, the coding sequence SSRs showed most variation at the higher taxonomic levels. This is expected, as presumably there is less tolerance for mutation in the functional coding sequence region than in untranslated regions. For this same reason the coding sequence SSRs are also highly transferable.

To summarize the Scott *et al.* (2000) study, from the small number of primers tested in grape it appeared that 3′ untranslated region SSRs were more variable at the lower taxonomic levels (more related), the 5′ untranslated region SSRs have intermediate variability, and the coding sequence SSRs had the potential to be the most useful for differentiation at the higher taxonomic level. The availability of a selection of SSR sources (untranslated regions versus coding sequence) may be a valuable way to target microsatellite choice to studies of appropriate accessions.

The comparison of a standard grape-enriched microsatellite library (in this case the *Vitis* microsatellite consortium (VMC) SSRs) with the EST SSR libraries (Scott *et al.*, 2000) showed an obviously higher frequency of SSRs (up to 47%) in the VMC compared with 2.5% in the EST-derived SSRs. The SSRs generated from ESTs were, however, more representative of all repeat motifs than the enriched SSRs (Table 15.2). This is a direct result of the biased protocols for SSR enrichment. The proportion of $(GA)_n$ motifs in the EST-derived SSRs was much higher than the proportion of $(CA)_n$ repeat motifs. This characteristic is found in many plant species (Gupta *et al.*, 1996). Powell *et al.* (1996) reported the $(AT)_n$ repeat to be the predominant motif in plants, this is apparently not the case in *Vitis*. As in other studies (Gupta *et. al.*, 1996), $(AAG)_n$ was the most common trinucleotide repeat in the EST SSRs; however, $(AAT)_n$, which is also often abundant in other plant species (Gupta *et al.*, 1996), is apparently not common in *Vitis*. The average repeat length was longer in the VMC-enriched SSRs, in comparison with EST-derived SSRs.

Grape EST-derived microsatellites are, therefore, a viable source of polymorphic, highly transferable SSRs for mapping, identity and research applications in *Vitis*. Although only a small number of SSRs were tested in this study, their utility was illustrated, and the continuation of the work will provide approximately 1000 more usable *Vitis* SSRs from the grape EST project (Ablett *et al.*, 1998).

Table 15.2. Comparison of the characteristics of expressed sequence tag (EST)-derived simple sequence repeats (SSRs) with those obtained through genomic enrichment (VMC repeat data based on the clones from which primers were designed) (Scott *et al.* 2000).

	EST SSRs	Enriched genomic SSRs
Frequency in library (%)	2.5	47
Repeat types within each length class (%)	4.3 CA/GT	0.9 CA/GT
	19.6 AT/TA	0 AT/TA
	67.4 CT/GA	97.3 CT/GA
	8.7 compound dinucleotide	1.8 compound dinucleotide
	2.5 TTA/AAT	0 TTA/AAT
	2.6 CAA/GTT	0 CAA/GTT
	3.8 CGT/GCA	0 CGT/GCA
	5.1 CAT/GTA	0 CAT/GTA
	9 CAG/GTC	40 CAG/GTC
	10.3 CAC/GTG	6.7 CAC/GTG
	11.5 CCT/GGA	0 CCT/GGA
	12.8 CGG/GCC	26.7 CGG/GCC
	41 CTT/GAA	13.3 CTT/GAA
	1.4 compound trinucleotide	13.3 compound trinucleotide
Average repeat length	7.9	20.2

Table 15.3. Comparison of the advantages and disadvantages of the range of protocols available for microsatellite isolation.

Microsatellite isolation source	Time/labour costs for development	Polymorphism	Transferability	Ability to derive large numbers	Bias in repeat types	Reliance on prior research
Genomic library	High	High	Moderate	Moderate	Low	None
Enriched library	Moderate	High	Moderate	Easy	High (protocol dependent)	None
BAC/YAC library	High	High	Moderate	Moderate	Low	Moderate
cDNA library	High (unless sequenced)	Moderate	Moderate/high	Moderate	Moderate[a]	Moderate
Public databases	Low	Moderate/high[b]	Moderate/high[b]	Moderate[c]	Moderate[b]	High
EST databases	Low	Moderate	Moderate/high	Moderate[c]	Moderate[a]	High
Related species	Low	Moderate	Dependent on original protocol	Difficult	Dependent on original protocol	High

[a]Bias due to repeat types which occur in gene regions not due to bias resulting from the protocol.
[b]Dependent on the sequence availability and types within the database.
[c]Dependent on total amount of sequence data available in the database.
BAC, bacterial artificial chromosome; YAC, yeast artificial chromosome; EST, expressed sequence tag.

Conclusions

Expressed sequence tag libraries are a useful approach for the fast development of large numbers of useful microsatellites in those species for which EST databases are available. There are also many other valid approaches to SSR derivation and these are outlined in Table 15.3. The ultimate selection of approach depends on the time and research resources available to the user, and on the research outcome requirements.

References

Ablett, E.M., Lee, L.S. and Henry, R.J. (1998) Analysis of grape ESTs. VIIème Symposium International sur la Genetique et l'Amelioration de la Vigne. Montpellier, France.

Akkaya, M.S., Bhagwat, A.A. and Cregan, P.B. (1992) Length polymorphisms of simple sequence repeat DNA in soybean. *Genetics* 132, 131–139.

Becker, J. and Heun, M. (1995) Barley microsatellites: allele variation and mapping. *Plant Molecular Biology* 27, 835–845.

Bowers, J.E., Dangl, G.S., Vignani, R. and Meredith, C.P. (1996) Isolation and characterization of new polymorphic simple sequence repeat loci in grape (*Vitis vinifera* L.). *Genome* 39, 628–633.

Broun, P. and Tanksley, S.D. (1996) Characterization and genetic mapping of simple repeat sequences in the tomato genome. *Molecular and General Genetics* 250, 39–49.

Brown, S.M., Hopkins, M.S., Mitchell, S.E., Senior, M.L., Wang, T.Y., Duncan, R.R., Gonzalez-Candelas, F. and Kresovich, S. (1996) Multiple methods for the identification of polymorphic simple sequence repeats (SSRs) in sorghum [*Sorghum bicolor* 9L. Moench]. *Theoretical and Applied Genetics* 93, 190–198.

Bryan, G.J., Collins, A.J., Stephenson, P., Orry, A., Smith, J.B. and Gale, M.D. (1997) Isolation and characterisation of microsatellites from hexaploid bread wheat. *Theoretical and Applied Genetics* 94, 557–563.

Chen, H., Pulido, J.C. and Duyk, G.M. (1995) MATS: a rapid and efficient method for the development of microsatellite markers from YACs. *Genomics* 25, 1–8.

Cho, Y.G., Ishii, T., Temnykh, S., Chen, X., Lipovich, L., McCouch, S.R., Park, W.D., Ayres, N. and Cartinhour, S. (2000) Diversity of microsatellites derived from genomic libraries and GenBank sequences in rice (*Oryza sativa* L.). *Theoretical and Applied Genetics* 100, 713–722.

Cregan, P.B., Mudge, J., Fickus, E.W., Marek, L.F., Danesh, D., Denny, R., Shoemaker, R.C., Matthews, B.F., Jarvik, T. and Young, N.D. (1999) Targeted isolation of simple sequence repeat markers through the use of bacterial artificial chromosomes. *Theoretical and Applied Genetics* 98, 919–928.

Davies, K.P. and Maddox, J.F. (1997) Genetic linkage mapping of three anonymous ovine EST microsatellites: KD101, KD103 and KD721. *Animal Genetics* 28, 453–461.

Delseny, M., Cooke, R., Raynal, M. and Grellet, F. (1997) The *Arabidopsis thaliana* cDNA sequencing projects. *Federation of European Biochemical Societies* 405, 129–132.

Depeiges, A., Goubely, C., Lenoir, A., Cocherel, S., Picard, G., Raynal, M., Grellet, F. and Delseny, M. (1995) Identifcation of the most represented repeated motifs in *Arabidopsis thaliana* microsatellite loci. *Theoretical and Applied Genetics* 91, 160–168.

Edwards, K.J., Barker, J.H.A., Day, A., Jones, C. and Karp, A. (1996) Microsatellite libraries enriched for several microsatellite sequences in plants. *Biotechniques* 20, 758–760.

Gupta, P.K., Balyan, H.S., Sharma, P.C. and Ramesh, B. (1996) Microsatellites in plants: a new class of molecular markers. *Current Science* 70, 45–54.

Haddad, L.A., Fuzikawa, A.K. and Pena, S.D.J. (1997) Simultaneous detection of size and sequence polymorphisms in the transcribed trinucleotide repeat D2S196E (EST00493). *Human Genetics* 99, 796–800.

Ito, T., Smith, C.L. and Cantor, C.R. (1992) Sequence-specific DNA purification by triplex affinity capture. *Proceedings of the National Academy of Sciences USA* 89, 495–498.

Kijas, J.M.H., Fowler, J.C.S., Garbett, C.A. and Thomas, M.R. (1994) Enrichment of microsatellites from the *Citrus* genome using biotinylated oligonucleotide sequences bound to streptavidin-coated magnetic particles. *Biotechniques* 16, 657–662.

Kijas, J.M.H., Fowler, J.C.S. and Thomas, M.R. (1995) An evaluation of sequence tagged microsatellite site markers for genetic analysis within *Citrus* and related species. *Genome* 38, 349–355.

Kostia, S., Varvio, S.-L., Vakkari, P. and Pulkkinen, P. (1995) Microsatellite sequences in a conifer, *Pinus sylvestris*. *Genome* 38, 1244–1248.

Lagercrantz, U., Ellegren, H. and Andersson, L. (1993) The abundance of various polymorphic microsatellite motifs differs between plants and vertebrates. *Nucleic Acids Reseach* 21, 1111–1115.

Ma, Z.Q., Röder, M. and Sorrells, M.E. (1996) Frequencies and sequence characteristics of di-, tri-, and tetra-nucleotide microsatellites in wheat. *Genome* 39, 123–130.

Meyer, E., Wiegand, P., Rand, S.P., Kuhlmann, D., Brack, M. and Brinkmann, B. (1995) Microsatellite polymorphisms reveal phylogenetic relationships in primates. *Journal of Molecular Evolution* 41, 10–14.

Milbourne, D., Meyer, R.C., Collins, A.J., Ramsay, L.D., Gebhardt, C. and Waugh, R. (1998) Isolation, characterisation and mapping of simple sequences repeat loci in potato. *Molecular and General Genetics* 259, 233–245.

Miyao, A., Zhong, H.S., Monna, L., Yano, M., Yamamoto, K., Havukkala, I., Minobe, Y. and Sasaki, T. (1996) Characterization and genetic mapping of simple sequence repeats in the rice genome. *DNA Research* 3, 233–238.

Mörchen, M., Cuguen, J., Michaelis, G., Hänni, C. and Saumitou-Laprade, P. (1996) Abundance and length polymorphism of microsatellite repeats in *Beta vulgaris* L. *Theoretical and Applied Genetics* 92, 326–333.

Ostrander, E.A., Jong, P.M., Rine, J. and Duyk, G. (1992) Construction of small-insert genomic DNA libraries highly enriched for microsatellite repeat sequences. *Proceedings of the National Academy of Sciences USA* 89, 3419–3423.

Panaud, O., Xiuli, C. and McCouch, S.R. (1995) Frequency of microsatellite sequence in rice (*Oryza sativa* L.). *Genome* 38, 1170–1176.

Powell, W., Machray, G.C. and Provan, J. (1996) Polymorphism revealed by simple sequence repeats. *Trends in Plant Science* 7, 215–222.

Refseth, U.H., Fangan, B.M. and Jakobsen, K.S. (1997) Hybridization capture of microsatellites directly from genomic DNA. *Electrophoresis* 18, 1519–1523.

Ruyter-Spira, C.P., de Koning, D.J., van der Poel, J.J., Crooijmans, R.P.M.A., Dijkhof, R.J.M. and Groenen, M.A.M. (1998) Developing microsatellite markers from cDNA, a tool for adding expressed sequence tags to the genetic linkage map of chicken. *Animal Genetics* 29, 85–90.

Sasaki, T., Song, J.Y., Kogaban, Y., Matsui, E., Fang, F., Higo, H., Nagasaki, H., Hori, M., Miya, M., Murayamakayano, E., Takiguchi, T., Takasuga, A., Niki, T., Ishimaru, K.,

Ikeda, H., Yamamoto, Y., Mukai, Y., Ohta, I., Miyadera, N., Havukkala, I. and Minobe, Y. (1994) Toward cataloguing all rice genes – large-scale sequencing of randomly chosen rice cDNAs from a callus cDNA library. *Plant Journal* 6, 615–624.

Scott, K.D., Eggler, P., Seaton, G., Rossetto, M., Ablett, E.M., Lee, S.L. and Henry, R.J. (2000) Analysis of SSRs derived from grape ESTs. *Theoretical and Applied Genetics* 100, 723–726.

Smulders, M.J.M., Bredemeijer, G., Rus-Kortekaas, W., Arens, P. and Vosman, B. (1997) Use of short microsatellites from database sequences to generate polymorphisms among *Lycopersicon esculentum* cultivars and accessions of other *Lycopersicon* species. *Theoretical and Applied Genetics* 97, 264–272.

Wang, X. and Bowen, B. (1998) A progress report on corn genome projects at pioneer hi-bred. In: *Plant and Animal Genome VI Conference*, San Diego, California.

Wang, Z., Weber, J.L., Zhang, G. and Tanksley, S.D. (1994) Survey of plant short tandem DNA repeats. *Theoretical and Applied Genetics* 88, 1–6.

Westman, A.L. and Kresovich, S. (1998) The potential for cross-taxa simple-sequence repeat (SSR) amplification between *Arabidopsis thaliana* L. and crop brassicas. *Theoretical and Applied Genetics* 96, 272–281.

White, G. and Powell, W. (1997) Cross-species amplification of SSR loci in the Meliaceae family. *Molecular Ecology* 6, 1195–1197.

Yamamoto, K. and Sasaki, T. (1997) Large-scale EST sequencing in rice. *Plant Molecular Biology* 35, 135–144.

Chapter 16
Plant DNA Extraction

R.J. HENRY

Centre for Plant Conservation Genetics, Southern Cross University, Lismore, Australia

Introduction

The extraction of DNA from plants is the starting point for genotype analysis. The approach to preparation of DNA from plants is determined by the species, the type of tissue or sample available and the analysis required on the DNA. For example, extraction of DNA from the leaf of a cereal for analysis using a robust test may require a very different technique from that required for isolation of DNA from the bark of a tree for amplified fragment length polymorphism (AFLP) analysis.

This chapter considers the range of techniques available for plant DNA extraction in relation to the wide variety of plant genotyping applications. The overall process is defined in Fig. 16.1. Plant samples may be collected and stored, DNA may be extracted immediately or analysed directly. Stored plant samples may be extracted and the DNA either stored prior to analysis or subjected to immediate analysis.

Plant sample collection and storage

The process of plant DNA isolation may be considered to start when the plant is sampled in the field or laboratory. Appropriate harvesting and storage are essential. Labelling is also an important issue especially when large numbers of samples are being analysed.

Sampling of small pieces of plant leaves for DNA has been achieved using plastic pipette tips to punch a small disc from the leaf (Berthomieu and Meyer, 1991); tissue may then be squashed on to a nylon membrane (Langridge *et al.*,

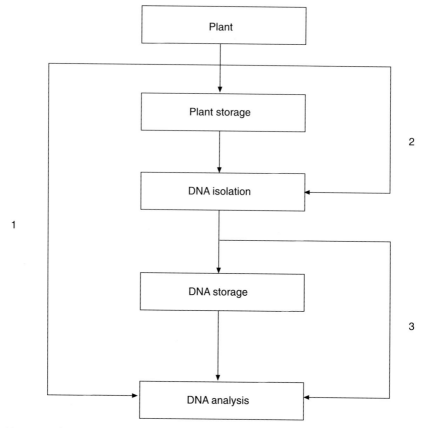

Fig. 16.1. Plant DNA analysis. **1** Direct DNA analysis avoids DNA isolation; **2** Immediate DNA isolation avoids tissue storage; **3** Immediate DNA analysis avoids DNA storage.

1991). This approach allows large numbers of samples to be processed and stored. Samples are often frozen for storage. The sample must remain frozen until DNA extraction as thawing of frozen tissue often results in degradation of the DNA. Alternatively, the DNA may be extracted from fresh material or tissue that has been dried for storage.

Analysis of fresh tissue

Fresh samples may be extracted immediately or stored for hours or days at appropriate low temperatures before extraction. Refrigeration (4°C if possible) usually allows extraction to be delayed for several days. Storage conditions that do not encourage the growth of fungi in these tissues are desirable. Alcohol may be used to help preserve some tissues for analysis (Flournoy *et al.*, 1996). Analysis of fresh tissue is probably the preferred option for DNA extraction in most cases.

Analysis of dry tissue

Longer-term storage of dry tissue is possible for many types of tissue (Savolainen et al., 1995; Fu et al., 1998). The simplest approach for the sampling of most leaves is to place the fresh leaf in a paper bag and store in an air-conditioned environment. Any office filing cabinet can allow storage of large numbers of samples (Thomson and Henry, 1993). Forced drying can result in tissue disruption and degradation of DNA. Conditions that produce the best quality herbarium samples also result in good DNA preservation. Dry seeds may also be used in many studies (McDonald et al., 1994). The analysis of dry samples may be useful when samples for a single study are to be collected over an extended period. Drying allows extraction when the sampling is complete.

Analysis of ancient DNA

The analysis of ancient DNA may be valuable in studies of evolution (Soltis and Soltis, 1993; Cano, 1996). Special procedures may be required to recover the degraded DNA in these samples and to avoid contamination (Yang, 1997; Yang et al., 1997; Schlumbaum and Jacomet, 1998).

Analysis of bark

Leaf samples may be very difficult to collect from large trees making bark an attractive source of DNA. DNA extraction from bark may be much more difficult than extraction from leaves in some species.

Analysis of plant pathogens

Diagnosis of plant disease often requires the preparation of plant samples for the analysis of DNA from pathogens. Methods have been devised for analysis of fungi (Bonito et al., 1995; Niepold and Schober-Butin, 1997), bacteria (Mahuku and Goodwin, 1997; Llop et al., 1999; Toth et al., 1999), viruses and phytoplasmas (Gibb and Padovan, 1994; Green and Thompson, 1999).

Analysis of pollen

Analysis of pollen DNA may be of great value in population genetics and gene mapping. Specific methods have been developed for working with single pollen grains (Petersen et al., 1996; Krabel et al., 1998).

Analysis of foods and other plant products

Genotyping of the plant may be of interest in foods, especially in establishing the accuracy of labelling or in food processing to match genotypes to specific products. Techniques for DNA extraction have been developed for many products. Yeast but not plant DNA has been recovered from beer (Hotzel et al., 1999). Identification of grain products during processing may be based upon DNA analysis (Henry et al., 1997).

Extraction of DNA from plant material

The steps that may be included in plant DNA preparation are defined in Fig. 16.2. Components of the process may include isolation of specific tissue, grinding (or other mechanical disruption), extraction into solution, solvent purification and precipitation.

Components of DNA extraction solutions

- *Buffer.* Buffers are used to control the pH of the extraction solution.
- *Salts.* Salts may influence the solubility of DNA and other molecules in the extraction.
- *Chelating agents.* Chelating agents such as EDTA bind metal ions in the extraction solution.
- *Detergents.* Detergents help to disrupt tissues. Many different detergents have been employed in plant DNA extraction protocols.
- *Phenolic binding agents.* PVP is often included to protect against phenolics (Kim et al., 1997). Citric acid has been added to extracts to prevent the formation of polyphenolics (Singh et al., 1998).
- *Other.* Activated charcoal has been used to bind contaminants in the isolation of DNA from coffee, rubber, cassava and banana (Vroh-Bi et al., 1996).
- *Enzymes.* The most common enzyme use in DNA isolation procedures is ribonuclease to remove contaminating RNA.
- *Solvents.* Water is the solvent used in almost all protocols.

Tissue disruption

- *Mechanical.* Physical disruption of tissue allows DNA to be released into the extraction solution. Excess mechanical disruption risks shearing of the DNA. Ideally only the minimum mechanical action necessary to release the DNA from the tissue is used. Tough plant tissues may be difficult targets for the isolation of high molecular weight DNA.

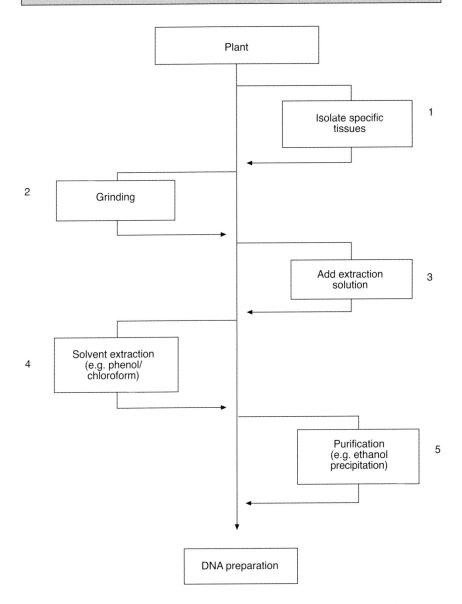

Fig. 16.2. DNA Isolation. **1** Separates tissues high in contaminants (selects tissue rich in DNA); **2** Breaks up tissue to release DNA; **3** Provides medium for DNA isolation; **4** Removes proteins and other contaminants; **5** Purifies DNA.

- *Heating.* Heating, especially at high temperatures (e.g. boiling), contributes to tissue disruption.
- *Microwave.* Microwave heat is an option for rapid extractions (Saini et al., 1999).

High throughput extraction methods

Extraction of large numbers of samples is possible using automation. Processing in 96-well microtitre plates may increase throughput (Dilworth and Frey, 2000). These procedures are of value in plant breeding applications (Gu *et al.*, 1995). Further automation of plant DNA extraction methods can be expected in the future as the extraction step is rate limiting in many DNA-based analyses in plants.

Plant DNA storage

Pure DNA free from the risk of biological contamination can be stored for long periods of time at room temperature. Low temperature storage is recommended for long-term storage, especially of valuable samples.

Purification of DNA

A variety of methods are available for DNA purification. Generally it is better to devise a method to isolate DNA of the required purity directly from the plant sample. Protocols resulting in the isolation of DNA with specific undesirable impurities should be avoided. In some cases it may be necessary to purify DNA in an extract for specific applications.

Requirements of different genotyping protocols

Analysis of microsatellites (simple sequence repeats) requires relatively simple DNA extraction protocols (Virk *et al.*, 1999) because the PCR involved is relatively robust. In contrast, random amplified polymorphic DNA (Boiteux *et al.*, 1999) and AFLP procedures are influenced by the quantity and quality of the DNA preparation requiring very uniform and usually more complex DNA isolation procedures. Preparation of plant DNA for sequencing is another common requirement (Nickrent, 1994). Applications involving the analysis of large numbers of samples such as studies of diversity in large germplasm collections (Virk *et al.*, 1996) require special, high throughput DNA extraction protocols.

Examples of DNA extraction methods

The literature contains very large numbers of DNA extraction protocols. Many of these are slight variations of other procedures (Graham *et al.*, 1994). Some tissues are especially difficult targets for DNA extraction (Steenkamp *et al.*, 1994) because of the presence of materials (e.g. polysaccharides, secondary metabolites) that react with DNA or co-purify with DNA during extraction. The method outlined by Graham and Henry (1997) is widely used. The following method is a good option for general application.

Standard method (Henry, 1997)

1. Grind sample to a fine powder in liquid nitrogen using a mortar and pestle.
2. Transfer the tissue to a tube of suitable size and add 1 ml of extraction buffer (2% (w/v) hexadecyltrimethylammonium bromide (CTAB), 100 mM Tris–HCl, 1.4 M NaCl, 20 mM EDTA) per gram of tissue.
3. Mix by gentle inversion and heat at 55°C for 20 min.
4. Centrifuge at 15,000 × g for 5 min.
5. Transfer the supernatant to a fresh tube and add one volume of chloroform:isoamyl alcohol (24:1).
6. Mix by gentle inversion for 2 min and centrifuge at 15,000 × g for 20 s.
7. Transfer the upper aqueous phase to a fresh tube and add 1/10 volume of 7.5 M ammonium acetate and two volumes of ice-cold ethanol.
8. Mix by gentle inversion and place in a freezer at −20°C for 60 min.
9. Centrifuge at 15,000 × g for 1 min and discard the supernatant.
10. Wash the pellet twice with 70% (v/v) ethanol, mixing each time by gentle inversion.
11. Dry the DNA in a desiccator.
12. The DNA may be resuspended in an appropriate volume of TE buffer (10 mM Tris–HCl, 1 mM EDTA, pH 8.0) or solution appropriate for the application intended.

Many applications do not require a procedure of this complexity.

Simple procedures for use in PCR Screening

The ultimate method is one that completely eliminates the DNA extraction step and allows direct analysis of DNA in plant tissues. This approach has been found to be successful in limited cases. Very small amounts of plant tissue may be introduced directly into PCR protocols. Some form of processing is usually required or advantageous. Klimyak *et al.* (1993) heated in alkali (40 μl of 0.25 M NaOH for 30 s on a boiling water bath), the solution was neutralized by adding 40 μl of 0.25 M HCl and 20 μl of 0.5 M Tris–HCl, pH 8.0, 0.25% (v/v) Nonidet P-40 and boiling for 2 min before use in PCR. In this procedure, the time of heating may require optimization for each species or type of tissue. This method has been applied to the analysis of barley tissues (Clancy *et al.*, 1996).

Most rapid methods have involved the preparation of an extract to add to the PCR because of the difficulty of sampling small amounts of tissue for direct introduction into the PCR and because of the presence of inhibitory substances in the plant tissue. Steiner *et al.* (1995) developed a protocol that involved heating the ground sample for 20 min at 90°C in 10 mM Tris–HCl, pH 8.0, 312.5 mM EDTA, 1% sodium lauryl sarkosyl and 1% polyvinylpolypyrrolidone.

Chunwongse et al. (1993) heated half seeds at 50°C for 1 h in 200 μl of 10 mM Tris–HCl, pH 8.0, 50 mM KCl, 1.5mM $MgCl_2$, 0.01% gelatin, 0.45% Nonidet P-40, 0.45% Tween and 12–48 μg of proteinase K. The temperature was increased to 100°C for 5 min before use of the supernatant in PCR.

The best method is usually the simplest method that gives a reliable result. The following protocol was developed as a general approach to DNA preparation.

Simple method (Thomson and Henry, 1995; Henry, 1997)

1. Place a small piece (0.5–1 mg works well for most tissues) of the plant tissue in a microfuge tube.
2. Add 20 μl of 100 mM Tris–HCl (pH 9.5), 1 M KCl, 10 mM EDTA.
3. Heat at 95°C for 10 min.
4. Add 1 μl directly to the PCR.

These extracts usually require dilution with water or PCR buffer before use in PCR. They can often be stored for several months at room temperature.

Removal of specific contaminants

Specific contaminants can be removed from tissues in which they cause problems. A mixture of glycoside hydrolases has been used to remove polysaccharides (Rether et al., 1993). Ion exchanges may also be used (Marechal-Drouard and Guillemaut, 1995).

Estimation of the quantity of DNA

The concentration of DNA in an extract can be measured in several ways. The A_{260} of the preparation is a good measure of concentration if the DNA is relatively pure (A_{260} = 1 for 50 pg ml^{-1} DNA). Impure preparations contain other material absorbing in the UV range. Gel electrophoresis on agarose allows estimation of both quality and quantity. Fluorometic methods for DNA estimation are also available. These are likely to be more accurate than UV-based methods, especially for impure samples.

Estimation of the quality of DNA

The purity of DNA can be estimated from the UV absorbance (A_{260}/A_{280} = 1.8 for pure DNA). Electrophoresis can be used to assess the size of DNA molecules in extracts. Digestion with restriction enzymes is often used to test for DNA purity and suitability for use in recombinant DNA protocols.

Isolation of DNA from mitochondria and chloroplasts

The isolation of DNA from organelles requires the prior isolation of organelles. Sucrose gradient centrifugation may be used to separate specific organelles. DNA may then be easily isolated from the organelle preparation. Isolation of organelles may also help to avoid specific contaminants (Scott and Playford, 1996). Simple procedures for analysis of chloroplast DNA are available (Fangan *et al.*, 1994).

Acknowledgements

Research by the author in this area has been supported by the Grains Research and Development Corporation and the Cooperative Research Centre for Molecular Plant Breeding.

References

Berthomieu, P. and Meyer, C. (1991) Direct amplification of plant genomic DNA from leaves and root pieces using PCR. *Plant Molecular Biology* 17, 555–557.

Boiteux, L., Fonseca, M. and Simon, P. (1999) Effects of plant tissue and DNA purification method on randomly amplified polymorphic DNA-based genetic fingerprinting analysis in carrot. *Journal of the American Society for Horticultural Science* 124, 32–38.

Bonito, R.D., Elliott, M. and Jardin, E.D. (1995) Detection of an arbuscular mycorrhizal fungus in roots of different plant species with the PCR. *Applied and Environmental Microbiology* 61, 2809–2810.

Cano, R., (1996) Analysing ancient DNA. *Endeavour* 20, 162–167.

Chunwongse, J., Martin, G. and Tanksley, S. (1993) Pre-germination genotypic screening using PCR amplification of half-seeds. *Theoretical and Applied Genetics* 86, 694–698.

Clancy, J., Jitkov, V., Han, F. and Ullrich, S. (1996) Barley tissue as direct template for PCR: a practical breeding tool. *Molecular Breeding* 2, 181–183.

Dilworth, E. and Frey, J. (2000) A rapid method for high throughput DNA extraction from plant material for PCR amplification. *Plant Molecular Biology Reporter* 18, 61–64.

Fangan, B., Stedje, B., Stabbetorp, O., Jensen, E. and Jakobsen, K. (1994) A general approach for PCR-amplification and sequencing of chloroplast DNA from crude vascular plant and algal tissue. *Biotechniques* 16, 484–494.

Flournoy, L., Adams, R. and Pandy, R. (1996) Interim and archival preservation of plant specimens in alcohols for DNA studies. *Biotechniques* 20, 657–660.

Fu, R.Z., Wang, J., Sun, Y.R. and Shaw, P.C. (1998) Extraction of genomic DNA suitable for PCR analysis from dried plant rhizomes/roots. *Biotechniques* 25, 797–801.

Gibb, K. and Padovan, A. (1994) A DNA extraction method that allows reliable PCR amplification of MLO DNA from 'difficult' plant host species. *Genome Research* 4, 56–58.

Graham, G. and Henry, R. (1997) Preparation of fungal genomic DNA for PCR and RAPD analysis in fingerprinting methods based upon arbitrarily primed PCR. In: Micheli, M.R. and Bova, R. (eds) *Springer Laboratory Manual*, Vol. 29–34. Springer, Heidelberg, pp. 29–34.

Graham, G.C., Mayers, P. and Henry, R.J. (1994) A simplified method for the preparation of fungal genomic DNA for PCR and RAPD analysis. *Biotechniques* 16, 48–50.

Green, M. and Thompson, D. (1999) Easy and efficient DNA extraction from woody plants for the detection of phytoplasmas by polymerase chain reaction. *Plant Disease* 83, 482–485.

Gu, W., Weeden, N., Yu, J. and Wallace, D. (1995) Large-scale, cost-effective screening of PCR products in marker-assisted selection applications. *Theoretical and Applied Genetics* 91, 465–470.

Henry, R.J. (1997) *Practical Applications of Plant Molecular Biology*. Chapman & Hall, London.

Henry, R.J., Ko, H.L. and Weining, S. (1997) Identification of cereals using DNA-based technology. *Cereal Foods World* 42, 26–29.

Hotzel, H., Muller, W. and Sachse, K. (1999) Recovery and characterization of residual DNA from beer as a prerequisite for the detection of genetically modified ingredients. *European Food Research and Technology* 209, 193–196.

Kim, C., Lee, C., Shin, J., Chung, Y. and Hyung, N. (1997) A simple and rapid method for isolation of high quality genomic DNA from fruit trees and conifers using PVP. *Nucleic Acids Research* 25, 1086–1086.

Klimyak, V., Carroll, B., Thomas, C. and Jones, J. (1993) Alkali treatment for rapid preparation of plant material for reliable PCR analysis. *Plant Journal* 3, 493–494.

Krabel, D., Vornam, B. and Herzog, S. (1998) PCR based random amplification of genomic DNA from single pollen grains of trees. *Journal of Applied Botany – Angewandte Botanik* 72, 10–13.

Langridge, U., Schwatt, M. and Langridge, P. (1991) Squashes of plant tissue as a substrate for PCR. *Nucleic Acids Research* 19, 6954.

Llop, P., Caruso, P., Cubero, J., Morente, C. and Lopez, M. (1999) A simple extraction procedure for efficient routine detection of pathogenic bacteria in plant material by polymerase chain reaction. *Journal of Microbiological Methods* 37, 23–31.

Mahuku, G. and Goodwin, P. (1997) Presence of *Xanthomonas fragariae* in symptomless strawberry crowns in Ontario detected using a nested polymerase chain reaction (PCR). *Canadian Journal of Plant Pathology* 9, 366–370.

Marechal-Drouard, L. and Guillemaut, P. (1995) A powerful but simple technique to prepare polysaccharide-free DNA quickly and without phenol extraction. *Plant Molecular Biology Reporter* 13, 26–30.

McDonald, M., Elliot, L. and Sweeney, P. (1994) DNA extraction from dry seeds for RAPD analyses in varietal identification studies. *Seed Science and Technology* 22, 171–176.

Nickrent, D., (1994) From field to film: rapid sequencing methods for field-collected plant species. *Biotechniques* 16, 470–475.

Niepold, F. and Schober-Butin, B. (1997) Application of the one-tube PCR technique in combination with a fast DNA extraction procedure for detecting *Phytophthora infestans* in infected potato tubers. *Microbiology Research* 152, 345–351.

Petersen, G., Johansen, B. and Seberg, O. (1996) PCR and sequencing from a single pollen grain. *Plant Molecular Biology* 31, 189–191.

Rether, B., Delmas, G. and Laouedj, A. (1993) Isolation of polysaccharide-free DNA from plants. *Plant Molecular Biology Reporter* 11, 333–337.

Saini, H., Shepherd, M. and Henry, R.J. (1999) Microwave extraction of total genomic DNA from barley grains for use in PCR. *Journal of the Institute of Brewing* 105, 185–190.

Savolainen, V., Cuenoud, P., Spichiger, R., Martinez, M., Crevecoeur, M. and Manen, J. (1995) The use of herbarium speciments in DNA phylogenetics: evaluation and improvement. *Plant Systematics and Evolution* 197, 1–4.

Schlumbaum, A. and Jacomet, S. (1998) Coexistence of tetraploid and hexaploid naked wheat in a neolithic lake dwelling of central Europe: evidence from morphology and ancient DNA. *Journal of Archaeological Science* 25, 1111–1118.

Scott, K. and Playford, J. (1996) DNA extraction technique for PCR in rainforest plant species. *Biotechniques* 20, 974–975.

Singh, R., Singh, M. and King, R. (1998) Use of citric acid for neutralizing polymerase chain reaction inhibition by chlorogenic acid in potato extracts. *Journal of Virological Methods* 74, 231–235.

Soltis, P. and Soltis, D. (1993) Ancient DNA: prospects and limitations. *New Zealand Journal of Botany* 31, 203–209.

Steenkamp, J., Wild, I., Lourens, A. and Helden, P.V. (1994) Improved method for DNA extraction from *Vitis vinifera*. *American Journal of Enology and Viticulture* 45, 102–106.

Steiner, J., Poklemba, C., Djellstrom, R. and Elliott, L. (1995) A rapid one-tube genomic DNA extraction process for PCR and RAPD analyses. *Nucleic Acids Research* 23, 2569–2570.

Thomson, D. and Henry, R. (1993) Use of DNA from dry leaves for PCR and RAPD analysis. *Plant Molecular Biology Reporter* 11(3), 202–206.

Thomson, D. and Henry, R. (1995) Single-step protocol for preparation of plant tissue for analysis by PCR. *Biotechniques* 19, 394–400.

Toth, I., Hyman, L. and Wood, J. (1999) A one step PCR-based method for the detection of economically important soft rot Erwinia species on micropropagated potato plants. *Journal of Applied Microbiology* 87, 158–166.

Virk, P., Ford-Lloyd, B., Jackson, M. and Newbury, H. (1996) Use of RAPD for the study of diversity within plant germplasm collections. *Heredity* 74, 170–179.

Virk, P., Pooni, H., Syed, N. and Kearsey, M. (1999) Fast and reliable genotype validation using microsatellite markers in *Arabidopsis thaliana*. *Theoretical and Applied Genetics* 98, 462–464.

Vroh-Bi, I., Harvengt, L., Chandelier, A., Mergeai, G. and Jardin, P.D. (1996) Improved RAPD amplification of recalcitrant plant DNA by the use of activated charcoal during DNA extraction. *Plant Breeding* 115, 205–206.

Yang, H. (1997) Ancient DNA from pleistocene fossils: preservation, recovery and utility of ancient genetic information for quaternary research. *Quaternary Science Reviews* 16, 1145–1161.

Yang, H., Golenberg, E. and Shoshani, J. (1997) Proboscidean DNA from museum and fossil specimens: an assessment of ancient DNA extraction and amplification techniques. *Biochemical Genetics* 35, 165–179.

Chapter 17

Collection, Reporting and Storage of Microsatellite Genotype Data

N. HARKER

Centre for Plant Conservation Genetics, Southern Cross University, Lismore, Australia

Introduction

The development and use of molecular markers in the investigation of plant genetics for varietal identification, genetic map development and marker-assisted selection is widely reported (Henry, 1994; Becker and Heun, 1995; Powell *et al.*, 1996; McCouch *et al.*, 1997; Röder *et al.*, 1998; Garland *et al.*, 1999). In particular, microsatellites or simple sequence repeats are considered useful because of their highly polymorphic nature and relative abundance in the genome (Wang *et al.*, 1994). Recent advances in scientific instrumentation, in particular the improvements in electrophoresis and analysis systems, have facilitated the development and use of molecular markers by allowing high throughput, semi-automated analysis (Ziegle *et al.*, 1992; Wenz *et al.*, 1998). The increasing range of instruments available for analysis of genotype data raises issues relating to accuracy and reliability that need to be considered. Further, as a consequence of the increased amounts of genotype data created by the availability of high throughput systems, there is an increased need for better data management methods. In many cases manual control of the data sets cannot be achieved to any satisfactory degree. This chapter addresses issues for consideration in the collection, reporting and storage of microsatellite data. Discussion focuses on issues in scoring accuracy and the effects of different instrumentation available for data generation. A simple yet functional database for assisting in the management of marker data is described.

Microsatellite separation and scoring of alleles

Genotype data are routinely generated by the employment of radioactively or fluorescently labelled primers or deoxynucleotide triphosphates in PCR. Separation of products is carried out by denaturing polyacrylamide gel electrophoresis or by capillary electrophoresis. Detection of the separated products is via exposure to autoradiograph film, scanning with phospho- or fluoro-imagers or through the laser-induced fluorescence detection system in the electrophoresis unit as is the case with the ABI systems for fluorescent products.

Increased market competitiveness in instrumentation for molecular biology has meant that there are now numerous systems available for high throughput generation and sizing of genotype data (e.g. ABI Prism 3700 DNA Analyser, ABI model 373A or 377 DNA Sequencer, Molecular Dynamics MegaBACE 1000 DNA Sequencing System and Genotyping System and the Corbett Research GS2000). Sizing of alleles is usually carried out with software that utilizes a series of external or internal size standards such as ABI GENESCAN ANALYSIS 3.1 (PE Applied Biosystems), ABI GENOTYPER (PE Applied Biosystems) or GENE PROFILER (Scanalytics, Inc.). The advantage of software programs such as GENOTYPER and GENE PROFILER is that they offer the ability to automate allele calling and tabulate data that can be exported into a MICROSOFT EXCEL-compatible spreadsheet. This can significantly increase the ease of data handling. These programs can make the data sets accessible by many end stream users without the need for purchasing the specialized software.

As with any scientific method, the question of accuracy and reproducibility is vital. The nature of microsatellites, in conjunction with the variety of systems used for analysis, raises a number of issues relevant to the accuracy and reliability in scoring and reporting of microsatellite data. For purposes of comparing results it is vitally important that data are accurately reported between research groups and in collaborative efforts between development and implementation groups.

Non-templated nucleotide addition by Taq DNA polymerase

The addition of a non-templated nucleotide, usually adenosine, to the 3′ end of DNA fragments by *Taq* DNA polymerase is a well-documented occurrence (Clark, 1988). This modification of fluorescently labelled PCR fragments can contribute to the mis-labelling of alleles scored by GENESCAN or GENOTYPER (Smith *et al.*, 1995) (Fig. 17.1). These genotyping errors increase the requirement for manual editing to ensure that the same peak is scored in each case. Inconsistent labelling of real peaks and 'plus A' peaks will be indicated by loci scored with a difference of one base pair. Since the addition of an adenosine is shown to be inconsistent and incomplete between and within microsatellites (Smith *et al.*, 1995; Fishback *et al.*, 1999) (Fig. 17.2), problems in reporting the data between collaborating groups can be created if variation occurs with respect to which peak is scored.

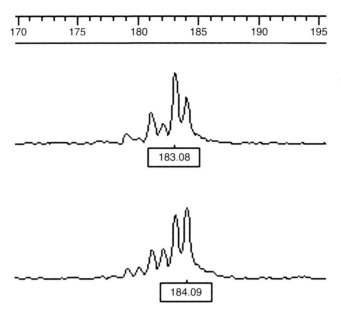

Fig. 17.1. Different levels of 'plus A' addition that can result in allele scores differing by one base, a genotyping error if manual checking is not performed.

Smith *et al.* (1995) proposed two alternative thermal cycling protocols for reducing the effects of 'plus A', favouring the addition of the non-templated adenosine or amplification of the 'true' allele. It has also been shown that a decrease in the amount of primer in association with an increase in the amount of *Taq* DNA polymerase can result in consistent addition of the 'plus A' adenosine (Fishback *et al.*, 1999). Ginot *et al.* (1996) reported the use of the proof reading enzyme *Pfu* in addition to *Taq* both during and after PCR, or T4 polymerase after PCR. It was found that *Pfu* polymerase did not exhibit an absolute ability to remove the non-templated adenosine added by the *Taq* DNA polymerase but that T4 polymerase was successful at removing the 'plus A' addition. Brownstein *et al.* (1996) demonstrated that the addition of a 'pigtail' (6 or 7 bp of particular sequence) to the 5' end of the reverse primer favoured complete 'plus A' addition.

Several approaches have been developed for dealing with the potential genotyping errors created by 'plus A' addition. The relative benefits of adopting any approach should be weighed by the cost of additional or alternative reagents (e.g. proof reading enzymes) and increased time requirements (e.g. longer PCR, post-PCR sample handling) compared with the potential cost of genotyping errors and intensive manual checking. However, if laboratories adopt alternative strategies for dealing with the 'plus A' addition of adenosine, the reporting and comparison of genotype data between groups may be affected.

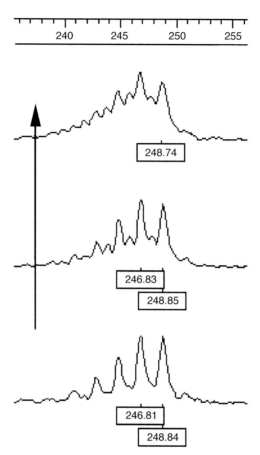

Fig. 17.2. Variable levels of 'plus A' addition within a microsatellite PCR across three different template DNA samples (the arrow indicates increasing levels of 'plus A').

Stutter in dinucleotide repeat microsatellites

PCR products amplified from microsatellite loci, especially dinucleotide repeats, can demonstrate a characteristic known as stutter, shadow banding or slippage (Weber and May, 1989). This manifests as a series of amplification products that are a multiple of the repeat number smaller than the true peak (Fig. 17.3). It is generally thought that stutter peaks are the product of slippage by the DNA polymerase during the PCR (Tautz, 1989). Murray *et al.* (1993) reported that stutter peaks are characterized by 2 bp deletions in the repeat region (for a dinucleotide CA repeat) generated during the PCR.

The presence of stutter peaks introduces issues for consistent size calling both within and between users utilizing allele calling software programs such as GENOTYPER. Although these programs are designed to recognize stutter peaks

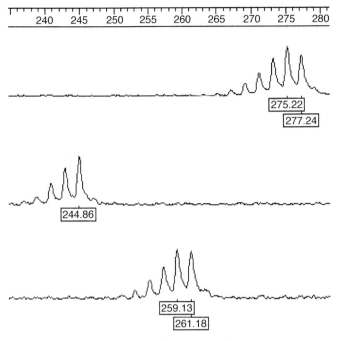

Fig. 17.3. Microsatellite dinucleotide repeat demonstrating how stutter peaks can complicate the process of allele scoring.

and therefore not call allele sizes for these bands, inconsistency with respect to which peak is called in the stutter pattern across samples (Fig. 17.3) has been observed. This can lead to the identification of a false polymorphism, if manual checking of peak calling is not performed. As recognized in the discussion of 'plus A' peaks, variable scoring of stutter peaks can affect the accuracy in reporting of allele sizes between groups. Although stutter peaks can be problematical for this reason, they are useful for distinguishing true alleles from PCR artefacts (Schwengel et al., 1994).

Allele scoring discrepancy in genotyping instruments

The instruments most commonly employed for genotype analysis utilize either slab gel polyacrylamide electrophoresis or capillary electrophoresis. Both electrophoresis systems are designed to provide semi-automated high throughput electrophoresis of PCR products. Fluorescence-based systems utilize the ability to detect multiple fluorescent dyes simultaneously to increase the number of loci that that can be detected at any one stage (Ziegle et al., 1992). Multiplexing is possible both by dye colour and by size (Schwengel et al., 1994). By functioning in stringent denaturing conditions, high temperatures and in the presence of denaturants (Rosenblum et al., 1997), both systems offer the ability to detect single nucleotide differences. In addition, capillary electrophoresis has the advantages of rapid DNA separation, high sensitivity, automatic sample loading and elimination of the need to pour gels (Wenz et al., 1998).

In terms of throughput, single capillary systems such as the ABI 310 Genetic Analyser are capable of electrophoresing a single sample at any one time (usually about 30 min per sample). However, the Molecular Dynamics MegaBACE and the ABI 3700 capillary electrophoresis systems have increased the throughput capacity of capillary electrophoresis by allowing electrophoresis of 96 samples simultaneously.

Despite the advantages in automation that modern electrophoresis systems offer, it has been documented in the literature and observed in this laboratory that sizing discrepancies for microsatellites occur between capillary electrophoresis, slab gel electrophoresis and sequencing, and can significantly influence the interpretation of results (Mansfield et al., 1996; Bruland et al., 1999; Williams et al., 1999). For example, Bruland et al. (1999) and Williams et al. (1999) reported that sizing of disease-causing alleles for Huntington's disease (a CAG triplet repeat expansion) during capillary electrophoresis (ABI 310) and slab gel electrophoresis (ABI 373/377/^{32}P) could result in an inaccurate estimate of the true number of repeats relative to the sequence determined size, altering the interpretation of results and subsequent diagnosis in patients.

At the Centre for Plant Conservation Genetics, in our investigations of wheat microsatellites, we have primarily dealt with dinucleotide repeats. The tendency of the ABI 310 Genetic Analyser to under size repeat DNA fragments is also observed. In an investigation of 19 microsatellites scored across a selection of Australian wheat varieties by both capillary electrophoresis (ABI 310 Genetic Analyser) and slab gel electrophoresis (ABI 373 or 377 Sequencing Systems), 14 microsatellites are underscored by the ABI 310 relative to the slab gel systems in a range of 2–6 bp (average across varieties). Five microsatellites scored similarly (within 0.5 bp) between capillary electrophoresis and slab gel electrophoresis. No pattern was observed between the discrepancies and type and size of repeat or fluorescent label.

Several theories have been proposed to explain the sizing discrepancies observed between the different electrophoresis systems. The mobility anomalies observed are thought to be sequence related (Rosenblum et al., 1997; Wenz et al., 1998) and a result of higher order structure formed by microsatellite DNA (Kiba and Baba et al., 1999). Therefore, incomplete denaturation and the different qualities of the separation matrices can also influence the rate of migration (Rosenblum et al., 1997; Wenz et al., 1998). The different chemistry of the various fluorescent labels is also proposed as a mechanism which may influence mobility (Mansfield et al., 1996; Wenz et al., 1998) although Haberl and Tautz (1999) suggested that the mobility discrepancies they observed were not a result of the different fluorescent dyes.

Bruland et al. (1999) and Williams et al. (1999) were able to show that by generating their own internal size standards derived from CAG repeat fragments of known length, the mobility discrepancy could be overcome. Mansfield *et al.* (1996) showed that a calculated 'normalization factor' derived from confirmed allele sizes could be applied to improve sizing accuracy. Haberl and Tautz (1999) also recommended that at least one allele from every locus should be sequenced allowing a correction factor to be applied to other alleles at the same locus. This would allow comparison of alleles identified by different studies utilizing different equipment. The assigning of arbitrary names rather than absolute allele sizes would also be a potential method for alleviating the difficulties that would arise from the comparison of allele sizes. Kline *et al.* (1997) reported an interlaboratory evaluation of microsatellite analysis. In the analysis of four microsatellites across seven samples in 34 laboratories, it was found that comparable results were obtained when arbitrary allele names were used instead of scored allele sizes. In the comparison of called allele sizes, scores differed by up to 5 bp.

Establishing a protocol

Several issues have been discussed which highlight areas of potential allele calling and reporting errors both within a laboratory and also between collaborating laboratories. There are two conclusions to be reached from these issues. One is that although there are excellent systems for high throughput semi-automated electrophoresis and automated allele sizing software available, there is still a requirement for intensive manual checking of the data to ensure consistency in the calling of allele sizes. Approaches to handling both 'plus A' and stutter peaks must be contemplated if accurate sharing of data is to occur. Secondly, in collaborative efforts or in reporting of called allele sizes, it is important always to remember that different electrophoresis and analysis software programs may result in sizing discrepancies. As long as groups involved in collaborative efforts are collectively aware of the potential for error in allele comparisons, then protocols can be established to ensure valid results are available for comparison. Haberl and Tautz (1999) appropriately recommended that sizing and electrophoresis conditions should be specified whenever results are reported.

Genotype data storage

Put simply, 'a database is an organized collection of information' (Schwartz, 1998). The term database commonly refers to a computer system that utilizes a database program or database management system to control access to and structure of the data files (Rob and Coronel, 1993; Kroenke, 1998).

If consideration is given to the volume of data generated by molecular biology projects, it quickly becomes clear that a paper-based data storage system has obvious disadvantages. Schwartz (1998) details the relative comparisons between paper and computer data storage systems. In summary, maintenance of paper records can be time-consuming and costly, including filing, sorting and editing of records. In a computer database system, the advantage is the ease with which data can be manipulated; the task of adding, deleting or modifying records is a simple one, finding and sorting records is easy (Schwartz, 1998). Many database programs assist in the data entry process by including features such as short cut keys and auto-entry options. Further, data already stored in files generated by compatible programs can be easily imported into the new database eliminating the need to re-enter the data.

Importantly in scientific research, collaborative efforts at an international level occur frequently. The use of database programs that can easily display data via the Internet utilizes a valuable and convenient resource to facilitate the sharing of data. Data stored in the database can be shared not only within the research group but also between research groups while still allowing the maintenance of confidentiality by establishing user access passwords. The value in this is reflected by the large number of publicly available sequence databases (Canhos et al., 1996).

There are numerous software systems available for creating and developing databases. In this chapter, the establishment of a relatively simple yet functional database, which may assist some groups to increase their ability to maintain and share large amounts of data generated by their projects, is described.

The Australian National Wheat and Barley Molecular Marker Programs microsatellite databases as an example of a functional database system

The Australian National Wheat and Barley Molecular Marker Programs (NWMMP and NBMMP) microsatellite databases were designed to offer a simple and effective method for the storage of data generated by the microsatellite component of the two programmes. These projects aim to provide overall coordination of wheat and barley molecular marker development in Australia.

The microsatellite components of these programmes are primarily involved in the development, optimization and screening of microsatellites across Australian varieties, and also in the generation of genetic maps for a number of significant mapping populations. As a consequence of this work, a large amount of genotypic data has been collected. In establishing the microsatellite databases, the aim was not only to increase the manageability of the data load but also to create a database that could be readily shared via the Internet to promote access to the data for collaborators who are geographically quite distant. Further, it is intended that these data will ultimately be made available to all Australian researchers and plant breeders in the hope that it will facilitate plant breeding programmes by providing access to data regarding useful genetic markers for the detection of quality and disease resistance traits in Australian varieties.

The databases are created in FILEMAKER PRO 4.0 (Filemaker, Inc.). This program was chosen by the programme leaders as it has two features, among many others, which make it an attractive tool for receiving and displaying the data generated by these projects:

1. FILEMAKER PRO 4.0 is compatible with MICROSOFT EXCEL, meaning that data already entered into an EXCEL spreadsheet can be directly imported into the microsatellite databases. This is particularly useful as many allele scoring software programs such as ABI GENOTYPER are capable of generating an EXCEL spreadsheet containing the allele sizes, reducing the manual handling of data required at the point of data entry. Without this option, data entry could be a tedious, potentially error-prone process requiring many hours at a computer terminal.

2. FILEMAKER PRO 4.0 incorporates the capacity for instant or custom web publishing, making access to the data from remote locations possible.

Data stored in the wheat and barley microsatellite databases

The primary step in establishing a simple yet effective database is to determine what data will need to be stored. Further, if data are to be included from collaborating groups, it may be important to establish at an early stage a protocol or submission form detailing the types of data to be provided to the database manager for entering.

The NWMMP and NBMMP microsatellite databases are relational databases. Relational databases are structured in a certain fashion, where data are separated into individual related files, linked by a common field (Schwartz, 1998). This aims to keep large sets of data broken down into smaller, more manageable files. Data stored in a related file can be easily viewed in any file with which a relationship is established. This not only saves space, but may also eliminate errors associated with the entry of the same data multiple times, improving the integrity of the data (Kroenke, 1998). The type of data that is stored in the NWMMP microsatellite database and their division into the files of the relational database are seen in Table 17.1. The common field linking the individual data sets is, in this case, the microsatellite name. Essentially, the data are maintained in small, manageable files while still allowing all the data to be browsed on a single screen if desired.

Searching the NWMMP microsatellite database

Through the use of CLARIS HOME PAGE 3.0 (Filemaker, Inc.), the establishment of custom report pages to display data via the Internet in a fashion that best suits the end users of the database is made possible. Many software programs are available which make generation of pages for the Internet possible with little or

Table 17.1. Data stored in the NWMMP microsatellite database and the division of these data into relational files linked by the common field, microsatellite name. All data can be browsed on a single page within the database for ease of access.

Wheat file	Library file	Primer file	PCR file	Variety file	Mapping file	Contact file
Microsatellite name	Microsatellite name	Microsatellite name	Microsatellite name	Microsatellite name	Microsatellite name	Microsatellite name
Chromosome	Variety	Primer name	No. of cycles	Variety	Polymorphic in crosses?	Name
Owner of sequences	Clone name	Sequence	MgCl	Accession no.	Is it mapped?	Institution
Public availability	Clone sequence	Tm	Reaction volume	Size of product	No. of loci	Address
	Type of repeat	Primer length	PCR technique	Method of scoring		Phone
	Size of product		PCR profile	Number of varieties tested		Email
			Quality of PCR product	Number of alleles identified		

no knowledge of the language required for such programming. CLARIS HOME PAGE 3.0 was chosen because of its in-built ability to recognize the language used in FILEMAKER PRO 4.0.

The search pages for the NWMMP microsatellite database have been created with the intention of meeting the needs of the end users, presenting data in an appropriate fashion with emphasis on remaining user friendly and easy to search. Currently, there are a number of ways in which the data can be searched, which in turn controls the type of summary report presented while always offering the user the flexibility to browse the entire data set. The summary reports are designed to present particular aspects of the data in a table that reduces the need for the user to extract data manually from the larger data set.

1. *General microsatellite summary*: presents microsatellites with the chromosome name, number of varieties screened and number of alleles identified in a table format. This report is designed to provide details on chromosome location and polymorphism levels.

2. *Crosses summary*: a summary report table indicating microsatellites that have been tested on both the parents of the population specified in the search criteria, indicating if that microsatellite is polymorphic between those two parents and whether it has been mapped on the doubled haploid population.

3. *Variety summary*: reports which microsatellites have been tested on that variety and the expected PCR product size along with the method of scoring.

4. *Varietal identification summary*: this search is designed to allow the user to identify a variety on the basis of testing carried out with a specific microsatellite marker and the allele size identified. A report is returned that shows all varieties that meet the criteria specified.

In each of the summary reports, each microsatellite name represents a link to the comprehensive set of data about that microsatellite. From that location, information about specific allele sizes, mapping data and PCR conditions can be found. The number of search pages and summary reports that can be created is unlimited although each search simply presents a different way to search the complete data set.

The template developed for the NWMMP and NBMMP microsatellite databases can be easily modified for other species. Regardless of the species studied, similar types of information will be collected in a microsatellite study, meaning that data fields are already established. Further, this template could be easily modified for the storage of information from other marker types such as restriction fragment length polymorphisms, amplified fragment length polymorphisms and expressed sequence tags.

The linking of these NWMMP and NBMMP microsatellite databases with data sets containing genetic map information and traits of interest provides the mechanism by which plant breeding programmes can be provided with the information to identify useful genetic markers for quality and disease resistance traits in Australian wheat varieties.

Summary

A number of important issues relating to the collection, reporting and storage of genotype data have been discussed. Techniques for reducing the impact of some of the issues have been reviewed. In the scoring of microsatellite alleles it is clear that some degree of manual checking is always required. It is important to recognize the potential for errors to occur in the handling of genotype data and to establish suitable protocols from the outset of any project or collaborative effort. Such steps should ensure a more accurate and reliable approach to the reporting and sharing of information both within and between groups. A simple and functional database for the handling of microsatellite genotype data has been described that can easily be transferred to other species or marker types and which should be useful for the facilitation of plant breeding programmes.

Acknowledgements

The contribution of Anne McLauchlan to investigations of sizing discrepancies in capillary electrophoresis is gratefully acknowledged. Funding for the microsatellite programme and development of the NWMMP and NBMMP databases is provided by the Grains Research and Development Corporation.

References

Becker, J. and Heun, M. (1995) Barley microsatellites: allele variation and mapping. *Plant Molecular Biology* 27, 835–845.

Brownstein, M.J., Carpten, J.D. and Smith, J.R. (1996) Modulation of non-templated nucleotide addition by *Taq* DNA polymerase: primer modifications that facilitate genotyping. *Biotechniques* 20, 1004–1010.

Bruland, O., Almqvist, E.W., Goldberg, Y.P., Broman, H., Hayden, M.R. and Knappskog, P.M. (1999) Accurate determination of the number of CAG repeats in the Huntington disease gene using a sequence-specific internal DNA standard. *Clinical Genetics* 55, 198–202.

Canhos, V.P., Manfio, G.P. and Canhos, D.A.L. (1996) Networking the microbial diversity information. *Journal of Industrial Microbiology* 17, 498–504.

Clark, J.M. (1988) Novel non-templated nucleotide addition reactions catalyzed by prokaryotic and eukaryotic DNA polymerases. *Nucleic Acids Research* 16, 9677–9686.

Fishback, A.G., Danzmann, R.G., Sakamoto, T. and Ferguson, M.M. (1999) Optimization of semi-automated microsatellite multiplex polymerase chain reaction systems for rainbow trout (*Oncorhynchus mykiss*). *Aquaculture* 172, 247–254.

Garland, S.H., Lewin, L., Abedina, M., Henry, R. and Blakeney, A. (1999) The use of microsatellite polymorphisms for the identification of Australian breeding lines of rice (*Oryza sativa* L.). *Euphytica* 108, 53–63.

Ginot, F., Bordelais, I., Nguyen, S. and Gyapay, G. (1996) Correction of some genotyping errors in automated fluorescent microsatellite analysis by enzymatic removal of one base overhangs. *Nucleic Acids Research* 24, 540–541.

Haberl, M. and Tautz, D. (1999) Comparative allele sizing can produce inaccurate allele size differences for microsatellites. *Molecular Ecology* 8, 1347–1350.

Henry, R.J. (1994) Molecular methods for cereal identification. *Australasian Biotechnology* 4, 150–152.

Kiba, Y. and Baba, Y. (1999) Unusual capillary electrophoretic behaviour of triplet repeat DNA. *Journal of Biochemical and Biophysical Methods* 41, 143–151.

Kline, M.C., Duewer, D.L., Newall, P., Redman, J.W., Reeder, D.J. and Richard, M. (1997) Interlaboratory evaluation of short tandem repeat triplex CTT. *Journal of Forensic Science* 42, 897–906.

Kroenke, D.M. (1998) *Database Processing – Fundamentals, Design and Implementation*, 6th edn. Prentice Hall, Englewood Cliffs, New Jersey.

Mansfield, E.S., Vainer, M., Enad, S., Barker, D.L., Harris, D., Rappaport, E. and Paolo, F. (1996) Sensitivity, reproducibility, and accuracy in short tandem repeat genotyping using capillary array electrophoresis. *Genome Research* 6, 893–903.

McCouch, S.R., Chen, X., Panaud, O., Temnykh, S., Xu, Y., Cho, Y., Huang, N., Ishii, T. and Blair, M. (1997) Microsatellite marker development, mapping and applications in rice genetics and breeding. *Plant Molecular Biology* 35, 89–99.

Murray, V., Monchawin, C. and England, P.R. (1993) The determination of the sequences in the shadow bands of dinucleotide repeat PCR. *Nucleic Acids Research* 21, 2395–2398.

Powell, W., Machray, G.C. and Provan, J. (1996) Polymorphism revealed by simple sequence repeats. *Trends in Plant Science* 1, 215–222.

Rob, P. and Coronel, C. (1993) *Database Systems – Design, Implementation and Management*, Wadsworth Publishing Company, Belmont, California.

Röder, M.S., Korzun, V., Wendehake, K., Plaschke, J., Tixier, M.-H., Leroy, P. and Ganal, M.W. (1998) A microsatellite map of wheat. *Genetics* 149, 2007–2023.

Rosenblum, B.B., Oaks, F., Menchen, S. and Johnson, B. (1997) Improved single-strand DNA sizing accuracy in capillary electrophoresis. *Nucleic Acids Research* 25, 3925–3929.

Schwartz, S.A. (1998) *Filemaker Pro 4 Bible*. IDG Books Worldwide, Foster City, California.

Schwengel, D.A., Jedlicka, A.E., Nanthakumar, E.J., Weber, J.L. and Levitt, R.C. (1994) Comparison of fluorescence-based semi-automated genotyping of multiple microsatellite loci with autoradiographic techniques. *Genomics* 22, 46–54.

Smith, J.R., Carpten, J.D., Brownstein, M.J., Ghosh, S., Magnuson, V.L., Gilbert, D.A., Trent, J.M. and Collins, F.S. (1995) Approach to genotyping errors caused by non-templated nucleotide addition by *Taq* DNA polymerase. *Genome Research* 5, 312–317.

Tautz, D. (1989) Hypervariability of simple sequences as a general source for polymorphic DNA markers. *Nucleic Acids Research* 17, 6463–6471.

Wang, Z., Weber, J.L., Zhing, G. and Tanksley, S.D. (1994) Survey of plant short tandem DNA repeats. *Theoretical and Applied Genetics* 88, 1–6.

Weber, J.L. and May, P.E. (1989) Abundant class of human DNA polymorphisms can be typed using the polymerase chain reaction. *American Journal of Human Genetics* 44, 388–396.

Wenz, H.M., Robertson, J.M., Menchen, S., Oaks, F., Demorest, D.M., Scheibler, D.,

Rosenblaum, B.B., Wike, C., Gilbert, D.A. and Efcavitch, J.W. (1998) High-precision genotyping by denaturing polyacrylamide electrophoresis. *Genome Research* 8, 69–80.

Williams, L.C., Hegde, M.R., Herrera, G., Stapleton, P.M. and Love, D.R. (1999) Comparative semi-automated analysis of (CAG) repeats in the Huntington disease gene: use of internal standards. *Molecular and Cellular Probes* 13, 283–289.

Ziegle, J.S., Su, Y., Corcoran, K.P., Nie, L., Maynard, P.E., Hoff, L.B., McBride, L.J., Kronick, M.N. and Diehl, S.R. (1992) Application of automated DNA sizing technology for genotyping microsatellite loci. *Genomics* 14, 1026–1031.

Chapter 18

Commercial Applications of Plant Genotyping

L.S. LEE AND R.J. HENRY

Centre for Plant Conservation Genetics, Southern Cross University, Lismore, Australia

Introduction

Although genotyping of plant species and varieties using DNA analysis was developed initially as a research tool, it soon became apparent that the technology would prove very useful for commercially oriented activities. As a result, plant DNA analysis services have become available in recent times to address a variety of needs for commercial clients. The benefits and the limitations of DNA fingerprinting techniques that apply to research activities also come into play in the commercial context. This raises issues of the kinds of commercial problems that can be addressed by DNA genotyping, the suitability of the various methods that may be employed and the influences of clients' requirements.

Commercial applications

Commercial plant DNA analysis is conducted to address problems or questions with respect to:

- product quality assurance;
- breach of intellectual property rights;
- support for plant breeders' rights and patents;
- guiding plant breeding programmes; and
- forensic investigations.

Quality assurance

A major aspect of commercial DNA analysis is the gathering of data for quality assurance purposes. DNA analysis services are called upon to determine the genetic uniformity of crop seed batches and nursery plants, the genetic fidelity of prescribed plant varieties, and the presence of genetically modified plant material.

Genetic uniformity within a plant variety can be reasonably assured in vegetatively propagated plants, in seed-propagated self-fertilizing plants that have been backcrossed for an appropriate number of generations, in seed-propagated progeny of controlled crosses, and in obligate apomictic species. However, when seed propagation is carried out from open pollinated parents, considerable genetic variability can result in the progeny. This is a particular problem in the nursery industry but also occurs in some crop plants (tea tree, coffee and certain artichokes are examples with which we have dealt). When unacceptable levels of phenotypic variability are observed, DNA analysis services may be called upon to quantify the extent of the problem. These situations can occasionally pose a difficulty for the service provider because the client will usually be concerned about the extent of phenotypic variability which may not be concomitant with the level of genotypic variability revealed by the analysis.

Genetic fidelity or 'trueness to type' presents a problem when clients are not satisfied that the plant variety they have been sold is in accordance with their expectations. This may result either when seed of a prescribed variety is sold to a farmer for sowing, or when nursery plants are sold. Determination of fidelity will entail comparison of DNA profiles of a known specimen of the variety in question and the subject plants. The reliability of analysing plant DNA profiles to assess genetic uniformity and fidelity will be affected by the genetic relatedness of the contending samples. Very closely related accessions will prove difficult to differentiate and the analysis technique employed will influence the outcome.

Genetically modified plant product identification is an issue of interest to food processors. Routine testing of batches of such products as canola and soybean is carried out in order to assure food processors of the genetically modified organism (GMO) content of their consignments. This form of quality assurance is highly reliable because it usually entails PCR-derived profiles obtained using highly specific primers designed to detect known transgenes (Hurst *et al.*, 1999). The converse situation also arises where seed companies selling seed of genetically modified crops want testing performed to ensure that the appropriate transgene is still in place. The application of DNA genotyping for purposes of quality assurance is often employed in litigious situations when parties are in conflict over genetic material. Accordingly, attention to detail is vital in laboratory practices and thorough record keeping is necessary because of the possibility of the service providers' staff being called upon to testify in legal proceedings.

Breach of intellectual property rights

Many cases of commercial plant DNA genotyping are undertaken to seek proof of, or to provide defence against, allegations of breaches of intellectual property (IP) rights. The approach here is similar to that for determining fidelity of genotype as described above. It is a matter of comparing DNA profiles of the protected plant variety against those of the variety in question. The same issue of reliability of the analyses as confronts testing for fidelity also occurs. Indeed, it is more likely to be a problem in cases of IP contention because the plant varieties in contest will always be very similar, if not identical.

Support for plant breeders' rights (PBR) and patents

It is not acceptable to the relevant authorities to consider DNA profiles alone as proof of unique identity of a plant variety for purposes of PBR or patent protection. None the less, plant breeders are often keen to strengthen their applications for protection of plant varieties by presenting molecular data which corroborate their claims of the distinctness of their variety. This is particularly the case where there is only a small apparent phenotypic difference between the new variety and an extant one, or where the difference is difficult to observe, such as maturity time or growth rate. Molecular profiling has successfully demonstrated distinction between plant varieties which are morphologically similar (Graham *et al.*, 1994; Ko *et al.*, 1996; Scott *et al.*, 2000), and have subsequently been submitted for protection of breeders' rights.

The issue of the acceptability of molecular data for the purposes of plant variety protection is currently under detailed scrutiny by the International Union for the Protection of New Varieties of Plants through its 'Working Group on Biochemical and Molecular Techniques and DNA-Profiling in Particular'.

Guiding plant breeding programmes

DNA genotyping of progeny generated in breeding programmes greatly improves efficiency (Poulsen *et al.*, 1996)(Chapter 20, this volume). In most cases, such work is conducted in-house in the research agency conducting the breeding. However, some plant breeders have found it more cost effective to outsource this work to commercial laboratories which specialize in DNA genotyping technology, thereby enabling the breeders to focus on their core business. Plant breeders may use genotyping simply to confirm the hybrid identity of breeding lines or to check parentage to avoid costly errors in crossing. Major providers of such services include AgroGene S.A. (http://www.agrogene.com/product.htm) in France and PE AgGen Inc. (http://www.appliedbiosystems.com/ab/aggen/html/plantdnatesting.htm) in the USA.

Forensic investigations

The use of commercial plant DNA services to provide genotype analyses in relation to criminal cases is relatively uncommon. Typically, plant identification is the objective in criminal investigations and this can be undertaken by alternative means. The precise resolution of discrete plant varieties is usually not a requirement. However, linking of a suspect to a specific individual plant (e.g. a tree at a crime scene) may be valuable evidence. However, our laboratory has been approached to undertake work for clients who needed evidence pertaining to criminal charges. A case in point called for distinguishing between non-THC marijuana varieties and the narcotic counterparts.

Difference versus similarity

Regardless of the reason for which an analysis has been commissioned, the objective of the work will typically be to provide evidence for either difference between two or more plant samples, or similarity between samples. This is the fundamental issue which will dictate how the job will be approached and will have a strong bearing upon the likelihood of a 'successful' outcome.

To some extent the technology utilized will determine the chances of differentiating between two samples (for instance, amplified fragment length polymorphism (AFLP) has much greater capacity to discriminate than random amplified polymorphic DNA (RAPD), which in turn is more effective than isozyme analysis). Furthermore, plant accessions with a close genetic relationship will be more difficult to differentiate than those more distant. Moreover, it is generally much simpler to show genotypic differences than to demonstrate definitively that two samples are identical.

Whereas a single molecular polymorphism can be considered as proof of distinction between two samples, it is theoretically impossible to prove that two samples are identical without examining the entire genome, a prospect entirely beyond commercial consideration. Therefore, in cases where the client wishes to demonstrate identicalness between accessions, the service provider must resort to exhibiting similarity on the balance of probability. Sometimes this requires such a large amount of data that it is not cost effective to undertake such work. Legal cases built upon probability of similarity can be difficult to defend.

A difficulty with demonstrating similarity arises from the intrinsic genetic variability within the species in question. If the species has a narrow genetic base it is much more difficult to argue that a lack of polymorphism in a DNA profile (which represents only a portion of the genome) provides adequate evidence of identity than if the species has a broad genetic base. In cases where evidence of similarity is sought, consideration must be given to the statistical analysis that will be required to support the case most strongly.

Types of markers

Many of the techniques which are employed in DNA genotyping are subject to patent protection. For commercial exploitation of this technology, appropriate licence agreements must be secured between the service provider and the patent holder. A variety of techniques are available and most have been described in detail elsewhere in this volume. Each approach offers its distinct set of benefits and disadvantages. The choice of technique will depend upon the facilities and expertise of the laboratory, the nature of the client's requirements and their intended use of the data, the available budget for the work, presence or absence of pre-existing molecular marker information for the species in question and the anticipated genetic distance between the accessions. The relative merits of various types of molecular markers are summarized in Table 18.1.

Table 18.1. Advantages and disadvantages of various DNA fingerprinting methods.

Type	Advantages	Disadvantages
Isozymes	Inexpensive Co-dominant Expressed genes	The small number of isozymes available can lead to poor discrimination capacity between samples Poor resolution of electrophoretic bands Markers can be difficult to differentiate because of complex banding patterns of oligomeric isozymes and co-migration in electrophoresis
RAPD	Inexpensive Simple	Uncharacterized DNA which is amplified in RAPD provides no additional information, which is sometimes a useful property of other marker systems Low stringency of the PCR process used in RAPD can result in profile reproducibility problems Unknown gene function
AFLP	High discrimination Many markers	Expensive Complex procedure Unknown gene function
SSR	High discrimination Highly reproducible Co-dominant Characterized DNA	Complex procedure to establish initially for given taxa Expensive (unless using extant SSR primers) Taxa-specific, there are limitations to the transferability of SSR markers between taxa; only successful with closely related genera Unknown gene function (unless EST-derived)
RFLP	Highly reproducible Co-dominant Characterized DNA	Complex procedure Laborious
SNP	High discrimination Co-dominant	Complex to establish initially for given taxa Not in general use

RAPD, random amplified polymorphic DNA; AFLP, amplified fragment length polymorphism; SSR, simple sequence repeat; RFLP, restriction fragment length polymorphism; EST, expressed sequence tag; SNP, single nucleotide polymorphism.

The amount of work that can be undertaken and the selection of marker type will have a major impact upon the likelihood of a successful outcome to the commercial job. Success is generally measured in terms of the surety with which the laboratory can discriminate between samples. As explained above, if the client's objective is to procure sound evidence of similarity, this may require a considerable amount of testing.

The available funding to conduct the DNA analysis is often a major limitation which will dictate the amount and type of analysis performed. Commercial clients invariably wish to spend as little as possible on such testing and many are surprised at the expense of DNA analysis. Costs aside, other factors must also come into consideration.

Discriminating power

The discriminating power of the marker type used varies. Isozymes are generally coded by only a few alleles and have limited power of discrimination which, therefore, necessitates the use of a suite of such markers. Unfortunately there are relative few isozymes markers available. RAPDs have a lower power of discrimination than other DNA markers. However, the problem can be addressed by utilizing a large selection of the huge number of RAPD primers available, and unlike isozymes the same protocol can typically be employed for all. Marker types such as simple sequence repeats (SSRs) and AFLPs have much higher power of discrimination but are more complex and expensive to develop and apply. SSRs are particularly useful markers if they have already been characterized for the species in question or a closely related species. They are highly polymorphic, abundant, co-dominantly inherited and easy to process in the laboratory. However, the cost of characterizing SSRs *de novo* is considerable.

Although it is beyond the scope of most commercial DNA analysis services to accommodate the development of markers such as SSRs, some degree of development is often necessary. When working with major crop species for which considerable previous genotyping research has been conducted, costs can be contained because the work is routine. However, in our laboratory, many analysis requests involve plant species for which little, if any, molecular genotyping has been conducted. Accordingly, considerable developmental work must be undertaken to achieve satisfactory DNA extraction, to optimize PCR conditions and to identify a suite of polymorphic markers for the species (if not for the accessions in question). This developmental work constitutes a significant portion of the price of the service.

Genetic characteristics

In particular circumstances, the genetic nature of the available markers may be an important consideration. Where precise genotyping is required (for example in plant breeding situations or phylogenetic studies or for population genetics) co-dominant markers such as isozymes, SSRs or restriction fragment length polymorphisms (RFLPs) will be most suitable. Conversely, if the task simply requires discrimination between samples, non-co-dominant markers such as RAPDs or AFLPs are acceptable.

In certain genotyping jobs, such as those supporting PBR registrations, markers which are derived from the expressed genome are preferred; for example, isozymes and RFLPs. It has been argued by some PBR authorities that genotypic variation in non-coding regions is not a valid discriminator of plant varieties because there is no resultant phenotypic variation. Accordingly, markers which do not necessarily associate with coding regions of the genome are unacceptable in such situations (see for example Le Buanec, 2000).

Other issues

A variety of other factors will influence the appropriateness of particular marker types. An imperative of most commercial genotyping is to maximize efficiency and simplicity. Marker types which are complex to utilize, such as AFLPs and RFLPs, are often less attractive for practical application. For similar reasons, laboratories usually prefer to employ well-established technologies rather than less familiar procedures.

Reproducibility of results is essential in most genotyping applications. It is essential in commercial work where the client will always have a vested interest, and often a legal imperative, that the results are reliable. Poor reproducibility can be a problem with RAPDs, for example, due to low stringency of the PCR conditions employed.

Problems of electrophoretic band resolution in some systems can lead to difficulties in interpretation of results. Isozymes separated on starch gels may display diffuse bands, and SSRs on agarose gels often produce poor separation of bands. The fine-tuning of electrophoretic conditions and the utilization of more precise systems, such as polyacrylamide gels or capillary electrophoresis, can overcome these problems.

Another consideration may be the system's propensity to yield artefacts which are not associated with the genotypic profile of the accessions, but are a product of the chemistry or physics of the technique being employed. These artefacts can result in difficulties, and indeed errors, in interpreting gels. Stutter peaks in SSRs, primer dimers in RAPDs and ghost bands in isozymes are examples of these artefacts. The skilled and experienced laboratory technician will often be able to minimize artefacts or at least instantly recognize them when they occur.

Statistical analysis and experimental design

Statistics

The amount and type of statistical analysis of the data which will be employed will depend upon the objectives of the client. Analyses can range from none at all through to simple summary statistics, t-tests, analysis of variance and cluster analyses. A huge amount has been written about the vast array of such analyses and the point here is not to discuss these techniques, but to point out the importance of recognizing that such analysis may be required and that it is best referred to expert statisticians. An inappropriate analytical tool can produce misleading results.

Experimental design

Many commercial jobs which entail routine genotype analysis, such as testing for the presence of GMOs, can be undertaken with little regard to experimental design and replication of samples. However, the importance of sound experimental design should not be underestimated, particularly in 'one-off' genotyping cases where some level of procedure development is required to undertake the DNA analysis.

Replication of results improves reliability and confidence in the interpretation thereof and in the DNA analysis system employed. Replication can occur at a number of levels in the process, depending upon the situation:

- discrete plant samples (multiple accessions of each subject plant);
- discrete DNA samples from each individual plant sample;
- discrete analyses of the one DNA sample (repeat PCRs);
- pyramiding of replication (combinations of the above).

Sometimes replication may not be necessary, although it is usually advisable even if only as an internal check for quality assurance. The available budget often dictates the amount of replication that can be undertaken, as does the number of different accessions that require analysis.

Number of comparators (i.e. the number of different accessions) that are incorporated into the experiment design is partially related to the degree of replication. Typically, within a given budget, as the amount of replication increases the number of comparators will decline in order to contain the workload. Initially, the client's requirements will define the number of accessions to be compared. However, it is usually beneficial to include additional accessions beyond those which are the specific subject of the client's enquiry. Because commercial DNA genotyping often examines near-isogenic plant varieties (or indeed identical samples), the general ability of the selected markers to discriminate may come into question if no polymorphisms are detected among the subject plants. The inclusion of additional accessions, particularly ones known to be genetically distant from the subject accessions, provides a test of the discrimination capacity of the markers being used.

References

Graham, G.C., Henry, R.J. and Redden, R.J. (1994) Identification of navy bean varieties using random amplification of polymorphic DNA. *Australian Journal of Experimental Agriculture* 34, 1173–1176.

Hurst, C.D., Knight, A. and Bruce, I.J. (1999) PCR detection of genetically modified soya and maize foodstuffs. *Molecular Breeding*, 5, 579–586.

Ko, H.L., Henry, R.J., Beal, P.R., Moisander, J.A. and Fisher, K.A. (1996) Distinction of *Ozothamnus diosmifolius* (Vent.) DC genotypes using RAPD. *HortScience* 31(5), 858–861.

Le Buanec, B. (2000) DUS Testing: Phenotype vs genotype. UPOV Working Group on Biochemical and Molecular Techniques and DNA-Profiling in Particular, paper BMT/6/6, February 2000. Geneva, Switzerland.

Poulsen, D.M.E., Ko, H.L., van der Meer, J.G., Van de Putte, P.M. and Henry, R.J. (1996) Fast resolution of identification problems in seed production and plant breeding using molecular markers. *Australian Journal of Experimental Agriculture* 36, 571–576.

Scott, K.D., Ablett, E.M., Lee, L.S. and Henry, R.J. (2000) AFLP markers distinguish an early mutant of Flame Seedless grape. *Euphytica* 113, 245–249.

Further reading

Anon (1998) *Intellectual Property Reading Material*, WIPO Publication Number 476. World Intellectual Property Organization, Geneva, Switzerland.

Anon (1999–2000) International Union for the Protection of New Varieties of Plants (UPOV) *Working Group on Biochemical and Molecular Techniques and DNA-Profiling in Particular*.

Erbisch, F.H. and Maredia, K.M. (1998) *Intellectual Property Rights in Agricultural Biotechnology*. CAB International, Wallingford, UK.

Chapter 19
Non-gel Based Techniques for Plant Genotyping

R. KOTA

Centre for Plant Conservation Genetics, Southern Cross University, Lismore, Australia; and Plant Genome Research Centre, Institute for Plant Genetics and Crop Plant Research (IPK), Sachsen-Anhalt, Germany

Introduction

PCR amplification of DNA has proved to be an invaluable tool for researchers in diverse areas. Currently the most common way to evaluate the outcome of PCR amplification is to analyse the amplified products using gel electrophoresis once thermal cycling is complete. This process takes extra time and effort, and increases the chances of lab equipment or reagents being contaminated with amplified products as tubes must be opened following PCR to remove samples for analysis. Contamination is a particularly important problem when running the same primer/template systems repeatedly since a contaminated amplified product can serve as a template in subsequent experiments and hence, produce false-positive results. Consequently, attempts have been made to develop a means to detect the accumulation of PCR products, either as thermal cycling is occurring or following thermal cycling, without having to remove samples for processing. Monitoring reactions with fluorescence in a closed tube, homogeneous format reduces the time, effort and the risk of contamination involved with performing PCR analysis.

An ideal system for high throughput testing would involve a 'homogeneous PCR assay', that is one in which the processes of amplification and detection are formed simultaneously. Homogeneous assay methods can be characterized as specific or non-specific. Non-specific methods detect the presence or absence of the amplicon but provide no information about the amplified product. Examples of this type of method are the use of an intercalating agent (Wittwer *et al.*, 1997) or sunrise primers (Nazarenko *et al.*, 1997). These methods are prone to false positives in that undesired products such as primer dimers and other spurious amplicons can increase fluorescence.

Higuchi *et al.* (1992, 1993) developed the first homogeneous real-time method for detecting PCR products. They exploited the fact that fluorescence of the intercalator dye such as ethidium bromide increases in the presence of dsDNA. By monitoring this increased fluorescence as the dsDNA product accumulated during amplification, they could continuously follow the progress and kinetics of a PCR. Despite the obvious advantages of this intercalator dye system, its major drawback is its non-specificity; it cannot discriminate between the PCR product of interest and the common PCR 'noise' from mis-priming and primer–dimer artefacts. Although it is possible to minimize the amplification of side products by careful primer design and to prevent primer dimer accumulation, it is desirable for diagnostics to use specific methods that probe the amplification product. Examples of such methods include TaqMan (Livak *et al.*, 1995b) and molecular beacons (Tyagi and Kramer, 1996).

The probes employed in both molecular beacon and TaqMan technologies are based on the principle of fluorescence resonance energy transfer (FRET) (Stryer, 1978). During FRET, a donor fluorophore is excited by an external light source and emits light that is absorbed by a second, acceptor fluorophore. Again, this energy transfer only occurs when the two probes are hybridized and in close proximity. The acceptor fluorophore emits light of a different wavelength, which is filtered and measured. The amount of FRET signal is proportional to the amount of specific DNA product available for hybridization, thus the signal increases after each thermal cycle.

Holland *et al.* (1991) pioneered an alternative probe-based strategy for detecting specific PCR products. They exploited the 5′–3′ nucleolytic activity of *Taq* DNA polymerase and demonstrated that it could act on a probe specifically targeted to a PCR product during the amplification reaction. To accomplish this, they first designed a reporter oligonucleotide that would hybridize to a target PCR product. After chemically blocking the 3′ end to prevent it acting as a primer and labelling the 5′ end with ^{32}P, this reporter oligo was simply included with the normal PCR reagents. The reporter oligo annealed to the target sequence of the accumulating PCR product, forming a structure that could be cleaved by the 5′ nuclease activity of the polymerase as it extended the upstream PCR primer into the vicinity of the annealed oligo. The cleaved remains of the oligo were later separated and detected using thin layer chromatography. The dependence on hybridization and polymerization for oligo cleavage ensured that a signal was generated only if the intended amplification had occurred; an unhybridized probe is unaffected by the enzyme. Thus, the usefulness of this system is its ability to report the presence of a particular PCR product while avoiding background PCR noise. However, a major drawback was the significant post-PCR handling still required.

The 5′ nuclease assay of Holland *et al.* (1991) was enhanced by Lee *et al.* (1993) into a homogeneous assay for PCR amplification and specific product detection. They developed an improved reporter oligonucleotide design incorporating fluorescent dyes for signal detection. This made it possible to eliminate post-PCR processing, as Higuchi *et al.* (1992, 1993) had done, but with the added advantage that only specific amplification products were detected. It is continued development of the 5′ nuclease assay (Holland *et al.*, 1991; Lee *et al.*, 1993; Livak *et al.*, 1995a; Livak, 1996) that forms the basis of the TaqMan system.

The TaqMan system uses a fluorescent probe in addition to standard PCR components and controls. The fluorescent probes are labelled with two fluorescent dyes. One dye acts as a 'reporter' and one as a quencher. The reporter transfers fluorescent energy to the quencher by FRET when they are in close proximity. A signal is generated when the hybridized oligo probe is cleaved by the DNA polymerase and the dyes are separated. This separation negates FRET effects and the fluorescent signal emitted by the reporter dye measurably increases (Fig. 19.1). The probes are double-labelled with reporter (FAM or TET) and quencher (TAMRA) dyes at the 5′ and 3′ ends, respectively, and the cleavage of the probes causes an increase in fluorescence of the reporter dyes in solution.

Along with the development of the TaqMan system, alternative fluorescent probe-based methods such as molecular beacons have also been making an impact on the development of rapid non-gel based assays. Molecular beacons are single-stranded nucleic acid molecules that possess a stem-and-loop structure (Tyagi and Kramer, 1996). The loop portion of the molecule serves as a probe sequence that is complementary to a target nucleic acid. The stem is formed by the annealing of the two complementary arm sequences on either side of the probe sequence (Fig. 19.2). A fluorescent moiety is attached to the end of one arm and a non-fluorescent quenching moiety is attached to the end of the other arm. The fluorophore–quencher pairing represents a unique sys-

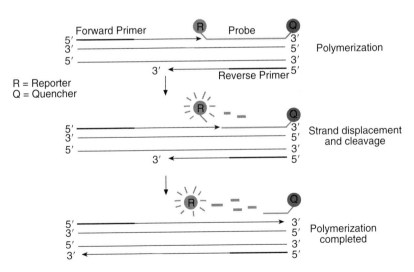

Fig. 19.1. Schematic representation of the TaqMan assay. The action of the *Taq* DNA polymerase on the hybridized fluorogenic probe is shown during one extension phase of the PCR.

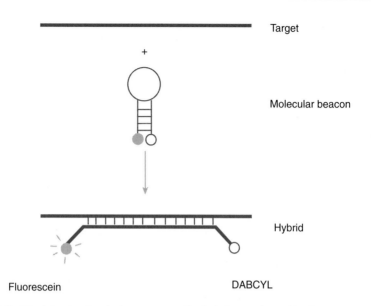

Fig. 19.2. Principle of molecular beacon assay. The hairpin stem formed by the complementary arm sequences cannot coexist with the double helix that is formed when the probe hybridizes to its target. Consequently, the molecular beacon undergoes a conformational change that forces the arm sequences apart and causes the fluorophore to move away from the quencher, thereby allowing fluorescence.

tem in which the energy received by the fluorophore is directly transferred to the quencher and dissipated as heat, rather than being emitted as light, resulting in the fluorophore being unable to fluoresce (Tyagi et al., 1998). When the molecular beacon encounters a target sequence, it forms a probe–target hybrid that is stronger and more stable than the stem hybrid. The probe undergoes spontaneous conformational reorganization which forces the arm sequence apart, separating the fluorophore from the quencher, and permitting the fluorophore to fluoresce (Bonnet et al., 1999). The power of molecular beacons lies in their ability to hybridize only to target sequences that are perfectly complementary to the probe sequence, hence permitting detection of single base differences (Tyagi et al., 1998).

Molecular beacons have several advantages over other systems for detecting the accumulation of PCR product using fluorescence. With molecular beacons, the presence of a distinct probe apart from the primers conveys an added level of specificity that other fluorescence-based real-time methods, such as those that use SYBR green or ethidium bromide, do not. Unless designed to tolerate mismatches, molecular beacons only report the presence of the completely complementary PCR product, assuring assay specificity.

The use of hairpin-shaped molecular beacons in PCR also provides several advantages over linear probes. It has been reported that the stem–loop structure of molecular beacons imparts an increased ability to discriminate single base-pair mismatches, compared with linear probes such as TaqMan probes (Livak, 1996). The hairpin shape of the molecular beacon makes mismatched probe–target hybrids less thermally stable than hybrids between the corresponding linear probes and their mismatched target. This means that molecular beacons are better suited for single nucleotide polymorphism (SNP) analysis than linear hydrolytic probes. Molecular beacons have also been demonstrated to be useful in discriminating between transgenic and non-transgenic cereals (Kota et al., 1999).

Guidelines for PCR probe and primer designs

In order to monitor PCR reactions successfully, fluorogenic probes should be designed so that they are able to hybridize to their targets at the annealing temperatures of PCR. Three main factors affect the performance of probes: (i) the efficiency of probe hybridization; (ii) the efficiency of reporter dye quenching; and (iii) the efficiency of probe cleavage by the polymerase. For efficient hybridization, the fluorogenic probe needs to have a melting temperature higher than that of the PCR primers used. This is because primer–template hybrids are quickly stabilized as they are extended by the DNA polymerase, whereas the fluorogenic probe is not extended and therefore not stabilized. The probe is blocked at its 3′ end so that it will not act as a primer. The higher Tm of the probes compensates for their relative instability. The recommended Tm of the probe is 70°C and compatible PCR primers should have a Tm approximately 10°C lower. Probe length is optimal between 20 and 25 nucleotides. Longer probes may display a reduced synthesis efficiency as they may form inter- or intramolecular structures that interfere with probe synthesis, cleavage, hybridization and DNA amplification.

Molecular beacons are generally 15–30 bp long, and the stem regions are 5–7 bp long. To monitor PCR successfully, molecular beacons need to be designed so that they hybridize to their targets at PCR annealing temperatures. At these temperatures, unhybridized beacons remain in their hairpin conformation and are non-fluorescent. The length and sequence of the stem should be designed so that the molecular beacon will melt at a temperature that is 7–10°C higher than the annealing temperature of the PCR primers. DNA-folding programs should be employed to estimate accurately the free energy of formation of the stem hybrid from which the stem melting temperature can be predicted. In general, GC-rich stems that are 5 bp in length exhibit melting temperatures between 55 and 60°C; GC-rich stems that are 6 bp long melt between 60 and 65°C, and GC-rich stems that are 7 bp in length melt between 65 and 70°C.

Just as there are important guidelines for PCR primer design, there are similar considerations for probe design. For instance, the G/C ratio can affect hybridization efficiency. For unknown reasons, probes with a higher proportion of C bases over G bases usually perform better. While the G+C content of complementary strands is identical, one strand often has more C bases and the other more G bases. Since probes can be designed for either strand of the PCR product, it is better to use the strand with more C bases as the probe. Other guidelines include avoiding runs of identical nucleotides, avoiding complementarity or overlap between the probe and primers, and avoiding a G at the 5' end of the probe.

Instrumentation

Initially instrumentation for the fluorogenic 5' nuclease assay consisted of a modified version of the model LS-50B luminescence spectrometer and custom PC-based software. The TaqMan LS-50B PCR detection system can read from either open microtitre plates or closed PCR tubes. However, with the development of new technologies, instruments dedicated to the fluorogenic 5' nuclease assay are now available: the ABI-7700 and 7200 Sequence Detectors.

The ABI Prism 7700 Sequence Detector system was designed for 'real-time' monitoring of the fluorogenic 5' nuclease assay by measuring the fluorescent signal as it is generated during thermal cycling. It integrates a thermal cycler, a laser light source, a fibre-optic network to distribute the laser light to, and the fluorescence signal from, each reaction tube, a cooled charged coupled device camera detector and a Macintosh computer with appropriate software. After collecting the signals from the 96-well plates sealed within the thermocycler block, the software first calculates the contribution of each component dye to the spectrum and normalizes the reporter signal against the standard ROX reference dye. Normalized peak signals are averaged and plotted against cycle number to produce 'amplification plots' of PCR products. The main uses for the 7700 are in quantitative PCR, very high throughput end-point detection, and PCR kinetics and optimization studies.

The ABI Prism 7200 detection system was designed for 'end-point' detection: it measures the fluorescent signal after thermal cycling has been completed. Like the 7700 system, detection is done from within sealed reaction tubes. The only post-PCR processing required is simply to transfer the reaction tubes from the 96-well thermal cycler to the automated sample drawer of the instrument. The 7200 is designed for routine pathogen detection and allelic discrimination assays.

An alternative to the ABI instruments, the LightCycler instrument is a rapid thermal cycler, combined with a microvolume fluorimeter utilizing high quality glass capillaries. Heating and cooling are controlled by alternating heated and ambient air as the medium for temperature transfer. This rapid cycling technique uses durable glass capillaries as reaction vessels. Because of the high surface area to volume ratio of these capillaries, they are highly efficient at transferring heat, thus permitting rapid cycling conditions. Furthermore, the glass capillary serves as an optical element for signal collection, piping the light and concentrating the signal at the tip of the capillary. The effect is efficient illumination and fluorescence monitoring of microvolume samples.

PCR requirements

Contamination issues cannot be overstated for the PCR. Even though amplification and detection are done within a sealed tube, standard reaction components for the fluorogenic 5' nuclease assay integrate the uracil N-glycosylase (UNG) carryover prevention strategy. In this strategy, uracil triphosphate is substituted for thiamine triphosphate during DNA polymerization. This allows PCR products from previous experiments to be differentiated from fresh template DNA in later experiments. UNG recognizes and enzymatically degrades the uracil-containing products before they can act as carryover templates in later experiments. UNG is not active at 55°C or above, so annealing temperatures of at least 55°C are used in the fluorogenic 5' nuclease assay.

For optimal PCR fidelity, standard reactions integrate a 'hot start' strategy (Erlich *et al.*, 1991). Hot starts overcome problems of non-specific primer annealing and extension that occur when reaction components are mixed at room temperature. Reliable hot starts can be achieved by using AmpliTaq Gold DNA polymerase (Birch *et al.*, 1996). This enzyme is activated only after exposure to elevated temperature. The polymerase is thus prevented from acting on reaction components until thermocycling begins.

Practical applications

SNP detection

SNPs are the class of DNA variants that are most common in the genome. Traditionally SNP discrimination has been done using techniques such as enzymatic restriction analysis of PCR products, single-stranded conformational polymorphism analysis or by direct sequencing. All the above-mentioned procedures require time and are labour intensive. It is now feasible routinely to differentiate between two forms of a gene differing by as little as single base using TaqMan or molecular beacon assays. These fluorogenic assays allow relatively high-throughput, high-quality genotyping of SNPs, with rapid development for individual SNPs of interest. The ability of fluorogenic assays to analyse SNPs has been exploited to develop rapid tests for SNP analysis. For instance, Piatek *et al.* (1998) employed this assay for detecting drug resistance in *Mycobacterium tuberculosis*. Similarly, Giesendorf *et al.* (1998) employed this assay to detect a point mutation in the *methylenetetrahydrofolate reductase* gene in humans. Although limited sequence information of many agronomically important plants is presently available, it is envisaged that these assays will have the potential to rapidly analyse SNPs in plants.

Quantitative studies

Quantitative gene expression analysis is often an integral part of determining gene function. However, many traditional methods employed to assess gene expression levels – including Northern analysis, dot blot analysis and RNase protection assays – are labour intensive, time consuming and require a large amount of sample for processing. For these reasons, RT–PCR is increasingly used for gene expression studies. However, as a quantitative method, it suffers from the problems inherent in PCR; due to the exponential nature of PCR amplification, small variations in reaction components and thermal cycling conditions can greatly influence the final yield of the amplification product (Livak, 1996). Also, during the later portion of the amplification reaction, PCR components can become limited, and reactions may reach a plateau phase in which no additional product is generated. Hence quantitation becomes intrinsically inaccurate if based on measurements of overall product yield at the end of the amplification reaction.

As a result, strategies such as competitive PCR have been developed in an attempt to improve the accuracy of end-point quantitative assessments of gene expression based on PCR (Larrick, 1997). However, competitive PCR is limited because an exogenous competitor must be synthesized for each gene being quantified. Also, PCR efficiencies of the competitor and wild-type products must be demonstrated to be equivalent; the technique is not amenable to high throughput because, typically, several competitor concentrations need to be run for each unknown. Fluorogenic assays such as TaqMan and molecular assays can be employed to monitor PCR amplification in real time. For instance, based on the molecular beacon assay, Stratagene has developed two RNA amplification kits that can be used to perform absolute or relative quantitation of gene expression. The molecular beacon β-actin and GAPDH detection kits are versatile and provide highly accurate quantitation. Similar kits are being developed to provide accurate quantitation of gene expression.

Conclusion

The nucleic acid sequence detection systems described here can automate the entire processes of PCR amplification, product detection, data processing and presentation of results. Using target-specific fluorogenic probes ensures that only amplification of the intended sequence is measured. The ability to detect PCR products provides detection sensitivity similar to gel electrophoresis but within a closed-tube format that decreases the risk of crossover contamination. The advantages of these assays over conventional gel-based assays include: (i) single-tube amplification and detection; (ii) fluorescence signal proportional to PCR product amplification; (iii) lower reagent and labour costs per genotype; (iv) genotype assignment based on clearly separated clusters of fluorescent values; and (v) simple, automatable assay implementation procedures. These fluorogenic assays provide new and powerful tools for a wide range of PCR-based diagnostics.

References

Birch, D.E., Kolmodin, L., Wong, J., Zangenberg, G.A., Zoccoli, M.A., McKinney, N., Young, K.K.Y. and Laird, W.J. (1996) Simplified hot start PCR. *Nature* 381, 445–446.

Bonnet, G., Tyagi, S., Libchaber, A. and Kramer, F.R. (1999) Thermodynamic basis of the enhanced specificity of structured DNA probes. *Proceedings of the National Academy of Sciences USA* 96, 6171–6176.

Erlich, H.A., Gelfand, D. and Sninsky, J.J. (1991) Recent advances in the polymerase chain reaction. *Science* 252, 1643–1651.

Giesendorf, B.A.J., Vet, J.A.M., Tyagi, S., Mensink, E.J.M.G., Trijbels, F.J.M. and Blom, H.J. (1998) Molecular beacons: a new approach for semi-automated mutation analysis. *Clinical Chemistry* 44, 482–486.

Higuchi, R., Dollinger, G., Walsh, P.S. and Griffith, R. (1992) Simultaneous amplification and detection of specific DNA sequences. *BioTechnology* 10, 413–417.

Higuchi, R., Fockler, C., Dollinger, G. and Watson, R. (1993) Kinetic PCR analysis: real-time monitoring of DNA amplification reactions. *BioTechnology* 11, 1026–1030.

Holland, P.M., Abramson, R.D., Watson, R. and Gelfand, D.H. (1991) Detection of specific polymerase chain reaction products by utilising the 5′ to 3′ exonuclease activity of *Thermus aquaticus* DNA polymerase. *Proceedings of the National Academy of Sciences USA* 88, 7276–7280.

Kota, R., Holton, T.A. and Henry, R.J. (1999) Detection of transgenes in crop plants using molecular beacon assays. *Plant Molecular Biology Reporter* 17, 363–370.

Larrick, J.W. (1997) *The PCR Technique: Quantitative PCR*. Bio-Techniques Books, Eaton Publishing, UK.

Lee, L.G., Connell, C.R. and Bloch, W. (1993) Allelic discrimination by nick-translation PCR with fluorogenic probes. *Nucleic Acids Research* 21, 3761–3766.

Livak, K.J. (1996) Quantitation of DNA/RNA using real-time PCR detection. P-E Applied Biosystems technology review, reference number 777902-001.

Livak, K.J., Marmaro, J. and Todd, J.A. (1995a) Towards fully automated genome-wide polymorphism screening. *Nature Genetics* 9, 341–342.

Livak, K.J., Flood, S.A.J., Marmaro, J., Giusti, W. and Deetz, K. (1995b) Oligonucleotides with fluorescent dyes at opposite ends provide a quenched probe system useful for detecting PCR product and nucleic acid hybridization. *PCR Methods and Applications* 4, 357–362.

Nazarenko, I.A., Bhatnagar, S.K. and Hohman, R.J. (1997) A closed tube format for amplification and detection of DNA based on energy transfer. *Nucleic Acids Research* 25, 2516–521.

Piatek, A.S., Tyagi, S., Pol, A.C., Telenti, A., Miller, L.P., Kramer, F.R. and Alland, D. (1998) Molecular beacon sequence analysis for detecting drug resistance in *Mycobacterium tuberculosis*. *Nature Biotechnology* 16, 359–363.

Stryer, L. (1978) Fluorescence energy transfer as a spectroscopic ruler. *Annual Review of Biochemistry* 47, 819–846.

Tyagi, S. and Kramer, F.R. (1996) Molecular beacons: probes that fluoresce upon hybridization. *Nature Biotechnology* 14, 303–308.

Tyagi, S., Bratu, D. and Kramer, F.R. (1998) Multicolour molecular beacons for allele discrimination. *Nature Biotechnology* 16, 49–53.

Wittwer, C.T., Herrmann, M.G., Moss, A.A. and Rasmussen, R.P. (1997) Continuous fluorescence monitoring of rapid cycle DNA amplification. *Biotechniques* 22, 130.

Chapter 20

Using Molecular Information for Decision Support in Wheat Breeding

H.A. EAGLES[1], M. COOPER[2], R. SHORTER[3] AND P.N. FOX[4]

[1] CRC for Molecular Plant Breeding, Department of Natural Resources and Environment, VIDA, Victoria, Australia; [2] School of Land and Food Sciences, The University of Queensland, Brisbane, Queensland, Australia; [3] CSIRO Tropical Agriculture, Indooroopilly, Queensland, Australia; [4] CIMMYT, Lisboa, Mexico

Today, the biological sciences are faced with an explosion of molecular data that will require the adoption of new strategies for data management, analysis and interpretation. The data generation processes that are available today, and currently being used to investigate the structure and function of genes and the architecture of traits, operate at rates that are orders of magnitude faster than our ability to convert the data into information that can be used by the plant breeder. While the conversion of the data into information is itself a daunting task, making the information available to breeders, to assist real-time decision making, is a major challenge for the application of molecular biosciences in plant breeding. The major private sector breeding programmes are already well advanced and have invested heavily in developing their bioinformatics programmes. However, even with their high levels of investment, genomics research is still generating data at a rate greater than computational biologists' abilities to manage and interpret them.

In most industries, which require decisions to be made based on incomplete information, powerful computer programs are used to predict outcomes of particular decisions. This allows the optimization of procedures and the minimization of costs. For example, in the mining industry, no mine of any consequence is ever developed without using powerful programs to predict the dimensions of the ore body from drilling data, and to model and predict the optimum method for accessing and mining the ore. In plant breeding, at least outside large multinational companies, the use of such software is in its infancy, but has the potential to be as useful and essential a tool as in the mining industry. A recent article in *The Economist* (1999), entitled 'Drowning in data', emphasized that 'biologists are confronted by a tidal wave of information. Unfortunately, few of them know how to swim'. We would argue that the data generated by many of the reasonably large molecular biosciences programmes around the world have yet to be converted into information. In the majority of cases these data are a long way removed from those who are in a position to act on the information they contain. This situation is clearly relevant to the use of molecular tools to improve the efficiency of plant breeding programmes. The issues of data taming, quality control, information extraction, management, access and use are central to the development of decision support systems for the use of molecular information in plant breeding.

In wheat breeding, as for other types of plant breeding, choice of parents, structure and method of selection in segregating generations, and the structure of the multienvironment (site–year combinations) are based on methods developed largely from experience, with limited opportunity to make data-driven choices based on quantitative predictions. While quantitative genetic theory provides a basis for the design of a breeding strategy and some predictive capability, it has limited application to many of the specific questions that are important to individual breeding programmes. As a consequence, usually less than 1% of crosses made in wheat breeding programmes ever produce a cultivar, although the international maize and wheat improvement center (CIMMYT) considers that by using the International Wheat Information System (IWIS) program, this could be raised to about 3% (Maarten van Ginkel, CIMMYT, 1998, personal communication). Quantitative predictions of parental combining ability and selection response have been theoretically possible for decades. However, before the advent of computers, many of the predictions were computationally too demanding to be used routinely. The availability of powerful computer hardware and software has changed that, as it has done in analogous industries. Large private sector breeding programmes now routinely use relatively sophisticated prediction procedures to assist decision making in their breeding programmes (e.g. Merrill, 1999).

In the public domain, software is being developed to integrate information from pedigrees, agronomic performance and molecular characteristics of wheat and other crops. This system can be used as the basis of a comprehensive decision support for plant breeders. In this chapter we will describe some of this software and its integration into a decision support framework for developing improved cultivars in the context of a farming system. We will concentrate on the elements relevant to the use of molecular markers, but will briefly mention other aspects to provide an overall perspective.

The challenge

The challenge is to develop software tools that will assist wheat breeders to make efficient *strategic* and *tactical* decisions while they continue to operate the breeding programme. Here strategic decisions relate to long-term aspects of the design and operation of the programme. Tactical decisions refer to short-term or medium-term decisions that influence what the breeder does within any one cycle or year.

Two examples of what would be considered strategic decisions are: (i) when should a breeding programme commence using molecular marker information to drive selection decisions? and (ii) when would it be advantageous to allocate the resources available to a breeding programme to the processes involved in developing molecular marker profiles of breeding lines in place of collecting additional phenotypic data from environments sampled as part of a multienvironment testing programme? Clearly these questions are related and part of a series of considerations involved in implementing marker-assisted selection strategies in a breeding programme. The answers to both questions are dependent on many variables, which differ in their ease of quantification. The breeder requires some measure of the relative merits of alternative breeding strategies for a range of operational procedures for the conduct of the programme. Answers to these types of questions can be obtained using several theoretical and simulation approaches for modelling the efficiency of breeding programmes.

For example, Lande (1992) used derivations from a theoretical trait model to quantify the efficiency of marker-assisted selection for a quantitative trait in a pedigree breeding programme (see Fig. 20.1a). In Fig. 20.1, derivations based on this theoretical model are depicted for a range of marker-assisted selection scenarios. It is observed that as the molecular markers account for larger proportions of the genetic variation for the trait (i.e. as P increases from 0.0 to 1.0), the selection response for marker-assisted selection increases. Also, in this example, as the emphasis is changed from sampling large numbers of F_2-derived individuals, with no or limited replication, to fewer F_2-derived individuals, with higher levels of replication (i.e. as the number of plants tested per line increases from 1 to 1000), the response from marker-assisted selection changes and the advantages of marker-assisted selection over phenotypic selection decrease. Cooper et al. (1999a,b) used a simulation approach to examine this marker-assisted selection scenario further (Fig. 20.1b). Where comparable genetic models are implemented for the quantitative trait, similar results are observed for the theory- and simulation-based approaches. An advantage of the simulation approach was that a wider range of genetic models could be examined. However, both approaches provide quantitative tools for evaluating the merits of marker-assisted selection in comparison with the situation where selection is based solely on phenotypic information (i.e. where $P = 0.0$ (Fig. 20.1a) and where RF = 0.5 Fig. 20.1b).

Whatever breeding strategy is in place, each year the breeder is faced with making many tactical decisions. Some of these may have longer-term implications for the breeding programme. For example, parent selection can have significant resource implications in the short and long term if specific parents are to be used in many crosses. To support tactical decisions the challenge is to develop software that will allow wheat breeders to:

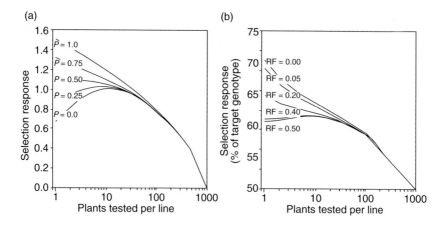

Fig. 20.1. Predictions of selection response for marker-assisted selection applied to a pedigree breeding scenario: (a) theoretical derivations based on selection among inbred lines plotted as a function of the number of plants tested per line and the proportion of total genetic variance (*P*) explained by molecular markers (the selection scenario was based on a total of 50,000 plants measured and the top 50 lines selected (Lande, 1992), and selection response is scaled in units of individual plant phenotypic standard deviations for the population as a whole); (b) simulation results for a quantitative trait controlled by 50 quantitative trait loci (QTLs) based on selection among inbred lines plotted as a function of the number of plants tested per line and the recombination frequency between flanking markers and the QTLs.

1. Predict the likelihood of outcomes from crossing particular parents. For example, assume that the breeder has 'Frame' as one adapted parent for mallee environments in southern Australia. 'Frame' has a high yield potential, is resistant to cereal cyst nematode, is the correct heading date and is tolerant to boron, but is only classified as Australian Premium White quality, a lower quality category. The breeder should be able to use the programme to search the database for parents which maintain the desirable features of 'Frame', but have a high probability of producing progeny which are also of superior quality (for example Australian Hard). The probability of obtaining desirable outcomes should be quantified.

2. Predict the outcomes for possible population types and selection procedures, for example F_2 progeny, BC_1, BC_2 and doubled haploids derived from F_1, and enable comparisons between alternative population types. Again, probabilities of obtaining desirable outcomes should be quantified. The likelihood of these outcomes will be influenced by the complexity of the genetic control of the traits to be manipulated. For example, the presence of epistasis will most probably reduce the chances of recovering improved genotypes, relative to the expectations based on non-epistatic models (Jensen *et al.*, 1999). However, the influences of these complexities will need to be considered where they are important components of the genetic architecture of the traits to be manipulated in the breeding programme.

3. Predict population sizes required for defined probabilities of outcomes. For example, if the breeder stipulates a 95% probability of obtaining a desirable line, a prediction of the number of individuals required in the F_2, or other defined generation, would be obtained. This is a relatively trivial exercise for simple genetic models, but becomes more complicated as the complexity of the genetic control of the traits increases.

Current software

Currently, software is available for parts of the requirement for an overall decision support system. Some of these are:

IWIS

This contains pedigree information on many wheat cultivars and germplasm accessions. It also contains variety performance data. It provides family trees, and can be used to calculate matrices of coefficients of parentage and provides easy access to trial data. This was developed by Dr Paul Fox and his colleagues at CIMMYT (Fox *et al.*, 1997).

QU-GENE (http://pig.ag.uq.edu.au/qu-gene/)

This is a platform for the quantitative analysis of genetic models. This consists of a genotype–environment system engine and application modules. It allows genetic predictions to be made from genetic models defined in terms of gene frequencies, types of gene action, and specified target populations of environments. QU-GENE was developed by Dr Mark Cooper and his colleagues at the University of Queensland (Podlich and Cooper, 1998) and has been used to evaluate and optimize components of conventional (Podlich *et al.*, 1999) and marker-assisted (Cooper *et al.*, 1999a,b, 2000) breeding strategies. A major motivation for using computer simulation tools to model breeding programmes is to deal with the specifics of individual breeding programmes and their germplasm. In many cases, traditional quantitative genetic theory cannot deal with this level of detail. Another advantage that can be gained by using simulation methodology is the capacity to deal with higher levels of complexity in genetic models for traits than is possible through classical theoretical approaches.

SUPERGENE

Several programs have been developed to integrate molecular marker, phenotypic and pedigree data, and provide a visual representation of the inheritance of alleles through pedigrees. One of these is SUPERGENE, which was developed by Boutin *et al.* (1995). An example of its application was to use restriction fragment length polymorphisms (RFLPs) to trace the flow of chromosome regions through the breeding history of soybeans in the USA (Lorenzen *et al.*, 1995). Another package of this type is GENEFLOW®, which was developed by Dr Edie Paul and Dr Susan McCouch at Cornell University (Edie Paul, New York, 1999, personal communication). Finally, the National Center for Genome Resources (http://www.ncgr.org/) is developing a series of databases and bioinformatics tools to link genome, molecular and biochemical information. Integration of molecular and phenotypic data using these tools and combining these data sources with the capacity to simulate breeding programmes is under investigation by several groups.

Integration

Cooper *et al.* (2000) discussed a framework for integrating these different types of programs to consider strategic questions in plant breeding. Here we describe a framework for a decision support system that would have a level of flexibility to consider elements of strategic and tactical decision support in a breeding programme (Fig. 20.2). This system utilizes the new International Crop Information System (ICIS) database, which is being developed by CIMMYT and collaborators to succeed IWIS. The examples discussed here are related to wheat breeding but could equally be developed for the other crops managed within ICIS.

It is critical that the data and information management system have a seamless interface with the data generation processes, that is, data on genes and allele variation, allele values, pedigree information and the allelic variants possessed by breeding lines and varieties, variety performance information and environmental characterization information for the trials that generated the performance data, molecular marker profiles of breeding lines and varieties, and molecular map relationships between molecular markers and genes. ICIS is being used to store data on pedigrees, genes, molecular markers and performance data for wheat varieties from information sources around the world. Thus, with its current structure, ICIS can be used to store many of the critical pieces of information used to quantify the performance of varieties within important target environments. Links to other databases can be established to enable relevant data transfer.

With access to the information contained in these database sources, 'breeder-friendly' software tools are required to enable the breeder to ask the relevant tactical and strategic questions of the databases. For the tactical questions, quick answers are required, whereas for the strategic questions longer time frames will be required to formulate answers. For both types of questions the breeder is likely to involve specific varieties in formulating the question. This can be achieved using the unique variety identifiers (GID) available within ICIS. Therefore, tactical questions, such as construction of a graphical genotype for

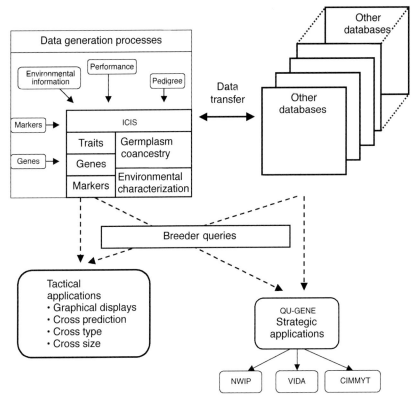

Fig. 20.2. Schematic representation of the components of a decision support system for a plant breeding programme. ICIS is the International Crop Information System, NWIP is the Northern Wheat Improvement Program simulation module of QU-GENE, VIDA is the Victorian Institute for Dryland Agriculture wheat breeding simulation module of QU-GENE, and CIMMYT is the wheat breeding simulation module for the International Center for Maize and Wheat Improvement.

the varieties 'Silverstar' (GID342030) and 'Frame' (GID196806), and highlighting the differences for the known stem, leaf and stripe rust resistance genes, become a relatively easy task when the relevant data are available in ICIS. A follow up tactical question may be: what molecular marker polymorphisms exist between these two varieties and how are they distributed across chromosomes? From there it is possible to examine the likelihoods of producing particular recombinant genotypes for different cross types and population sizes.

Equally, strategic questions can be asked, such as, what would be the efficiency of marker-assisted selection for a series of F_2- and BC_n-derived crosses for specific traits based on the parents 'Silverstar' and 'Frame'? The approach currently being investigated for these and other strategic questions is to transfer the available genetic model information for the selected genotypes from ICIS to the QU-GENE engine, construct an appropriate genotype–environment system model comprising a reference population of genotypes and a target population of environments, and then use the relevant QU-GENE breeding programme-specific modules to simulate the breeding scenarios. Graphical displays, such as those depicted in Fig. 20.1b can then be constructed to summarize the results of the simulation experiments.

The new ICIS database is planned to contain pedigree, varietal performance, gene and molecular marker information. Work is proceeding to link QU-GENE with ICIS to provide the foundation for the type of comprehensive decision support systems for wheat breeders we have described. In this scheme, ICIS is used as a mechanism for storing gene, genotype and environmental data to be used as the inputs to the QU-GENE engine. The QU-GENE engine can then provide a model of the genotype–environment system targeted by the breeding programme. QU-GENE application modules that are developed to represent the operation of specific breeding programmes can then simulate the likely outcomes of the breeding programmes for the target genotype–environment system. ICIS will include information on other cereal crops as well as wheat, but we have confined our discussion to wheat in this chapter.

Information on target environments, including farming systems, is required for the full characterization of the system. Methodology that can be used to obtain this information was discussed by Muchow *et al.* (1996), Cooper and Chapman (1996) and Cooper and Fox (1996). Software such as APSIM (McCowan *et al.*,1995) can handle some data of this type, but will not be discussed further in this chapter. Integration of such methodology and software into an overall decision support framework will facilitate improvements in breeding strategies to produce cultivars better adapted to evolving farming systems.

Simulation and prediction

The simplest form of genetic information to simulate should be for genes providing consistent phenotypes in easily defined target environments. For example, the cross 'Silverstar'/'Frame' is being extensively analysed at VIDA. This cross is expected to segregate at the $Sr9$ locus for resistance to stem rust, $Lr1$ and $Lr13$ loci for resistance to leaf rust, and the $Yr7$ locus for resistance to stripe rust. The phenotypes of progeny from this cross, in terms of resistance or susceptibility to each of these species of rust, can be readily defined for differentiating races of rust in environments conducive to expression. The chromosomal locations of these loci are known: $Sr9$, $Yr7$ and $Lr13$ are on the long arm of chromosome 2B, while $Lr1$ is on chromosome 5D (McIntosh *et al.*, 1995). Information on recombination frequencies of the linked loci is also known (McIntosh *et al.*, 1995). This cross is also expected to segregate for grain quality, which is influenced by combinations of different glutenin and gliadin proteins (Weegels *et al.*, 1996). For example, segregation is possible at the *Glu-D1* locus for the *a* and *d* alleles ('Silverstar' is actually heterogeneous at this locus) and at the *Glu-D3* locus for the *b* and *c* alleles. The *Glu-D1* locus is on the long arm of chromosome 1D, while the *Glu-D3* locus is on the short arm (Singh and Shepherd, 1988a,b).

Predictions of economically desirable combinations of these alleles are possible, but more difficult than for the rust phenotypes. Among other characters, this cross will also segregate for height, because, although both parents are semi-dwarf, they have different alleles at the *Rht1* and *Rht2* loci on chromosomes 4B and 4D. Given that the chromosome locations of these loci are known (McIntosh *et al.*, 1998), it is conceptually not difficult to calculate the probabilities of obtaining particular outcomes (for example a genotype combining the resistances of *Sr9g*, *Yr7*, *Lr1*, and *Lr13* with the *Glu-D1d* and *Glu-D3b* alleles and semi-dwarf height), and the population sizes necessary to have defined probabilities of obtaining them. In this simple example, desirable coupling phase linkages will keep the predicted population sizes relatively small. However, with undesirable repulsion linkages, the computational difficulty will increase rapidly with increasing numbers of linked loci, so that efficient algorithms for modelling recombination events will be required.

In the previous example, we considered genes with readily identified effects (qualitative traits) of economic importance. Using molecular markers, chromosome regions associated with quantitative traits of economic importance can be identified. An example was given by Fritz *et al.* (1995), who identified RFLP markers associated with grain yield and kernel weight in backcross populations derived from crosses between elite cultivars of wheat and *Triticum tauschii*. However, the RFLP markers explained a relatively small proportion of the total variation and, not surprisingly, their effects interacted with environment. Fritz *et al.* (1995) suggested that important regions of the genome might not have been covered by the RFLPs they used, and that duplicate effects of homeoalleles could have dampening effects. A recent workshop held at CIMMYT considered the status of molecular maps for quantitative trait loci (QTLs) associated with drought resistance. The presence of interactions between the QTLs and environment and epistatic interactions between QTLs was common for the traits examined. This suggests that predictions of outcomes will be complicated for these quantitative traits.

Molecular markers can be used to assist with gene introgression from a donor parent into the genome of a recipient line (Hospital and Charcosset, 1997). Usually, the donor will be an unadapted line possessing some desirable trait, while the recipient line will be an adapted cultivar. Often, the objective is to introgress the genes determining the desirable trait, while recovering the remaining genotype of the recipient line as rapidly and cheaply as possible. Molecular markers can be particularly effective for background selection, where no knowledge is required of the activities of genes on the marked chromosome regions.

The concept of selection for the background genotype can be extended to crosses involving relatively well-adapted parents, especially with varying degrees of relationship. In Australia, most cultivars are related and carry varying proportions of common chromosome segments (McCormick et al., 1995; Paull et al., 1998). This allows them to be classified into family groups. Often, wheat breeders making crosses to combine desirable characteristics can choose among many potential parents. Using molecular marker information on the background genotype would allow predictions of the extent of variation likely from particular crosses, or backcrosses, and thus allow rational choices to be made on which parents to use, and the population sizes required for defined outcomes.

Molecular marker storage and utilization

We regard the input of molecular marker data covering a high proportion of the wheat genome as essential to assist predictions by breeders. One way this could happen is to utilize database modules of ICIS containing two types of information.

1. *Gene location data.* This would be developed from mapping populations and would contain information on locations of gene loci on the wheat chromosomes. At this stage there is merit in the use of simple sequence repeats (SSRs) (Hearne et al., 1992) and as many other qualitative trait loci as possible. The qualitative trait loci would be as comprehensive as possible, and especially include loci controlling traits of importance to the breeder, such as the glutenin and gliadin proteins (*Glu* and *Gli* loci), types of starch, resistances to rust (*Sr*, *Lr* and *Yr* loci), height characteristics (*Rht* loci), presence and absence of awns, etc. The positions of many of these loci on the linkage map are known (McIntosh et al., 1998), and would be revised as more information becomes available.

We have suggested SSRs because of their potential polymorphisms in wheat, and because they provide a highly informative, co-dominant, PCR-based assay (Powell et al., 1996). When appropriate, other types of molecular markers could be included. This may be through anonymous markers located on the map and linked to the gene of interest or preferably through a marker located within the sequence of the target gene itself. We suggest using the best estimate of each locus position, ignoring the fact that recombination varies among crosses. Detection of recombination frequencies will vary among crosses for several reasons including: (i) the level of coancestry between the parents of the crosses; and (ii) differences in the frequency of recombination events due to structural features of the genome. In addition, position estimates are also subject to stochastic error in a similar way to other biological data.

2. *Genotype profile data.* This would contain information on the alleles present at each gene locus for each genotype (cultivar, breeding line, germplasm accession, etc.). This would be the larger of the two components of the database considered here.

The link between the two modules is provided by the gene locus, and the program should be able to create a linkage map of the full genome of any genotype. This means that alleles of a particular gene would be allocated to loci on the linkage map. Levels of polymorphism and the map for the two genotypes would provide the basis for prediction of the outcomes of segregating generations. This would require simulation of meiosis, including crossing over, and the generation of distributions of resulting genotypes. The probability of particular combinations, designated by the plant breeder running the program, should be possible for F_2 and backcross generations, and from this a prediction of population sizes required to identify these combinations. As mentioned previously, we envisage the SSR information as important for determining population sizes to recover proportions of designated parents. For example, in the earlier example with 'Frame', 'Silverstar' might be identified as a complementary parent. The breeder might then designate 90% 'Frame' and 10% 'Silverstar' for one run, 80% 'Frame' and 20% 'Silverstar', etc. Predictions of population types and progeny sizes would be extremely valuable for designing an efficient breeding programme. The outcomes will depend on degree of relatedness of the parents and on linkage relationships between desirable and undesirable alleles.

Pedigree information

The storage and retrieval of pedigree information can be handled through IWIS. This methodology has been the basis for developing ICIS, and can provide information to the wheat breeder on relatedness of particular genotypes using coefficients of parentage.

Linkage between molecular marker and pedigree information

Linkage of molecular marker and pedigree data in a database of the IWIS type would be through the unique genotype identifier, and would be able to provide the type of visual output from a program like SUPERGENE. This should include the inheritance of alleles through pedigrees. Furthermore, degree of relatedness can be calculated using molecular marker and other genotype data.

Linkages to QU-GENE

QU-GENE provides the simulation platform. Linkages between QU-GENE and ICIS are being developed (Cooper *et al.*, 2000). Under the structure suggested earlier in this chapter, the molecular data would be accessed, and maps of specific genotypes would be generated, in the molecular marker component of ICIS. These data would be transferred to the simulation platform of QU-GENE, which would then be used to examine the merits of alternative breeding strategies and generate probabilities of specific outcomes.

While some forms of target environment data can already be used in QU-GENE, further work is proceeding to broaden the range of characterizations that can be accommodated (e.g. Chapman *et al.*, 2000a,b,c). For complete utilization, a determination of the traits, genes or QTLs for specific adaptation to different types of target environment will be required. This work will continue to evolve as we improve our understanding of the genetic architecture of the traits manipulated in breeding programmes. However, a very useful system could operate, in the first instance, with minimal information of this type.

Conclusion

Computer programs can be developed to simulate plant breeding programmes and will provide a much more rational basis for a wheat breeder to determine the outcomes of crossing and selection strategies. Examples based on the QU-GENE software were demonstrated by Podlich and Cooper (1998), Podlich *et al.* (1999) and Cooper *et al.* (1999a,b). An important component of this system will be provided by molecular marker data which will allow simulation of meiosis and gene segregation across the entire genome. Parts of the system are already operational. For acceptance by wheat breeders, the system must be easy to use and provide simple visual output of the type required to make practical breeding decisions.

Acknowledgements

We thank Dr Edie Paul for sending information on GENEFLOW® and Dr Maarten van Ginkel for information on the rate of success from crosses in the CIMMYT wheat breeding programme.

References

Boutin, S.R., Young, N.D., Lorenzen, L.L. and Shoemaker, R.C. (1995) Marker-based pedigrees and graphical genotypes generated by supergene software. *Crop Science* 35, 1703–1707.

Chapman, S.C., Cooper, M., Butler, D. and Henzell, R.G. (2000a) Genotype by environment interactions affecting grain sorghum. I. Characteristics that confound interpretation of hybrid yield. *Australian Journal of Agricultural Research* 51, 197–207.

Chapman, S.C., Cooper, M., Hammer, G.L. and Butler, D. (2000b) Genotype by environment interactions affecting grain sorghum. II. Frequencies of different seasonal patterns of drought stress are related to location effects on hybrid yields. *Australian Journal of Agricultural Research* 51, 209–221.

Chapman, S.C., Hammer, G.L., Butler, D.G. and Cooper, M. (2000c) Genotype by environment interactions affecting grain sorghum. III. Temporal sequences and spatial patterns in the target population of environments. *Australian Journal of Agricultural Research* 51, 223–233.

Cooper, M. and Chapman, S.C. (1996) Breeding sorghum for target environments in Australia. In: Foale, M.A., Henzell, R.G. and Kneipp, J.F. (eds) *Proceedings of the Third Australian Sorghum Conference*. AIAS Occasional Publication 93, Australian Institute of Agricultural Science, Melbourne, Australia, pp. 173–187.

Cooper, M. and Fox, P.N. (1996) Environmental characterization based on probe and reference genotypes. In: Cooper, M. and Hamer, G.L. (eds) *Plant Adaptation and Crop Improvement*. CAB International, IRRI and ICRISAT, Wallingford, UK, pp. 529–547.

Cooper, M., Podlich, D.W. and Fukai, S. (1999a) Combining information from multi-environment trials and molecular markers to select adaptive traits for yield improvement of rice in water-limited environments. In: O'Toole, J.C., Ito, O., O'Toole, J. and Hardy, B. (eds) *Genetic Improvement of Rice for Water-limited Environments. Proceedings of the Workshop on Genetic Improvement of Rice for Water-Limited Environments, 1–3 Dec. 1998, Los Baños, Philippines*. International Rice Research Institute, Los Baños, Philippines, pp. 13–33.

Cooper, M., Podlich, D.W. and Jensen, N.M. (1999b) Modelling breeding programmes: genes, phenotypes and breeding strategies. In: Williamson, P., Banks, P., Haak, I., Thompson, J. and Campbell, A. (eds) *Proceedings of the Ninth Assembly of the Wheat Breeding Society of Australia*. Organising Committee of the Australian Wheat Breeding Society, The University of Southern Queensland, Toowoomba, 27 September – 1 October, pp. 121–128.

Cooper, M., Podlich, D.W. and Chapman, S.C. (2000) Computer simulation linked to gene information databases as a strategic research tool to evaluate molecular approaches for genetic improvement of crops. In: Ribaut, J.-M. and Poland, D. (eds) *Molecular Approaches for the Genetic Improvement of Cereals for Stable Production in Water-Limited Environments*. A Strategic Planning Workshop held at CIMMYT, El Batan, Mexico, 21–25 June 1999, pp. 162–166.

The Economist (US) (1999) Drowning in data. 351, p. 93.

Fox, P.N., Magaña, R.I., Lopez, C., Sanchez, H., Herrera, R., Vicarte, V., White, J.W., Skovmand, B. and Mackay, M.C. (1997) *The International Wheat Information System (IWIS), Version 2*, CD. CIMMYT, Mexico.

Fritz, A.K., Cox, T.S., Gill, B.S. and Sears, R.G. (1995) Marker-based analysis of quantitative traits in winter wheat × *Triticum tauschii* populations. *Crop Science* 35, 1695–1699.

Hearne, C.M., Ghosh, S. and Todd, J.A. (1992) Microsatellites for linkage analysis of genetic traits. *Trends in Genetics* 8, 288–294.

Hospital, F. and Charcosset, A. (1997) Marker-assisted introgression of quantitative trait loci. *Genetics* 147, 1469–1485.

Jensen, N.M., Podlich, D.W. and Cooper, M. (1999) The influence of crossing strategy on genetic progress from pedigree breeding. In: Langridge, P., Barr, A., Auricht, G., Collins, G., Granger, A., Handford, D. and Paull, J. (eds) *Proceedings of the 11th Australian Plant Breeding Conference* Vol. 2 *Contributed Papers, Adelaide, 19–23 April*, CRC for Molecular Plant Breeding, Adelaide, pp. 83–84.

Lande, R. (1992) Marker-assisted selection in relation to traditional methods of plant breeding. In: Stalker, H.T. and Murphy, J.P. (eds) *Plant Breeding in the 1990s*. CAB International, Wallingford, UK, pp. 437–451.

Lorenzen, L.L., Boutin, S., Young, N., Specht, J.E. and Shoemaker, R.C. (1995) Soybean pedigree analysis using map-based molecular markers: I. Tracking RFLP markers in cultivars. *Crop Science* 35, 1326–1336.

McCormick, K.M., Panozzo, J.F. and Eagles, H.A. (1995) Starch pasting properties and genetic relationships of wheat cultivars important to Victorian wheat breeding. *Australian Journal of Agricultural Research* 46, 861–871.

McCowan, R.L., Hammer, G.L., Hargraves, J.N.G., Holzworth, P.D. and Freebairn, D.M. (1995) APSIM: a novel software system for model development, model testing and simulation in agricultural systems research. *Agricultural Systems* 48, 1–17.

McIntosh, R.A., Wellings, C.R. and Park, R.F. (1995) *Wheat Rusts: an Atlas of Resistance Genes*. CSIRO Publications, Melbourne, Australia.

McIntosh, R.A., Hart, G.E., Devos, K.M., Gale, M.D. and Rogers, W.J. (1998) Catalogue of gene symbols for wheat. *Proceedings of the 9th International Wheat Genetics Symposium*, Vol. 5, *2–7 August 1998, Saskatoon, Canada*. University Extension Press, University of Saskatchewan, Canada.

Merrill, R.E. (1999) Research information management at Pioneer Hi-Bred International Inc. In: Langridge, P., Barr, A., Auricht, G., Collins, G., Granger, A., Handford, D. and Paull, J. (eds) *Proceedings of the 11th Australian Plant Breeding Conference, Additional Papers, Adelaide, 19–23 April*, CRC for Molecular Plant Breeding, Adelaide.

Muchow, R.C., Cooper, M. and Hamer, G.L. (1996) Characterizing environmental challenges using models. In: Cooper, M. and Hamer, G.L. (eds) *Plant Adaptation and Crop Improvement*. CAB International, IRRI and ICRISAT, Wallingford, UK, pp. 349–364.

Paull, J.G., Chalmers, K.J., Karakousis, A., Kretschmer, J.M., Manning, S. and Langridge, P. (1998) Genetic diversity in Australian wheat varieties and breeding material based on RFLP data. *Theoretical and Applied Genetics* 96, 435–446.

Podlich, D.W. and Cooper, M. (1998) QU-GENE: a simulation platform for quantitative analysis of genetic models. *Bioinformatics* 14, 632–653.

Podlich, D.W., Cooper, M. and Basford, K.E. (1999) Computer simulation of a selection strategy to accommodate genotype-by-environment interactions in a wheat recurrent selection programme. *Plant Breeding*, 118, 17–28.

Powell, W., Morgante, M., Andre, C., Hanafey, M., Vogel, J., Tingey, S. and Rafalski, A. (1996) The comparison of RFLP, RAPD, AFLP and SSR (microsatellite) markers for germplasm analysis. *Molecular Breeding* 2, 225–238.

Singh, N.K. and Shepherd, K.W. (1988a) Linkage mapping of genes controlling endosperm storage proteins in wheat: genes on the short arms of group 1 chromosomes. *Theoretical and Applied Genetics*, 75, 628–641.

Singh, N.K. and Shepherd, K.W. (1988b) Linkage mapping of genes controlling endosperm storage proteins in wheat: genes on the long arms of group 1 chromosomes. *Theoretical and Applied Genetics*, 75, 642–650.

Weegels, P.L., Hamer, R.J. and Schofield, J.D. (1996) Functional properties of wheat glutenin. *Journal of Cereal Science* 23, 1–18.

Chapter 21
Application of DNA Profiling to an Outbreeding Forage Species

J.W. FORSTER[1], E.S. JONES[1], R. KÖLLIKER[1,2], M.C. DRAYTON[1], M.P. DUPAL[1], K.M. GUTHRIDGE[1] AND K.F. SMITH[3]

[1] Plant Biotechnology Centre, Agriculture Victoria, La Trobe University, Bundoora, Victoria, Australia; [2] Swiss Federal Research Station for Agroecology and Agriculture, Zürich, Switzerland; [3] Pastoral and Veterinary Institute, Agriculture Victoria, Hamilton, Victoria, Australia

Introduction

Forage and pasture species provide the basis for grazing agricultural systems on a global scale. The production of meat, dairy products and pelt derivatives such as wool and leather from sheep, cattle and other herbivores is of high economic importance. In Australia, the pastoral industries have been estimated to generate products of gross value c. AU\$10 billion year^{-1} (Australian Bureau of Statistics, 2000). The most important herbage species are pasture grasses (*Poaceae*) and legumes (*Leguminosae*). In addition, several species of grass are significant for turf and amenity purposes, supporting a global seed export market valued at c. US\$419 million year^{-1} (Le Buanec, 1997).

The most valuable pasture grasses in temperate regions include the members of the two closely related genera *Lolium* and *Festuca*. The predominant species are perennial ryegrass (*Lolium perenne* L.), Italian ryegrass (*Lolium multiflorum* Lam.), meadow fescue (*Festuca pratensis* Huds.) and tall fescue (*Festuca arundinacea* Schreb; Holmes, 1980). Other grasses such as cocksfoot (*Dactylis glomerata* L.), Kentucky bluegrass (*Poa pratensis* L.), phalaris (*Phalaris aquatica* L.) and the brome grasses such as smooth brome grass (*Bromus inermis* L.) show more limited usage. Warm season and tropical pasture grass species include members of the genera *Brachiaria* and *Paspalum* as well as buffelgrass (*Pennisetum ciliare* L.).

The key pasture legumes include lucerne (*Medicago sativa* L.) and clovers of the genus *Trifolium* (white clover (*Trifolium repens* L.), red clover (*Trifolium pratense* L.) and subterranean clover (*Trifolium subterraneum* L.)). Minor species include bird's foot trefoil (*Lotus corniculatus*) and sweet clover (*Melilotus alba*). Members of the genera *Stylosanthes*, *Macroptilium* and *Centrosema* are cultivated as tropical pasture legumes.

The pasture grasses and legumes show a wide range of variation in breeding system, annual/perennial behaviour and ploidy levels. Among the pasture and turfgrasses and their wild relatives, inbreeding (autogamous), outbreeding (allogamous), vegetatively propagated and apomictic species have been described. The most important of the temperate forages are cross-pollinated. Grasses such as perennial ryegrass and tall fescue are obligate outbreeders, with a gametophytic self-incompatibility system controlled by two genes designated S and Z (Cornish *et al.*, 1979). White clover, red clover and lucerne are also allogamous species. White clover shows a gametophytic self-incompatibility system controlled by a single gene designated S (Attwood, 1940), similar to that of red clover, while lucerne shows a partial self-incompatibility system (Barnes *et al.*, 1972). As a consequence of the outbreeding habit, both natural ecotypes and synthetic populations of these species are likely to be highly genetically heterogeneous. Strategies for genetic diversity analysis based on DNA profiling must address this issue and allow the quantification of variation within and between populations.

Evaluation of genetic variation for outbreeding forage species is important for the processes of cultivar identification and seed purity analysis, the ecological analysis of pasture populations and the selection of genetically divergent parents for genetic mapping studies. This chapter describes the implications of current breeding methods for cross-pollinated forage crops, the strategies for analysis that may be employed and the current and future genotyping technology.

Cultivar development in outbreeding forage species

Classical breeding of forage species is based on the generation of synthetic populations (Vogel and Pedersen, 1993). For perennial ryegrass, the following process has been employed: evaluation of suitable base populations of 2000–5000 genotypes, selection of *c.* 200 genotypes which are separated into one or more groups based on time of reproductive maturity, open pollination (polycrossing) within maturity groups, selection of superior 'mother plants' based on half-sib analysis, polycrossing of these selected individuals (generally in groups of seven to ten parents) to produce a synthetic 1 (Syn 1) population, all followed by further cycles of multiplication to produce certified seed. This produces a cultivar with a relatively non-restricted genetic base. A similar process using four to six parental clones from the synthetic 1 population produces a cultivar with a relatively restricted genetic base. In polyploid species such as tall wheatgrass (*Thinopyrum ponticum*) and lucerne, between 50 and 100 parental clones may be used during synthetic cultivar formation (Bray and Irwin, 1999; Smith, 2000). These breeding methods inevitably lead to genetic heterogeneity within and between varieties that may be morphologically similar. The large number of cultivars available for the main forage species compounds the problem of accurate discrimination between genetically diverse varieties. Some 720 perennial ryegrass cultivars are eligible for Organization for Economic Cooperation and Development certification (Anon, 1999), while a world checklist of white clover lists 326 cultivars (Caradus and Woodfield, 1997).

Under International Union for the Protection of New Varieties of Plants (UPOV), plant breeders' rights (PBR) are established based on criteria of distinctiveness, uniformity and stability. As most new forage cultivars have been selected for improved total or seasonal forage yield, which are not permitted as traits to describe varietal distinctiveness, the majority of cultivars are distinguished on the basis of a number of morphological traits. Flowering time, plant height, growth habit, flag leaf length, leaf colour, awns, ploidy, seedling florescence and leaf length and width are all listed in the UPOV guidelines for *Lolium*. The environment may influence the expression of many of these characters, some of which are only expressed late in the life cycle of the plant. The criteria of uniformity and stability over generations of seed multiplication are assessed by appropriate field trial experiments. The possibility of genetic drift over generations for a cross-pollinating species, along with potential pollen contamination from nearby grass populations, clearly poses problems for these assessments.

Forage grass and legume species are propagated by the production and subsequent distribution of seed. Seed purity is an important requirement for certification, particularly for grass breeders. Contamination of seed batches may occur through pollen contamination from adjacent field plots, seed mixing from other varieties during the harvesting and processing stages of seed multiplication, or contamination with seed of resident naturalized populations. These problems are acute when production is highly localized geographically, as in the Willamette Valley of Oregon, USA ('the grass seed capital of the world'). Production of *c.* 60% of the world's cool-season grass seed is performed in Oregon (Hampton, 1991). Apart from contamination between cultivars, contamination of perennial ryegrass with seed from annual varieties of Italian ryegrass (*L. multiflorum* Lam. var. *westerwoldicum*) is a serious problem that limits the persistence and agronomic performance of perennial ryegrass varieties. The potential contamination of white clover seed crops by plants arising from hard, buried seed has been identified as a major threat to the white clover seed production industry (Hampton *et al.*, 1987).

The methods used to discriminate between cultivars should ideally be uninfluenced by environmental factors and be detectable at all stages of the plant life cycle. The use of DNA profiling data has the potential to support PBR registration strongly and determine the purity of seed lots. The application of molecular marker data to assist the establishment of breeders' rights is the remit of a working group on biochemical and molecular techniques and DNA profiling convened by UPOV. Such data are acceptable as a supplementary character for PBR registration within Australia (Morell *et al.*, 1995). Molecular marker data will allow the evaluation of genetic distinctiveness for new cultivars as well as the assessment of genetic stability over time. DNA profiling is also likely to be of key importance in the establishment of essential derivation. The international breeders' rights organization ASSINSEL has defined an essentially derived variety (EDV) as one which is clearly distinct from the initial variety (IV) but conforms to the IV in its essential characteristics and is predominantly derived from the IV. For a successful EDV claim, evidence of phenotypic conformity and genetic similarity is required.

Strategies for analysis of genetic variation in heterogeneous populations

Analysis of individual genotypes

In order to evaluate within and between population variation, samples of individual genotypes from cultivars and ecotypes must be selected for analysis. The number of individual genotypes required will be influenced by several criteria. The genetic structure of the population determines the extent of genetic variability. A cultivar may have a wide genetic base (with genetically distant parents) or a narrow genetic base (with genetically similar parents). The genetic base may also be categorized as non-restricted (with a large number of parents) or restricted (with a small number of parents). The level of variability will be a major factor in determining the number of individuals necessary to characterize the cultivar effectively. Low population variability, arising from a narrow or restricted base, will require a small number of individuals. Conversely, high intrapopulation variability is expected of wide and non-restricted base cultivars, and larger numbers of individuals may be required for genotypic analysis. It is, of course, possible to derive a restricted base cultivar with high genetic variability by choosing highly divergent parental clones.

The optimal sampling size in order to assay genetic variation within cross-pollinated populations has been determined in the context of germplasm conservation (Crossa, 1989). Statistical analysis indicates that the optimal sample size largely depends on the frequency of the least common allele or genotype. In addition, for multi-allelic loci, the sample size is determined more by the frequency of rare alleles than allele number. Alleles at frequencies of more than 10% may be effectively sampled by 40 individuals, while 100 individuals are required for allele frequencies of 5%. A selection of 300–400 genotypes is necessary for allele frequencies close to 1%. Bottleneck effects due to small sample numbers lead inevitably to loss of rare alleles (Crossa *et al.*, 1992), suggesting a typical value of 100 interpollinating individuals to maintain a given accession.

For the comparison of different cultivars and populations, the genetic structure of each population and their degree of relatedness will be important. Close bred varieties which share a high proportion of common alleles will be more difficult to discriminate than those based on genetically divergent parents, and this effect will again be influenced by the restricted or non-restricted structure of the synthetic population.

The cost and complexity of the analysis process is another major consideration that limits the numbers of genotypes that may be effectively handled. For molecular markers, the cost of DNA extraction and subsequent manipulation for large sample sizes may prove prohibitive. Marker assays based on Southern hybridization are lengthy compared with PCR, providing another constraint. In practice, between 10 and 70 individuals have been used in studies of perennial ryegrass cultivars (Huff, 1997; Barker and Warnke, 1998). Numbers of individual markers have varied according to the detection system, from c. 30–100 for dominant random amplified polymorphic DNA (RAPD) loci to 20–60 for co-dominant restricted fragment length polymorphism (RFLP) loci. The number of markers required for discrimination of individuals will increase in proportion to the degree of genetic similarity.

Analysis of bulked samples

In order to analyse larger numbers of genotypes, bulking or pooling strategies may be employed. These are particularly important for the screening of gene bank accessions and for the large-scale analysis of cultivar identity and seed batch purity. Bulking methods for molecular markers were originally developed for use with members of segregating populations (Michelmore *et al.*, 1991). Bulking may occur at the level of plant material or purified genomic DNA template. The optimal bulk size has been estimated in several studies. Small replicated bulks of three to five individuals are optimal in order to detect rare alleles and effectively measure genetic variation within populations (Gilbert *et al.*, 1999; Kraft and Säll, 1999). In order to allow cultivar identification by producing diagnostic profiles, rare alleles are preferentially eliminated by dilution in larger bulk samples. Studies based on RAPD analysis (Williams *et al.*, 1990) have determined the relevant bulk size to be seven for lucerne (Yu and Pauls, 1993) and 20 for red clover (Kongkiatngam *et al.*, 1996). Sweeney and Danneberger (1994) used 30 bulked individuals to distinguish populations of perennial ryegrass, while Liu *et al.* (1994) analysed 20 bulked individuals per ecotype of the forage and turfgrass seashore paspalum (*Paspalum vaginatum* Swartz.). Based on the experience of these studies, a value of 20 genotypes per population seems appropriate for most forage species.

Previous studies of genetic diversity in outbreeding forage species

Forage grasses

The earliest use of molecular markers to detect genetic variation in forage grasses was based on protein polymorphisms. A number of studies have demonstrated the potential for use of isozyme polymorphism to discriminate between cultivars of *Lolium* (Hayward and McAdam, 1977; Gilliland *et al.*, 1982) and a catalogue of allele frequency at three isozyme loci has been formed in *Lolium* (Lallemand *et al.*, 1991). However, it was only possible to separate a large number (355) of *Lolium* cultivars into 23 groups on the basis of allele frequency at the *PGI-2* locus, and a set of 160 cultivars into 40 groups when *PGI-2* and *ACP-1* allele frequencies were considered together (Booy *et al.*, 1993). Other studies have identified cultivar-based variation in the banding pattern of storage proteins in seed samples when analysed by SDS-PAGE (Ferguson and Grabe, 1986; Krishnan and Sleper, 1997).

Most studies to date using DNA-based markers have concentrated on RAPD analysis. Huff (1997) used RAPDs to characterize 18 perennial ryegrass populations, including related crosses, derived selections and unrelated ecotypes. Ten individuals from each population were analysed for the presence of 33 polymorphic RAPD markers. No population-specific markers were observed in this study, but analysis of molecular variance allowed the relationships between populations to be determined. Closely related populations could not be reliably separated, and turfgrass varieties were shown to be derived from a narrow genetic base. Barker and Warnke (1998) describe the analysis of the minimum numbers of individual plants and RAPD markers required to characterize perennial ryegrass populations effectively. RAPDs were used to detect genetic contamination from volunteer seedlings in seed production plots when changing cultivars. Seventy plants were analysed from each group. Once again, no cultivar-specific RAPD markers were detected. The frequency distributions for selected bands were determined, showing that the discriminatory power is correlated with the frequency bias of individual bands.

Kölliker *et al.* (1999) analysed genetic variability with RAPDs within and between cultivars of three forage species: perennial ryegrass, cocksfoot and meadow fescue. Twenty-eight individuals were selected from each of three cultivars from each species, and were analysed using 104 polymorphic RAPD markers. Genetic variability was lower in meadow fescue than in the other species, and is apparently related to management practices in permanent pastures (Kölliker *et al.*, 1998). RAPDs have also been used to identify perennial ryegrass varieties for cultivar identification purposes by De Loose *et al.* (1994), Sweeney and Danneberger (1994), Wiesner *et al.* (1995) and Posselt and Bolaric (1999). The use of RAPDs generated from seed bulks was evaluated by Sweeney and Danneberger (1997). Eight of eleven perennial ryegrass test cultivars were distinguished, providing the basis for seed lot certification, although the purity of genomic DNA template was a significant constraint in this study.

RFLP analysis has been used to a lesser extent than RAPD. RFLP in the ribosomal DNA intergenic spacer and coding sequence regions was used to distinguish ten unique patterns among 35 perennial ryegrass cultivars (Warpeha *et al.*, 1998), while mitochondrial DNA RFLP (Sato *et al.*, 1995) revealed variation within single *L. perenne* cultivars. Low copy nuclear DNA sequences derived from a *Pst*I genomic library were used to detect RFLP in nine turf and seven forage tall fescue cultivars (Xu *et al.*, 1994). Twenty plants from each cultivar were selected and analysed with 24 RFLP probes. Genetic variation within cultivars was high, 16 randomly chosen plants being sufficient to assay and maintain genetic diversity in these accessions. The separation of cultivars was consistent with pedigree information, with higher divergence among forage than turf cultivars.

Forage legumes

RAPDs have been used to analyse variation within and between white clover populations of permanent pastures in the north-eastern USA (Gustine and Huff, 1999). A high degree of genetic variation was detected within populations, with each of 18 populations showing significant differences but no consistent differences between the three states of origin. In red clover, two cultivars were analysed for genetic variation using morphological, isoenzyme and RAPD markers (Kongkiatngam *et al.*, 1995) in order to validate the RAPD technique, which was then used in conjunction with bulking to distinguish 15 red clover cultivars (Kongkiatngam *et al.*, 1996). A total of 55 polymorphic RAPD markers were used, with cultivar-specific bands detected in bulks with 13 of 14 primers tested.

RAPD markers were used on DNA bulked samples of lucerne by Yu and Pauls (1993) to estimate the genetic distance between heterogeneous populations. Based on these data, it was estimated that the use of ten primers in concert with ten individuals would suffice to distinguish up to 70 different cultivars. Ghérardi *et al.* (1998) analysed genetic variation from individual plants from eight natural and cultivated populations of *M. sativa* and *Medicago falcata*. Very few population-specific RAPD markers were identified, but genetic distance analysis indicated that *c.* 60 markers analysed over 20 individuals could distinguish *Medicago* germplasm. RFLP markers based on cDNA and gDNA clones were used to evaluate the approach to homozygosity under selfing in diploid lucerne (Brouwer and Osborn, 1997) and to correlate genetic distance, forage yield and heterozygosity in isogenic diploid and tetraploid lucerne populations (Kidwell *et al.*, 1994).

DNA profiling with AFLP and SSRP markers in outbreeding forage species

Background

RAPD and RFLP analysis in forage crops has given valuable information on the genetic structure of populations, optimum bulk sizes and the number of markers required for population discrimination. However, RAPD analysis is now relatively disfavoured due to problems with reproducibility (Jones *et al.*, 1997).

RFLP analysis is highly reproducible and informative but logistically complex for large numbers of samples. In future, analysis based on amplified fragment length polymorphism (AFLP) (Vos et al., 1995) and simple sequence repeat polymorphism (SSRP) (Rafalski et al., 1996) will be more important. Both methods are based on PCR analysis and are highly reproducible. AFLP produces mainly dominant markers with a high multiplex ratio (typically 80–100 loci), while SSRP detects single co-dominant loci. There have been few reports to date of the development of these systems for forage species. In forage legumes, AFLP profiles have been obtained and compared with RFLP data for 24 parental genotypes of two synthetic white clover varieties (Griffiths et al., 1999) and AFLP has been developed for genetic mapping in lucerne (Barcaccia et al., 1999). SSRs for forage legumes have so far only been reported in lucerne and other *Medicago* species (Diwan et al., 1997). For forage grasses, AFLPs have been used for genetic mapping by Bert et al. (1999), while small numbers of SSR markers have been reported for perennial ryegrass (Kubik et al., 1999) and seashore paspalum (Liu et al., 1995).

Current activities in our laboratory are aimed at the development and implementation of AFLP- and SSRP-based marker systems for the molecular breeding of perennial ryegrass and white clover. These species are frequently co-cultivated in a mixed sward system, and constitute the key species for dairy production in temperate regions. The ability to discriminate genotypes and populations has been demonstrated using AFLP systems, providing the basis for the implementation of SSR markers in future.

AFLP-based analysis of genetic diversity in perennial ryegrass

AFLP detection may be based on the generation of primary template using either *Eco*RI or *Pst*I as rare cutter enzymes. *Pst*I-generated AFLPs target demethylated regions of higher plant genomes due to the methylation sensitivity of the enzyme. This property may ensure closer association with genic regions than for *Eco*RI-generated AFLPs, which contain significant numbers of repetitive, methylated DNA sequences. Both systems are used for DNA profiling, but the *Pst*I AFLP system may be of special value as loci are likely to be less clustered within the genome. Putative association with agronomic genes will also assist the selection of divergent potential parental genotypes for trait-specific mapping crosses. Polymorphic AFLP loci may be mapped and evaluated for linkage with genes and quantitative trait loci (QTLs) for the relevant target trait.

Due to the methylation sensitivity of *Pst*I, variability between different templates may be observed, reflecting spatial or temporal variation in levels and specificity of genomic methylation. Organ-specific differences in AFLP profiles of wheat due to differences in DNA methylation have been previously reported (Donini et al., 1997). In order to test the reproducibility of both *Eco*RI and *Pst*I AFLPs, three independent template DNA samples were extracted from mature leaf tissue of three perennial ryegrass genotypes. Several different primer combinations were used for each rare cutter. High levels of reproducibility were seen for both enzyme systems. Some very minor variations were detected with certain primer pairs for each enzyme, but there was no consistent difference attributable to methylation effects (data not shown). The proportion of variable bands was not more than 2% of the polymorphic bands (K.M. Guthridge, unpublished). Extra bands may be attributable to contamination of template DNA, possibly due to the presence of the fungal endophyte *Neotyphodium lolii*, which varies in distribution within individual plants and between genotypes of perennial ryegrass and may be differentially represented in replicated DNA templates. The frequency of spurious bands is unlikely to have a significant impact on either diversity studies or AFLP mapping in perennial ryegrass.

AFLP profiles were generated from 12 different *L. perenne* genotypes covering a range of Australian, New Zealand and North African germplasm. Five $PstI+3/MseI+3$ primer combinations generated a total of 175 polymorphic bands. The data were analysed to generate a phenogram (Fig. 21.1a) based on coefficient of genetic similarity (K.M. Guthridge, unpublished). Each of the genotypes was effectively discriminated, and pairs of plants from the cultivars 'Vedette' and 'Ellett' and the Tunisian ecotype T35 were positioned close to one another. However, individual genotypes from within the cultivar 'Yatsyn' were separated, indicating substantial within-population diversity. The North African accessions (from Algeria, Morocco and Tunisia) formed a single clade separate from the other genotypes, demonstrating the genetic divergence of these populations from cultivated perennial ryegrass varieties. The Mediterranean basin, specifically the Middle East, is the likely centre of origin for perennial ryegrass (Balfourier *et al.*, 1998).

The detection of AFLP diversity based on single genotypes was extended to individuals selected from several distinct populations. As part of a strategy for molecular marker mapping of QTLs for drought tolerance in perennial ryegrass, four populations were selected to provide potential parents for mapping crosses. Selection was based on existing phenotypic data. The North African ecotype is drought tolerant through the mechanism of summer dormancy, while 'Aurora', a UK cultivar derived from a Swiss ecotype, is drought sensitive and summer active. The Australian cultivar 'Victorian' has been derived from an ecotype locally adapted to low rainfall zones, but is summer active, while the New Zealand dairy cultivar 'Aries' is drought sensitive.

Six individuals from each population were profiled with two $EcoRI+3/MseI+3$ and three $PstI+3/MseI+3$ primer combinations to generate 184 polymorphic bands (63 and 121, respectively). The data were analysed to generate a phenogram based on coefficient of genetic similarity (Fig. 21.1b) and a three-dimensional principal coordinate analysis (PCA) plot (Fig. 21.1c). The dendrogram shows a separate cluster of four of the six North African genotypes, which are most distant from the other groups, confirming the divergent nature of this germplasm. 'Victorian' and 'Aurora' also form distinct groups, although there is some overlap between them, and two of the North African genotypes cluster with 'Victorian'. This provides evidence for substantial genetic variation

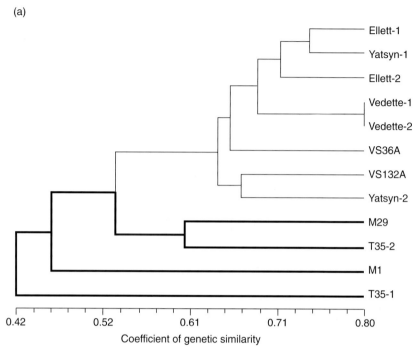

(a)

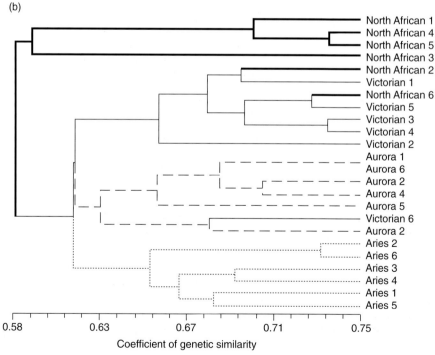

(b)

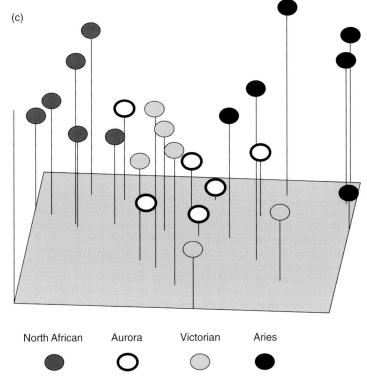

Fig. 21.1. UPGMA phenograms showing relationships between (a) (*opposite*) 12 *L. perenne* genotypes. The scale indicates the coefficient of genetic similarity based on the method of Nei and Li (1979). VS36 and VS132: selections from Australian cultivar 'Victorian'; 'Ellett', 'Yatsyn', 'Vedette': New Zealand cultivars (all normal lines); M1: Moroccan ecotype; T35: Tunisian ecotype; M29: selection from Algerian ecotype (all bold lines). (b) (*opposite*) Six genotypes from each of four populations of perennial ryegrass. The scale indicates the coefficient of genetic similarity based on the method of Nei and Li (1979). North African: Moroccan ecotype (bold lines); 'Victorian': Australian cultivar (normal lines); 'Aurora' UK cultivar (broken lines); 'Aries': New Zealand cultivar (dotted lines). (c) Three-dimensional principal coordinate analysis plot for six genotypes from each of four populations of perennial ryegrass.

within the ecotype. 'Aries' forms a separate clade. The 3-D PCA plot also clearly shows the clustering and relative genetic distance between the parental groups (K.M. Guthridge, unpublished).

AFLP analysis permits the detection of variation within perennial ryegrass populations and partial discrimination between ecotypes and cultivars. The overlap between cultivars probably reflects the broad and non-restricted genetic base of many varieties. Further studies on restricted base cultivars (Guthridge *et al.*, 2001) indicate that such populations may be effectively discriminated by AFLPs. The data also allow the selection of parental genotypes for the most genetically divergent crosses and identification of polymorphic loci for trait mapping, as demonstrated for RFLP analysis in lucerne (Kidwell *et al.*, 1999). The relative genetic divergence of North African genotypes ensures a high level of genetic polymorphism in crosses involving these plants.

AFLP profiling of inbred lines of white clover

Joyce *et al.* (1999) measured the genetic distance between inbred lines of white clover, based on selfing of plants with the rare S_f (self-fertile) allele at the self-incompatibility locus (Yamada *et al.*, 1989a,b). Twenty-two inbred lines were derived from pre-existing lines by at least six generations of single seed descent, belonging to four distinct groups designated H, J, R and S (Michaelson-Yeates *et al.*, 1997). A number of lines from within each group were selected and profiled using five RAPD primers, generating a dendrogram which showed substantial partitioning of the genotypes into their respective groups and produced a tentative ordering of genetic distance between groups.

In order to test the efficiency of AFLP analysis for the separation of closely related white clover germplasm, six inbred genotypes were selected from group H (H_1, H_2), group R (R_1, R_2), group J (J_1) and group S (S_1). The H_1 and H_2 genotypes were full sibs from seven generations of inbreeding, while the R_1 and R_2 genotypes were full sibs from six generations of inbreeding. AFLP profiles based on *Eco*RI as the rare cutter enzyme were derived with six *Eco*RI+3/*Mse*I+3 primer combinations (Fig. 21.2a). A total of 86 polymorphic bands were scored across the six genotypes (R. Kölliker, unpublished). Cluster analysis based on Euclidean distance (ε^2) assigned genotypes from the same heterosis groups to the same cluster (Fig. 21.2b). Two of the inbred lines from group H could not be separated in this analysis, unlike the R_1 and R_2 genotypes. The extra generation of single seed descent may contribute to an enhanced level of isogenicity in group H. The genetic distances between groups are partially congruent with the RAPD analysis, except that group H, rather than group J, is most divergent in the AFLP-based tree. The greater genetic distance between groups H and S in the AFLP analysis is more consistent with the observation of heterosis for dry weight in F_1 hybrids between these groups, based on a positive correlation between genetic distance and heterosis as described for maize (Smith *et al.*, 1990).

Cultivar identification by AFLP analysis in white clover

AFLP analysis has also been used to discriminate white clover cultivars, and to test the efficiency of bulking strategies in this species (Kölliker *et al.*, 2001). Three white clover cultivars ('Prestige', 'Prop' and 'Ladino') were analysed using 20 individual plants from each cultivar. Leaf material from these plants was bulked to represent each cultivar. For the cultivar 'Prop', three replicated bulked samples of the same individual plants were analysed to test reproducibility. Six

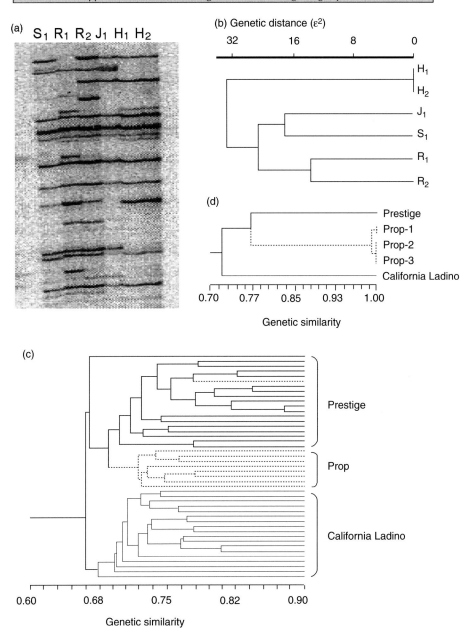

Fig. 21.2. (a) Typical amplified fragment length polymorphism banding pattern for six inbred white clover genotypes from heterosis groups H, J, R and S. (b) Cluster analysis based on Euclidean distance. Numbers indicate sister lines within a single heterosis group. UPGMA phenograms for (c) individual genotypes of cultivars 'Prestige' (bold lines), 'Prop' (broken lines) and 'California Ladino' (solid lines) and (d) bulked samples of cultivars 'Prestige', 'Prop' and 'California Ladino'. Results for three replicated 'Prop' bulks are shown. The scale indicates the coefficient of genetic similarity based on the method of Nei and Li (1979).

EcoRI+3/MseI+3 primer combinations produced 172 polymorphic bands. On average, 10% of the bands present in individual plants were absent in bulked samples of the respective cultivar. However, most of the bands missing in bulked samples appeared in individual samples at a frequency lower than 20%.

Genetic similarity was calculated for individual plants and bulked samples. Cluster analysis of individual plants resulted in a clear separation of the three cultivars, 'California Ladino' being the most divergent (Fig. 21.2c). Analysis of molecular variance revealed 84% of total variation within cultivars and 16% between cultivars. Cluster analysis of bulked samples revealed the same relationship between cultivars, as well as near identity between the replicated bulks (Fig. 21.2d). This analysis demonstrates that AFLPs may effectively discriminate white clover populations, and tissue bulking from 20 individuals is a viable strategy for large-scale screens of germplasm accessions.

Development of SSR markers for DNA profiling in perennial ryegrass

Although AFLPs provide a powerful system for DNA profiling, problems for implementation arise due to technical complexities in the assay, while the dominant nature of AFLP loci limits the information content. Simple sequence repeat (SSR) loci are co-dominant, multi-allelic and capable of simple implementation, providing the ideal marker system for outbreeding species. SSR markers may also be combined in multiplex assays. Our objective is to validate the genetic relationships detected by AFLP analysis with polymorphic SSR markers.

A limited number of SSRs (16) were isolated from perennial ryegrass by Kubik *et al.* (1999) through hybridization screening of 13,000 clones of a plasmid genomic library, of which six were used to develop polymorphic loci. This set was sufficient to discriminate between 11 perennial ryegrass genotypes. However, this number is likely to be insufficient for all DNA profiling applications, and is inadequate for genetic mapping purposes, for which 100–200 SSR loci are required. In order to develop a set of 200 efficiently amplified, polymorphic *L. perenne* SSR loci, the method of Edwards *et al.* (1996) was modified (Jones *et al.*, 2001). Two libraries were developed: the LPSSRH library (*c.* 50% enrichment) was selected for $[CA]_n$ repeats, while the LPSSRK library (*c.* 60% enrichment) was selected for a number of dinucleotide and trinucleotide repeat motifs. Large-scale SSR discovery was based on DNA sequencing of randomly selected clones. Across all libraries, 1853 clones were analysed, of which 859 (46%) contained SSRs. Redundancy accounted for 16% of the SSR clones, leaving 718 unique SSR clones for primer design. Truncation of SSR sequences, such that less than 25 bp of 5′- or 3′-flanking sequence is available for primer design, is a serious problem in enrichment libraries. The non-truncated SSR clones accounted for 51% of the total, such that 366 clones were available for primer design.

The LPSSRH library is largely (89%) composed of dinucleotide repeats, of which the most common motif type was $[CA]_n$. The LPSSRK library contains 52% dinucleotide repeats and 46% trinucleotide repeats, with 2% contributed by $n > 3$

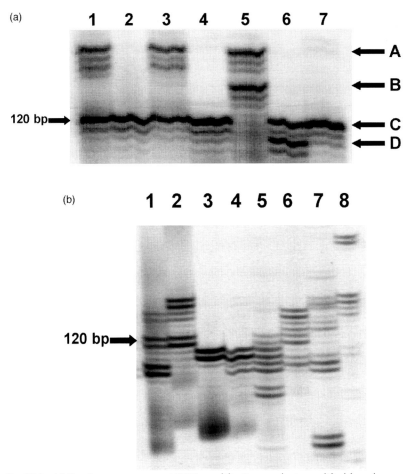

Fig. 21.3. (a) Simple sequence repeat (SSR) amplification products amplified from locus LPSSRXX0D7 from seven perennial ryegrass genotypes. The SSR structure in the cloned sequence is $[CA]_{10}$. 1 'Ellett'; 2 'Kangaroo Valley'; 3 'Grasslands Pacific'; 4 'Victorian'; 5 'Yatsyn'; 6 p150/112 heterozygous parent; 7 doubled haploid DH297. Four alleles are indicated from A (largest) to D (smallest). (b) SSR ortholocus detection among *Poaceae* species related to *Lolium perenne*. Amplified products were obtained for locus LPSSRXX0K21. The SSR structure in the cloned sequence is $[CA]_{10}$. 1 *L. rigidum*; 2 *L. multiflorum*; 3 *F. pratensis*; 4 *F. arundinacea*; 5 *F. rubra*; 6 *P. aquatica*; 7 *P. pratensis*; 8 *Avena sativa* (oat).

repeat types. Perfect, imperfect and compound repeats were observed (Weber, 1990). PCR primers were designed for 100 LPSSR loci and tested for amplification efficiency and polymorphism across panels of diverse genotypes. Amplification products of the expected size were produced by 81% of the primer pairs, of which 67% were polymorphic. A mean allele number of 3.5 was observed, with a range from two to seven alleles. A representative amplification profile is shown in Fig. 21.3a, in which four alleles are detected in seven genotypes.

Perennial ryegrass SSR loci show significant transfer to other related Poaceae taxa. Cross-amplification in other *Lolium* species is *c*. 70–80% efficient, and *c*. 40–60% efficient in members of the related genus *Festuca*, while lower levels of transfer are seen in other pasture and turfgrass species, as well as several cereals (Fig. 21.3b).

Development of SSR markers for DNA profiling in white clover

The method for enrichment library construction optimized for perennial ryegrass was used to construct $[CA]_n$-enriched libraries for white clover (Kölliker *et al.*, 2001). The levels of enrichment across six libraries varied from 30 to 90%. Sequence analysis of 1123 clones detected 793 SSR-containing clones (71%), with 12% redundancy, leaving 696 unique SSR clones. Almost 60% of the unique SSR sequences were non-truncated, providing 397 clones for primer design. The majority of the SSR clones (93%) contained dinucleotide repeats, of which 91% contained $[CA]_n$ motifs. The proportions of imperfect repeats were lower than for perennial ryegrass.

Primer pairs were designed for 117 TRSSR loci, of which 101 (86%) produced amplification products of the correct predicted size. Eighty-nine (88%) of the amplified TRSSR loci detected polymorphism in a screening panel of eight genotypes, including a fifth generation J group inbred and a fourth generation R group inbred, as described in 'AFLP profiling of white clover' above. A representative amplification profile is shown in Fig. 21.4a, in which six alleles are detected in seven genotypes. The relatively high genetic distance between the J and R groups predicted by RAPD and AFLP analysis led to the selection of individuals from these groups as inbred parents of a segregating F_2 reference mapping population for white clover (T.P.T. Michaelson-Yeates and M.T Abberton, personal communication). TRSSR locus polymorphism in this family has been evaluated, allowing the construction of a framework SSR genetic map. As for perennial ryegrass, the aim is to develop a set of *c*. 200 efficient, polymorphic SSR loci.

White clover SSR loci also show significant transfer to other related *Leguminosae* taxa. The most efficient cross-amplification is seen for other *Trifolium* species belonging to section 'Amoria' (Zohary, 1970), with lower level transfer to red clover (*T. pratense*) and subterranean clover (*T. subterraneum*). Transfer to other genera is most efficient for *Medicago* and *Melilotus*, which are classified together with *Trifolium* in the tribe *Trifoliae* of subfamily of the family *Leguminosae* (Williams, 1987). Cross-amplification in more distantly related genera such as *Lotus* and *Glycine* is relatively inefficient (Fig. 21.4b).

Conclusions

The cross-pollinated breeding system of the most important temperate forage species ensures that cultivars are genetically heterogeneous and provides a

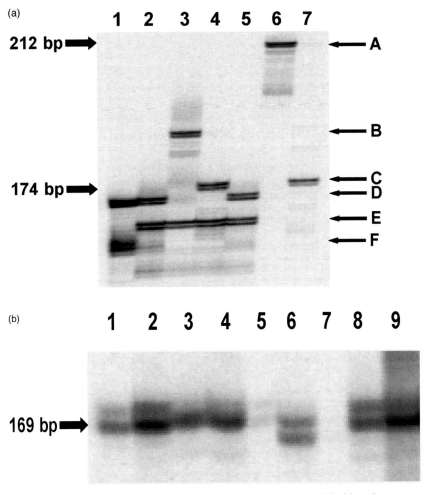

Fig. 21.4. (a) Simple sequence repeat (SSR) amplification products amplified from locus TRSSA01H11 from seven white clover genotypes. The SSR structure in the cloned sequence is $[CAAAA]_3[GAA]_{20}$. 1 Dusi 35; 2 Haifa 27; 3 Haifa 38; 4 Prop 16; 5 Prop 39, 6 I5J, 7 I4R. Six alleles are indicated from A (largest) to F (smallest). (b) SSR ortholocus detection among Leguminosae species related to *Trifolium repens*. Amplification products were obtained for locus TRSSRA02B09. The SSR structure in the cloned sequence is $[GT]_2GG[GT]_5$. 1 *T. pratense*; 2 *T. subterraneum*; 3 *Trifolium ambiguum* (Caucasian clover); 4 *Trifolium nigrescens* (putative diploid ancestor of white clover); 5 *Glycine max* (soybean); 6 *M. sativa*, 7 *L. corniculatus*, 8 *M. alba*, 9 *T. repens*.

requirement for DNA profiling strategies which assess within and between population genetic diversity. Analysis of individual genotypes and bulking strategies has proved successful for cultivar discrimination and germplasm screening. The majority of studies to date have involved RAPDs and RFLPs. However, AFLPs are capable of detecting variation between closely related genotypes in perennial ryegrass and white clover, as well as cultivar discrimination and the selection of parents for genetic mapping strategies. The large-scale isolation of SSR loci from perennial ryegrass and white clover now provides the means for highly efficient DNA profiling based on simple or multiplex assays. A subset of highly polymorphic loci from each species may be arranged into a series of marker panels multiplexed by a combination of different fluorochromes and size distributions. High throughput genotyping may then be performed by automated gel or capillary electrophoresis (Mitchell *et al.*, 1997). The observation of significant levels of transfer to other forage species enhances the value of these marker systems. SSR loci will provide the method of choice for DNA profiling in outbreeding forage species in the immediate future.

Acknowledgements

The authors thank Professors Michael Hayward and German Spangenberg for careful critical reading of this manuscript. The original research described here was supported by the CRC for Molecular Plant Breeding, Dairy Research and Development Corporation and the Department of Natural Resources and Environment, Victoria, Australia. R. Kölliker was the recipient of a Swiss National Science Foundation Fellowship.

References

Anon (1999) *List of Cultivars Eligible for Certification – 1999.* Organisation for Economic Co-operation and Development, Paris (http://www.oecd.org/agr/code/index.htm).

Attwood, S.S. (1940) Genetics of cross-incompatibility among self-incompatible plants of *Trifolium repens*. *Journal of the American Society of Agronomy* 32, 955–968.

Australian Bureau of Statistics (2000) *2000 Year Book Australia.* Commonwealth of Australia, Canberra.

Balfourier, F., Charmet, G. and Ravel, C. (1998) Genetic differentiation within and between natural populations of perennial ryegrass and annual ryegrass (*Lolium perenne* and *L. rigidum*). *Heredity* 81, 100–110.

Barcaccia, G., Albertini, E., Tavoletti, S., Falcinelli, M. and Vernonesi, F. (1999) AFLP fingerprinting in *Medicago* spp: its development and application in linkage mapping. *Plant Breeding* 118, 335–340.

Barker, R.E. and Warnke, S.E. (1998) Characterization of grass cultivars by use of DNA markers. In: Nakagawa, H. (ed.) *Proceedings of the International Workshop on Utilization of Transgenic Plants and Genome Analysis in Forage Crops.* National Grassland Research Institute, Nishinasuno, Japan, pp. 123–134.

Barnes, D.K., Bingham, E.T., Axtell, J.D. and Davis, W.H. (1972) The flower, sterility mechanisms and pollination control. In: Hanson, C.H. (ed.) *Alfalfa Science and Technology. Agronomy* 15, 123–141. American Society of Agronomy, Madison, Wisconsin.

Bert, P.F., Charmet, G., Sourdille, P., Hayward, M.D. and Balfourier, F. (1999) A high-density molecular map for ryegrass (*Lolium perenne*) using AFLP markers. *Theoretical and Applied Genetics* 9, 445–452.

Booy, G., van Dreven, F. and Steverink-Raben, A. (1993) Identification of ryegrass varieties (*Lolium* spp.) using allele frequencies of the PGI-2 and ACP-1 isozyme systems. *Plant Varieties and Seeds* 6, 179–196.

Bray, R.A. and Irwin, J.A.G. (1999) *Medicago sativa* L. (lucerne) cv. Hallmark. *Australian Journal of Experimental Agriculture* 39, 643–644.

Brouwer, D.J. and Osborn, T.C. (1997) Molecular marker analysis of the approach to homozygosity by selfing of diploid lucerne. *Crop Science* 37, 1326–1330.

Caradus, J.R. and Woodfield, D.R. (1997) World checklist of white clover varieties II. *New Zealand Journal of Agricultural Research* 40, 115–206.

Cornish, M.A., Hayward, M.D. and Lawrence, M.J. (1979) Self-incompatibility in ryegrass. I. Genetic control in diploid *Lolium perenne* L. *Heredity* 43, 95–106.

Crossa, J. (1989) Methodologies for estimating the sample size required for genetic conservation of outbreeding crops. *Theoretical and Applied Genetics* 77, 153–161.

Crossa, J., Jewell, D.C., Deutsch, J.A. and Taba, S. (1992) Gene action and the bottleneck effect in relation to sample size for maintenance of cross-pollinated populations. *Field Crops Research* 29, 225–239.

De Loose, M., Van Laecke, K., Depicker, A. and Van Bockstaele, E. (1994) Identification of ryegrass using DNA markers. *Proceedings of the Ninth Meeting of the EUCARPIA Section on Biometrics in Plant Breeding*. Wageningen, The Netherlands, pp. 222–223.

Diwan, N., Bhagwat, A.A., Bauchan, G.B. and Cregan, P.B. (1997) Simple sequence repeat DNA markers in lucerne and perennial and annual *Medicago* species. *Genome* 40, 887–895.

Donini, P., Elias, M.L., Bougourd, S.M. and Koebner, R.M.D. (1997) AFLP fingerprinting reveals pattern differences between template DNA extracted from different plant organs. *Genome* 40, 521–526.

Edwards, K.J., Barker, J.H.A., Daly, A., Jones, C. and Karp, A. (1996) Microsatellite libraries enriched for several microsatellite sequences in plants. *Biotechniques* 20, 758–759.

Ferguson, J.M. and Grabe, D.F. (1986) Identification of cultivars of perennial ryegrass by SDS-PAGE of seed proteins. *Crop Science* 26, 170–176.

Ghérardi, M., Mangin, B., Goffinet, B., Bonnet, D. and Huguet, T. (1998) A method to measure genetic distance between allogamous populations of lucerne (*Medicago sativa*) using RAPD molecular markers. *Theoretical and Applied Genetics* 96, 406–412.

Gilbert, J.E., Lewis, R.V., Wilkinson, M.J. and Caligari, P.D.S. (1999) Developing an appropriate strategy to assess genetic variability in plant germplasm collections. *Theoretical and Applied Genetics* 98, 1125–1131.

Gilliland, T.J., Camlin, M.S. and Wright, C.E. (1982) Evaluation of phosphoglucoisomerase allozyme electrophoresis for the identification and registration of cultivars of perennial ryegrass (*Lolium perenne*). *Seed Science and Technology* 10, 415–430.

Griffiths, A., Kidwell, K., Barrett, B., Williams, W. and Woodfield, D. (1999) Comparing AFLP and RFLP marker systems for diversity assessment in white clover. *Proceedings of the 11th Australian Plant Breeding Conference. Adelaide*. Vol. 2, pp. 210–211.

Gustine, D.L. and Huff, D.R. (1999) Genetic variation within and among white clover populations from managed permanent pastures for the northeastern USA. *Crop Science* 39, 524–530.

Guthridge, K.M., Dupal, M.D., Kölliker, R., Jones, E.S., Smith, K.F. and Forster, J.W. (2001) AFLP analysis of genetic diversity within and between populations of perennial ryegrass (*Lolium perenne* L.). *Euphytica* (in press).

Hampton, J.G. (1991) Temperate herbage seed production: an overview. *Journal of Applied Seed Production* 9 (Suppl.), 2–13.

Hampton, J.G., Clifford, P.T.P. and Rolston, M.P. (1987) Quality factors in white clover seed production. *Journal of Applied Seed Production* 5, 32–40.

Hayward, M.D. and McAdam, N.J. (1977) Isozyme polymorphism as a measure of distinctiveness and stability in cultivars of *Lolium perenne*. *Zeitschrift für Pflanzenzüchtung* 79, 59–68.

Holmes, W. (ed.) (1980) *Grass: its Production and Utilisation*. Blackwell Scientific Publications, Oxford.

Huff, D.R. (1997) RAPD characterization of heterogeneous perennial ryegrass cultivars. *Crop Science* 37, 557–564.

Jones, C.J., Edwards, K.J., Castiglione, S., Winfield, M.O., Sala, F., Vandewiel, G., Bredemeijer, B., Vosman, B., Matthes, M., Daly, A., Brettschneider, R., Bettini, P., Buiatti, M., Maestri, E., Malcevschi, A., Marmiroli, N., Aert, R., Volckaert, G., Rueda, J., Lincaero, R., Vasquez, A. and Karp, A. (1997) Reproducibility testing of RAPD, AFLP and SSR markers in plants by a network of European laboratories. *Molecular Breeding* 3, 381–390.

Jones, E.S., Dupal, M.D., Kölliker, R., Drayton, M.C. and Forster, J.W. (2001) Development and characterisation of simple sequence repeat (SSR) markers for perennial ryegrass (*Lolium perenne* L.). *Theoretical and Applied Genetics* 102, 405–415.

Joyce, T.A., Abberton, M.T., Michaelson-Yeates, T.P.T. and Forster, J.W. (1999) Relationships between genetic distance measured by RAPD-PCR and heterosis in inbred lines of white clover (*Trifolium repens* L.). *Euphytica* 107, 159–165.

Kidwell, K.K., Bingham, E.T., Woodfield, D.R. and Osborn, T.C. (1994) Relationships among genetic distance, forage yield and heterozygosity in isogenic diploid and tetraploid lucerne populations. *Theoretical and Applied Genetics* 89, 323–328.

Kidwell, K.K., Hartweck, L.M., Yandell, B.S., Crump, P.M., Brummer, J.E., Moutray, J. and Osborn, T.C. (1999) Forage yields of lucerne populations derived from parents selected on the basis of molecular marker diversity. *Crop Science* 39, 223–227.

Kölliker, R., Stadelmann, F.J., Reidy, B. and Nösberger, J. (1998) Fertilization and defoliation frequency affect genetic diversity of *Festuca pratensis* Huds. in permanent grasslands. *Molecular Ecology* 7, 1757–1768.

Kölliker, R., Stadelmann, F.J., Reidy, B. and Nösberger, J. (1999) Genetic variability of forage grass cultivars: a comparison of *Festuca pratensis* Huds., *Lolium perenne* L. and *Dactylis glomerata* L. *Euphytica* 106, 261–270.

Kölliker, R., Jones, E.S., Drayton, M.C., Dupal, M.D. and Forster, J.W. (2000) Development and characterisation of simple sequence repeat (SSR) markers for white clover (*Lolium perenne* L.). *Theoretical and Applied Genetics* 102, 416–424.

Kölliker, R., Jones, E.S., Jahufer, M.Z.Z. and Forster, J.W. (2001) Bulked AFLP analysis for the assessment of genetic diversity in white clover (*Trifolium repens* L.). *Euphytica* (in press).

Kongkiatngam, P., Waterway, M.J., Fortin, M.G. and Coulman, B.E. (1995) Genetic variation within and between two cultivars of red clover (*Trifolium pratense* L.): comparison of morphological, isoenzyme and RAPD markers. *Euphytica* 84, 237–246.

Kongkiatngam, P., Waterway, M.J., Coulman, B.E. and Fortin, M.G. (1996) Genetic variation among cultivars of red clover (*Trifolium pratense* L.) detected by RAPD markers amplified from bulk genomic DNA. *Euphytica* 89, 355–361.

Kraft, T. and Säll, T. (1999) An evaluation of the use of pooled samples in studies of genetic variation. *Heredity* 82, 488–494.

Krishnan, H.B. and Sleper, D.A. (1997) Identification of tall fescue cultivars by sodium dodecyl sulphate polyacrylamide gel electrophoresis of seed proteins. *Crop Science* 37, 215–219.

Kubik, C., Meyer, M.A. and Gaut, B.S. (1999) Assessing the abundance and polymorphism of simple sequence repeats in perennial ryegrass. *Crop Science* 39, 1136–1141.

Lallemand, J., Michaud, O. and Greneche, M. (1991) Electrophoretical description of ryegrass varieties: a catalogue. *Plant Varieties and Seeds* 4, 11–16.

Le Buanec, B. (1997) An overview of the world seed market. *International Herbage Seed Production Research Group Newsletter* December, 12–15.

Liu, Z.-W., Jarret, R.L., Duncan, R.R. and Kresovich, S. (1994) Genetic relationships and variation among ecotypes of seashore paspalum (*Paspalum vaginatum*) determined by random amplified polymorphic DNA markers. *Genome* 37, 1011–1017.

Liu, Z.-W., Jarret, R.L., Kresovich, S. and Duncan, R.R. (1995) Characterization and analysis of simple sequence repeat (SSR) loci in seashore paspalum (*Paspalum vaginatum* Swartz). *Theoretical and Applied Genetics* 91, 47–52.

Michaelson-Yeates, T.P.T., Marshall, A., Abberton, M.T. and Rhodes, I. (1997) Self-incompatibility and heterosis in white clover (*Trifolium repens* L.). *Euphytica* 94, 341–348.

Michelmore, R.W., Paran, I. and Kesseli, R.V. (1991) Identification of markers linked to disease resistance genes by bulked segregant analysis: a rapid method to detect markers in specific genomic regions by using segregating populations. *Proceedings of the National Academy of Sciences USA* 88, 9828–9832.

Mitchell, S.E., Kresovich, S., Jester, C.A., Hernandez, C.J. and Szewc-McFadden, A.K. (1997) Application of multiplex PCR and fluorescence-based, semi-automated allele sizing technology for genotyping plant genetic resources. *Crop Science* 37, 617–624.

Morell, M.K., Peakall, R., Appels, R., Preston, L.R. and Lloyd, H.L. (1995) DNA profiling techniques for plant variety identification. *Australian Journal of Experimental Agriculture* 35, 807–819.

Nei, M. and Li, W.H. (1979) Mathematical model for studying genetic variation in terms of restriction endonucleases. *Proceedings of the National Academy of Sciences USA* 76, 5269–5273.

Posselt, U.K. and Bolaric, S. (1999) Genetic diversity among cultivars and ecotypes of *Lolium perenne*. *Proceedings of the 22nd Fodder Crops and Amenity Grasses Section Meeting of Eucarpia*. St Petersburg, Russia.

Rafalski, J.A., Vogel, J.M., Morgante, M., Powell, W., Andre, C. and Tingey, S.V. (1996) Generating and using DNA markers in plants. In: Birren, B. and Lai, E. (eds) *Non-mammalian Genomic Analysis: a Practical Guide*. Academic Press, San Diego, pp. 75–135.

Sato, M., Yamashita, M., Kato, S., Mikami, T. and Shimamoto, Y. (1995) Mitochondrial DNA analysis reveals cytoplasmic variation within a single cultivar of perennial ryegrass (*Lolium perenne* L.). *Euphytica* 83, 205–208.

Smith, K.F. (2000) *Thinopyrum ponticum* (Podp.) (tall wheatgrass) cv. Dundas. *Australian Journal of Experimental Agriculture* 40, 119–120.

Smith, O.S., Smith, J.S.C., Bowen, S.L., Tenborg, R.A. and Wall, S.R. (1990) Similarities among a group of elite maize inbreds as measured by pedigree, F_1 grain yield, heterosis and RFLPs. *Theoretical and Applied Genetics* 80, 833–840.

Sweeney, P.M. and Danneberger, T.K. (1994) Random amplified polymorphic DNA in perennial ryegrass: a comparison of bulk samples vs. individuals. *HortScience* 29, 624–626.

Sweeney, P.M. and Danneberger, T.K. (1997) RAPD markers from perennial ryegrass DNA extracted from seeds. *HortScience* 32, 1212–1215.

Vogel, K.P. and Pedersen, J.F. (1993) Breeding systems for cross-pollinated forage grasses. *Plant Breeding Reviews* 11, 251–274.

Vos, P., Hogers, R., Bleeker, M., Reijans, M., Vandelee, T., Hornes, M., Frijters, A., Pot, J., Peleman, J., Kuiper, M. and Zabeau, M. (1995) AFLP – a new technique for DNA fingerprinting. *Nucleic Acids Research* 23, 4407–4414.

Warpeha, K.M.F., Capesius, I. and Gilliland, T.J. (1998) Genetic diversity in perennial ryegrass (*Lolium perenne*) evaluated by hybridization with ribosomal DNA: implications for cultivar identification and breeding. *Journal of Agricultural Science, Cambridge* 131, 23–30.

Weber, J.L. (1990) Human DNA polymorphism based on length variations in simple-sequence tandem repeats. In: Davies, K.E. and Tilghman, S.M. (eds) *Genetic and Physical Mapping*. Vol. I *Genome Analysis*. Cold Spring Harbor Laboratory Press, Cold Spring Harbor, New York.

Wiesner, I., Samec, P. and Nasinec, V. (1995) Identification and relationships of cultivated accessions from *Lolium–Festuca* complex based on RAPD fingerprinting. *Biologia Plantarum* 37, 185–195.

Williams, J.G.K., Kubelik, A.R., Livak, K.J., Rafalski, J.A. and Tingey, S.V. (1990) DNA polymorphisms amplified by arbitrary primers are useful as genetic markers. *Nucleic Acids Research* 18, 6531–6535.

Williams, W.M. (1987) White clover taxonomy and biosystematics. In: Baker, M.J. and Williams, W.M. (eds) *White Clover*. CAB International, Wallingford, UK, pp. 343–419.

Xu, W.W., Sleper, D.A. and Krause, G.F. (1994) Genetic diversity of tall fescue germplasm based on RFLPs. *Crop Science* 34, 246–252.

Yamada, T., Fukuoka, H. and Wakamatsu, T. (1989a) Recurrent selection programmes for white clover (*Trifolium repens* L.) using self-compatible plants. I. Selection of self-compatible plants and inheritance of a self-compatibility factor. *Euphytica* 44, 167–172.

Yamada, T., Higuchi, A. and Fukuoka, H. (1989b) Recurrent selection of white clover (*Trifolium repens* L.) using self-compatible factor. *Proceedings of the XVI International Grassland Congress*. Nice, France, pp. 299–300.

Yu, K. and Pauls, K.P. (1993) Rapid estimation of genetic relatedness among heterogeneous populations of lucerne by random amplification of bulked genomic DNA samples. *Theoretical and Applied Genetics* 86, 788–794.

Zohary, M. (1970) *Trifolium*. In: Davis, P.H. (ed.) *Flora of Turkey*, Vol. 3. Edinburgh University Press, Edinburgh, pp. 384–448.

Index

AAD *see* Arbitrarily amplified DNA
Aegelops
    speltoides 150
    squarrosa 150
    tauchii 90
AFLP *see* Amplified fragment length polymorphism
Agarose 16
Allele 3
Allele-specific oligonucleotides 5
Allele-specific PCR 184
Allelic richness 65
Alocasia 109
Amplified fragment length polymorphism 1, 19, 62, 66, 134, 161–162, 167, 196, 306
Anthurium 109
Araceae 109
Arbitrarily amplified DNA 29, 36
Array based hybridization 187
ASOs *see* Allele-specific oligonucleotides

Bacterial artificial chromosome libraries 196
Bark 241
Barley 21, 161
Bermudagrass 38
Botanic garden 63
Brachiaria 299
Bread wheat 18
Bulked samples 303

Cabernet Sauvignon 231

Caladium 109
Capillary gel electrophoresis 1
CAPS *see* Cut amplified polymorphic sequence
Capsicum 70
Carnation 40
Cayratia japonica 231
CCM *see* Chemical cleavage of mismatches
cDNA libraries 227
Centrosema 300
CGIAR *see* Consultative Group on International Agricultural Research
Chardonnay 231
Chemical cleavage of mismatches 6
Chemotype 100
Chloroplasts 247
CIAT 73
CIMMYT 84, 286
Cissus cardiophylla 231
Cluster analysis 122, 169
Co-dominant markers 95
Colocasia esculenta 109
Commercial applications 265
Consultative Group on International Agricultural Research 85
Core collections 71
Cross transferability 219
    in animals 213
    in plants 214
Cross-species amplification 212
Cross-species transfer of microsatellites 195
Cultivar identification 310

Cut amplified polymorphic sequence 50, 51
Cynodon 37
Cyrtosperma 109

Dactylis glomerata L. 299
DAF *see* DNA amplification fingerprinting
Data-mining 184
dCAPS 51
Denaturing high-performance liquid chromatography 188
Denaturing PAGE 16
DHPLC *see* Denaturing high-performance liquid chromatography
Dieffenbachia 109
Direct sequencing 4
Discula destructiva 39
Diversity 65
DNA
    amplification fingerprinting 30
    chip 7, 51, 75, 86
    extraction 114, 239
    ligase 48
    melting-based approaches 50
    storage 244
DNA-BankNet programme 61
Dry tissue 241

EMC *see* Enzyme mismatch cleavage
Enriched microsatellite libraries 114, 193
Enrichment by hybridization 201
Enzyme mismatch cleavage 6
Escherichia coli 193
Ex situ conservation 61–62
Expressed sequence tag 5, 24, 225
Expressed sequence tag databases 229

Festuca arundinacea 299
Floriculture 40
Fluorescence resonance energy transfer 186, 276
Foods 242
Forage grasses 303
Forage legumes 305

Forensic investigations 265
Fresh tissue 240
FRET *see* Fluorescence resonance energy transfer

Gene flow 99
Genetic diversity 64
Genetically modified organism 266
Genetics of wild populations 106
Genomic libraries 196, 226
Genomic mismatch scanning 54
Glu-D1 locus 293
Glycine 215
GMO *see* Genetically modified organism
GMS *see* Genomic mismatch scanning
Grapes 230
Grass seed 301
Grasses 299

Haplotype 3
Heteroduplex 6
Heterotic groups 87
High throughput extraction methods 244
Hordeum
    spontaneum 172
    spp. 161
    vulgare 172
HPLC 51
Hybridization 52, 185

ICIS *see* International Crop Information System
in situ conservation 73
Inbred lines 310
Indels 2, 181
Indica 70
Insertions/deletions *see* Indels
Intellectual property 40, 265
International Center for Maize and Wheat Improvement 286
International Crop Information System 84, 290
International Triticeae EST Cooperative 24
International Wheat Information System 286

InterSSR 113
IP-RP-HPLC 51
Isozyme 68, 111, 132
ITEC *see* International Triticeae EST Cooperative
IWIS *see* International Wheat Information System

Jaccard's coefficient 117
Japonica 70

Landraces 62
LD *see* Linkage disequilibrium
Legumes 299
Leguminosae 299
Lens 68
Linkage disequilibrium 189
Living collections 72
Lolium
    multiflorum Lam. 299
    perenne L. 299
Lotus corniculatus 300

Macroptilium 300
Maize 90, 147, 182
Maize microsatellites 2
MALDI-TOF MS *see* Matrix-assisted laser desorption/ionization time of flight mass spectrometry
Marker-assisted selection 48, 55
Matrix-assisted laser desorption/ionization time of flight mass spectrometry 8, 49
MB *see* Molecular beacon
Medicago sativa L. 299
Melaleuca alternifolia 95
Melitotus alba 300
Mendelian inheritance 87
Microarray 47, 62, 76
Microsatellite 1, 15, 97, 147, 161, 179, 225, 251
Microsatellite databases 258
Microsequencing 49
Microwave 243
Mini-hairpin 30
Minisequencing 10

Mismatch detection 188
Mismatching 32
Mitochondria 247
Molecular beacon 186, 278
MutS 49
Mycobacterium tuberculosis 281
Myrtaceae 95

Nei's index 65
Non-gel based techniques 275
Non-template A-addition 17
Nylon membranes 202

OLA *see* Oligonucleotide ligation assay
Oligonucleotide ligation assay 48
On-farm conservation 73
Oryza
    glumaepatula 70
    longistaminata 68
    nivara 70
    rufipogon 70
    sativa 68
Outbreeding 300
Outbreeding forage species 299

PAGE *see* Polyacrylamide gel electrophoresis
Paspalum 299
PBR *see* Plant breeders' rights
Pedigree 86, 295
Pennisetum ciliare L. 299
Peptide nucleic acids 51
Perennial ryegrass 312
Petunia 40
Phalaris aquatica L. 299
Phaseolus 74
Philodendron 109
Physical mapping of SSRs 18
Phytophthera colocasiae 109
Picea
    mariana 3
    rubens 3
PIG-tailing 17
Plant breeders' rights 267
Plant genetic resources 59, 63
Plant pathogens 241

Plant sample collection 239
PNA/DNA duplexes 51
PNAs *see* Peptide nucleic acids
Poaceae 129, 299
Pollen 241
Polyacrylamide gel electrophoresis 16, 121
Potato 60
Product quality assurance 265
Prunus africana 69
Purification of DNA 244
Pyrosequencing 186

QTL *see* Quantitative trait loci
Quality of DNA 246
Quantitative trait loci 61, 74
Quantity of DNA 246
Quercus
    macrocarpa 215
    petrea 215
    robur 215
QU-GENE 292

Random amplified polymorphic DNA 1, 19, 30, 62, 72, 110, 112, 134, 162, 304
RAPD *see* Random amplified polymorphic DNA
Restriction endonuclease-based assays 50
Restriction fragment length polymorphism 1, 19, 62, 72, 111, 133, 161, 167, 179, 304
RFLP *see* Restriction fragment length polymorphism
Ribosomal RNA genes 72
Rice 68
Rice genome project 25
Riesling 231

Saccharum 129
    barberi 130
    edule 130
    officinarum 130
    robustum 130
    sinense 130
    spontaneum 130

SBE *see* Single base extension
SCARs *see* Sequence characterized amplified regions
Scoring of alleles 252
Seed storage 61
Sequence characterized amplified regions 33
Simple sequence repeat 15, 62, 97, 113, 135, 161, 179–180, 211, 312
Single-base extension 49, 188
Single-nucleotide polymorphism 1, 3, 47, 62, 76, 179–180, 281
Single-strand conformation polymorphism 5
SNP *see* Single-nucleotide polymorphism
Software 289
Solanum tuberosum 60
Solid phase assays 48
Sorghum 129
Sourcing of SSR markers from related species 213
Southern hybridization 34
Soybean 75
SSCP *see* Single-strand conformation polymorphism
SSR *see* Simple sequence repeat
SSR-enriched libraries 16
Streptavidin-coated magnetic beads 201
Structure dependent cleavase 50
Stutter 17, 254
Stylosanthes 300
Sugarcane 129

TaqMan 276
Taro 109
Tea tree 95
Temperature modulated heteroduplex assay 188
Teosinte 91
Third Wave Technology 187
Tissue disruption 242
TMHA *see* Temperature modulated heteroduplex assay
Trifolium
    pratense L. 299
    repens L. 299
    subterraneum L. 300
Triticeae 91

Triticum
   aestivum 18, 147
   monococcum 147, 150
   turgidum 147
Turfgrass 37

Ultra-low temperature storage 61
Unweighted pairgroup method 66, 117, 309
UPGMA *see* Unweighted pairgroup method

Vitaceae 231
Vitis vinifera 231

Weighted pairgroup method 117
Wheat 21, 90, 147
Wheat breeding 285
White clover 314
Wide crosses 74
Wild wheat 147
WPGMA *see* Weighted pairgroup method

Xanthosoma 109

Zea 129
Zip code 49